J.M.
D0025133

Mineral licks, geophagy, and biogeochemistry of North American ungulates

ROBERT L. JONES
PROFESSOR, SOIL MINERALOGY AND ECOLOGY
UNIVERSITY OF ILLINOIS

HAROLD C. HANSON
PROFESSIONAL SCIENTIST, ILLINOIS NATURAL HISTORY SURVEY
CHAMPAIGN, ILLINOIS

The Iowa State University Press . Ames

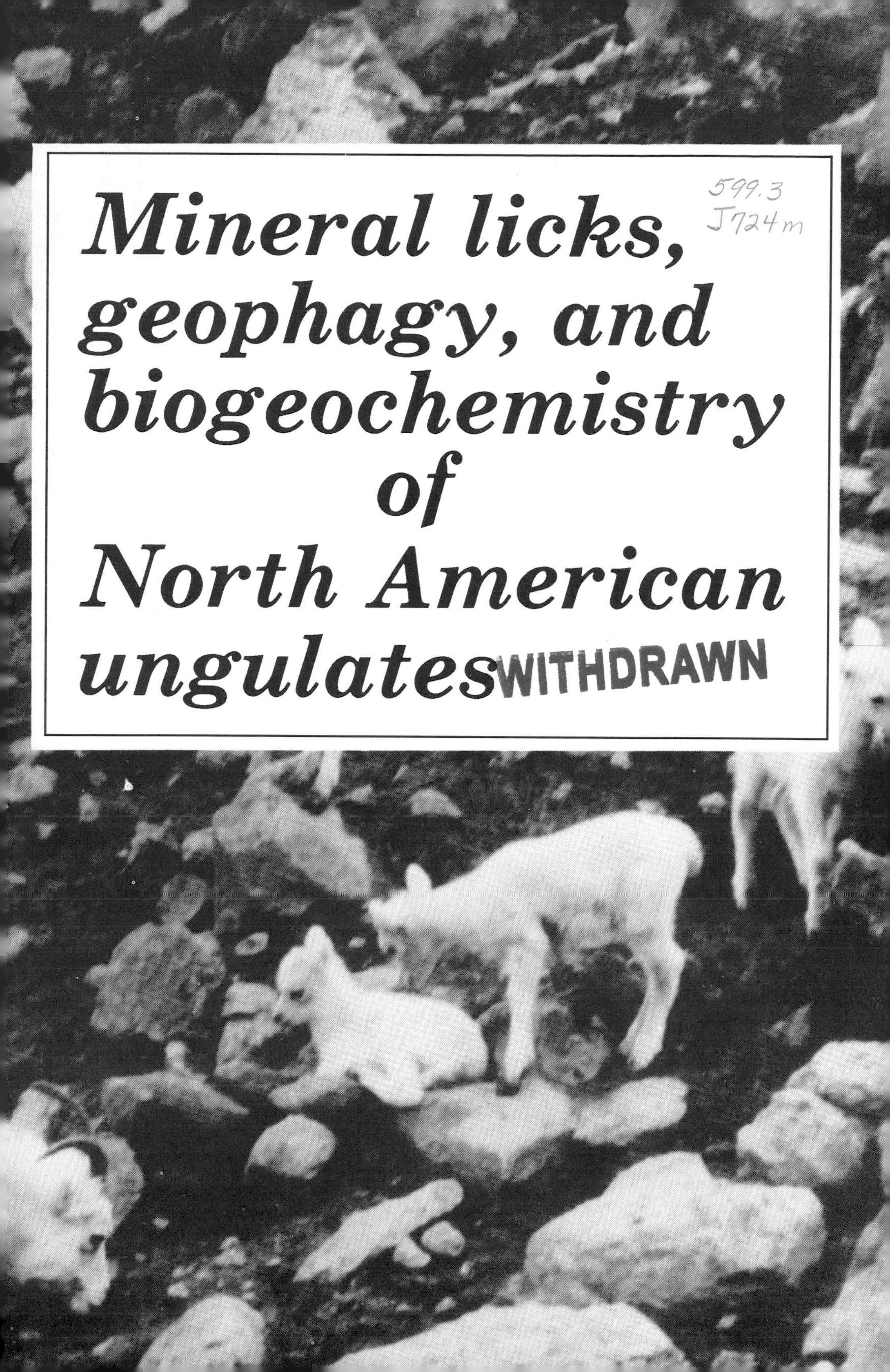

Mineral licks, geophagy, and biogeochemistry of North American ungulates

599.3
J724m
WITHDRAWN

Frontispiece: Dall sheep at Dry Creek Lick, Alaska Range
(Courtesy W. D. Heimer and S. W. Watson)

© 1985 The Iowa State University Press
All rights reserved

Composed by Compositors, Cedar Rapids, Iowa, and printed by The Iowa State University Press, Ames, Iowa 50010

First edition, 1985

Library of Congress Cataloging in Publication Data

Jones, Robert L. (Robert Lewis), 1936-
Mineral licks, geophagy, and biogeochemistry of North American ungulates.

Bibliography: p.
Includes index.
1. Ungulata—North America—Physiology. 2. Ungulatae—North America—Ecology. 3. Salt licks—North America. 4. Minerals in animal nutrition. 5. Mammals—North America—Physiology. 6. Mammals—North America—Ecology.
I. Hanson, Harold C. (Harold Carsten), 1917-
II. Title.
QL737.U4J66 1985 599.30413′097 84-10849
ISBN 0-8138-1151-1

CONTENTS

CAT May 20 '85

84-6810

5-2-83 MLS 20.30

PREFACE

IN 1912, George Shiras pointed out that there is scarcely an ungulate range without an earth lick or licks. In the course of preparing a discussion of the relationships between environmental sulfur and animal populations (Hanson and Jones 1976: Appendix 2), the chemical composition of mineral licks reported in the literature was reviewed. From this survey it became apparent that there was no consensus on the sought-after elements or what explicit physiological needs any given element fulfilled. Although from widely scattered areas of the continent, each of the previous studies was of a geographically parochial nature. It seemed this was the reason a coherent theory of mineral lick use had not emerged from these studies. Only when a comparative study was made of licks used by all ungulate species from all sectors of the continent and interpreted in the context of regional bedrock geology, soils, current knowledge of mineral metabolism, and biochemistry could a definitive assessment be made of the role of mineral licks in the life cycle of ungulates—and, as it turned out, certain aspects of the biology of herbivores in general.

Mineral licks can be regarded as a nutritional anomaly in the mineralogy of the ecosystem of an ungulate. To assess this character, *control sampling* of the surrounding terrain was necessary but was more than could be asked of cooperators who so graciously collected the lick samples. Furthermore, what constituted control samples? Were they to be taken at 5, 10, 50, or 5000 meters from a given lick? The annual range of many ungulates is extensive; it is measured in kilometers and often in thousands of vertical meters—hence, it can be argued convincingly that it is more important to understand lick chemistry in the context of regional bedrock geology and soil conditions than in the characteristics of the soils immediately outside the lick area. Consequently, the former strategy was followed to the extent that literature and maps were available.

The American bison outstandingly exemplifies the relation of ungulate range to mineral nutrition. In eastern North America, herds were small and relatively restricted in their movements because they were physiologically tied to mineral licks; on the Great Plains, their migrations encompassed great distances—a freedom of movement made possible by the base-rich soils that characterized their range. Thus, the lack of licks in the autecology of the bison on the short-grass prairies is indicative of the high levels of calcium, magnesium, sodium, and sulfur this environment affords.

It also became clear that not only would the bedrock substrate and soils have to be considered carefully in attempting to construct a secure theoretical framework for mineral lick use, but that (insofar as known) the whole nutrient chain of ungulates would have to be examined—food habits and associated geophagy; specific nutritional requirements, if known; and pathologies associated with mineral imbalances, particularly as exemplified by domestic ungulates. Perhaps our reviews—there being no conspectus available—of these aspects of ungulate biology will present the reader with an inordinate amount of material to digest before having the opportunity to review our own findings per se on the chemistry of the licks. Nevertheless, we were convinced that some attempt must be made to review the whole scope of the nutrient chain if a comprehensive understanding of "lick biology" was to be gained.

The importance of calcium and magnesium nutrition to phenotypic development in deer has been the subject of speculation since the late nineteenth century. Because of animal association with licks rich in these elements, we sought to relate size, fertility rates, and antler development of white-tailed deer with levels of calcium and magnesium in soil and/or type of bedrock over a wide geographic area. Our results give more support to the significance of relative levels of calcium and particularly magnesium in the ecosystems of deer and (by inference from the lick analyses) other native ruminants.

It would appear that mammalian herbivores generally confront unique nutritional challenges related not only to well-documented seasonal changes in the mineral content of vegetation but also to longer-lasting cyclical changes in the vegetation that now are poorly understood. The relationships of bedrock types and available soil nutrients to cycles in mineral content of vegetation will not be easy to evaluate in future investigations, but the geochemical contributions are real and (in some cases) of substantial magnitude. The present study illustrates some of the many fundamental relationships between the chemistry of rock types, soils, plants, and herbivore populations.

From our findings it became clear that in explaining the chemical and ecological characteristics of mineral licks and their relationships to ungulate ecology and physiology, we were evaluating nutritional problems that are related to all herbivorous animals. Magnesium was found to occupy a pivotal biochemical and physiological position in ungulate biology. We realized that, occupying an equivalent role in other mammalian herbivores, it could be one of the decisive factors in explaining small animal cycles as exemplified by snowshoe hares and lemmings. We were eminently rewarded from our explorations of the relevant literature.

Acknowledgments

MANY PEOPLE contribute to a study that is of the geographic breadth of this work. We are indebted to the following people who collected samples or assisted in their collection: *Alabama,* W. J. Hamrick; *Alaska,* O. E. Burris, W. E. Heimer, D. R. Klein, P. A. LeRoux, S. Linderman; *Alberta,* A. Bibaud, E. Bruns, J. Clark, W. Hall, P. C. Jacobson, G. Kemp, C. Klute, T. A. Kosinski, A. Lees, J. Olson, A. C. Paulsen, R. E. Salter, L. G. Sugden, Roy Trymble, P. Whyte, T. Woledge; *Arkansas,* H. H. Lingo, R. Taylor; *British Columbia,* D. Downing, J. Elliott, K. Fujino, R. McNab, Bryan Webster; *California,* J. DeForge, T. E. Ramsey, R. Weaver; *Colorado,* R. E. Keiss; *Georgia,* S. O. Drake, Jr., N. Nicholson, S. Pagans, W. Tillman; *Illinois,* G. B. Gill, J. W. Graber, R. R. Graber, G. Tripp; *Indiana,* H. P. Weeks, Jr.; *Kentucky,* H. Barber, L. D. Sharp, R. D. Smith; *Louisiana,* J. W. Farrar, D. Hall, R. B. Kimble, K. Martin, J. D. Newsom, C. Smith; *Manitoba,* F. Anderson, W. Antichow, D. Bigelow, R. Chartrand, L. Dubray, D. Maxwell, M. Pylypiw; *Maryland,* L. Roberts, J. Shugars; *Michigan,* R. H. Anderson, R. Bernard, C. Eacker, J. Greene, J. M. Scheidler; *Minnesota,* J. Lennartson; *Mississippi,* D. E. Temple; *Missouri,* W. Porath; *Montana,* L. Deist, B. May; *North Dakota,* S. D. Fairaizl; *Northwest Territories,* E. Hiscock, S. J. Miller, G. W. Scotter; *Nova Scotia,* H. C. Mowatt; *Ohio,* F. Brown, M. Budzik, R. W. Donohoe, M. W. McClain, C. McKibben, W. Parker; *Ontario,* C. A. Elsey, D. G. Fraser, J. D. Roseborough, H. R. Timmerman; *Oregon,* E. C. Meslow, R. R. Opp; *Pennsylvania,* W. E. Drake, J. S. Jordan, S. Liscinsky; *Quebec,* S. Georges, J. P. Lebel; *South Carolina,* W. Schrader; *Tennessee,* K. Garner; *Vermont,* L. E. Garland; *Virginia,* O. A. Burkholder, M. Carpenter, G. Dalton, S. E. Funkhouser, J. Huffer, G. H. Souder; *West Virginia,* D. Gilpin, K. Rhodes, J. J. Roy, J. Ruckel, R. W. Sharp, G. Strawn; *Wyoming,* R. H. Johnson, D. Lockman, B. Robbins.

We are indebted to the following scientists who provided geological soils or biological data: R. W. Arnold, K. Huffman, P. D. Karns, L. B. Keith, J. D. McFarland III, H. Mayland, H. L. Motto, B. L. Penzhorn, D. E. Pope, K. R. Robertson, W. I. Segars, H. P. Weeks, Jr., and L. P. Wilding. B. Springer and E. A. Anderson patiently and carefully typed the several versions of this manuscript. The staff of the libraries of the University of Illinois assisted in the use of the enormous collection of documents in their care.

We gratefully acknowledge financial support for publication by the Foundation for North American Wild Sheep.

Partial financial support was provided by the graduate college of the University of Illinois. T. R. Peck of the agronomy department graciously made available certain analytical instruments. M. R. Pence was a careful and talented analyst. J. S. Lohse did much of the figure drafting. The comments of a number of reviewers have been very helpful and we gratefully acknowledge the following: G. C. Fahey, Jr., J. J. Hassett, R. A. Hunt, J. D. Jones, K. McSweeney, C. M. Nixon, G. C. Sanderson, H. Schultz, H. P. Weeks, Jr., and P. E. Zollman.

Responsibility for the conclusions rests with the authors.

Mineral licks, geophagy, and biogeochemistry of North American ungulates

. . . But all went listlessly after the Wise One, whose calm decision really inspired confidence. When far below the safety-line, the leader began to prick up her ears and gaze forward. Those near her also brightened up. They were neither hungry nor thirsty, but their stomachs craved something which they felt was near at last. A wide slope ahead appeared, and down it a white streak. Up to the head of this streak the Wise One led her band. They needed no telling; the bank and all about was white with something that the Sheep eagerly licked up. Oh, it was the most delicious thing they had ever tasted! It seemed they could not get enough; and as they licked and licked, the dryness left their throats, the hotness went from eye and ear, the headache quit their brains, their fevered itching skins grew cool and their stomachs sweetened, their listlessness was gone, and all their nature toned. It was like a most delicious drink of life-giving cordial, but it was only *common salt.*

This was what they had needed—and this was the great healing Salt-lick to which the leader's wisdom had been their guide.

Ernest Thompson Seton
Lives of the Hunted

CHAPTER ONE

Historical Perspective

NO INTENSIVE EFFORT was made to trace the use of the word *lick,* as defined here, to its possible European origins or to find synonyms. Nevertheless, many of the nutritional benefits ascribed to animals and their behavior that are associated with licks are traceable to European thought and observations. French (1945) credits the influential German biologist, Bunge (1873), with implanting in zoological literature the importance of sodium chloride in lick use. Jakle cited Imlay (1792), who implied that *lick* was a term used by hunters in colonial America. In Imlay's words, "A salt spring is called a 'Lick,' from the earth about them being furrowed out in a most curious manner by the buffalo and deer, which lick the earth on account of the saline particles with which it is impregnated" (Jakle 1969:688). The association of sodium chloride with many licks led to the assumption that all licks were sodium chloride licks. The writings of Bunge, and of New World pioneers who settled the Appalachian region that had salt licks with brines, led to the conclusion in eastern North America that salt licks were synonymous with concentrations of sodium chloride. However, many biologists, particularly in western North America, were strained to explain that sodium chloride was the principal attractant in licks that were essentially devoid of it. Many biologists discounted sodium chloride as an important factor in salt licks.

Dominant role ascribed to sodium in licks

In the minds of many biologists and people interested in natural history, salt licks are just that—*licks*, i.e., places in the landscape where animals concentrate at certain times to satisfy their need for sodium chloride. Because of the dominant role of sodium in the physiology of animals, biologists believe licks provide sodium rather than chloride ions.

Only recently has the physiological role of the chloride ion been receiving adequate research attention. According to French's review (1945), Bunge was chiefly responsible for focusing the attention of biologists on ruminants' need for common salt. Bunge (1873:104–5) cited a quote from Ledebour's *Riese durch das Altai-Gebirge* in which attention was drawn to caves licked from salt-bearing slates by native cervids in the Altai Mountains of Asia. It is intriguing that Bunge should have cited behavior of ungulates in the mountains of western Mongolia when the deer of Germany also displayed geophagy. Bunge's reference to the Altai Mountains recently was cited again by Blair-West et al. (1968:922).

Weeks's study (1974) of deer licks in southwestern Indiana is perhaps the most intensive to date. Weeks analyzed food items eaten by white-tailed deer throughout the year as well as the earth from licks that deer in the study area used. In addition, he evaluated the morphology of their adrenal glands and noted their behavior and the intensity of their use of licks. He concluded that "sodium is the universal attractant in ungulate use of mineral licks" (p. 352). Weeks hypothesized that high levels of potassium in spring forage, particularly grasses, and the succulent condition of herbage at this season created conditions for decreased efficiency of tubular resorption of sodium in the kidney and a diarrhetic condition that further contributed to sodium loss.

However, some biologists who have carried out less intensive studies of geophagy have discounted the importance of sodium. French (1945), working in west Africa, found difficulty in ascribing importance to sodium. Cowan and Brink (1949), in a survey of 29 lick sites in the Rocky Mountains of British Columbia and Alberta, also discounted sodium chloride as the major attractant for the wide variety of native ruminants using these ranges. Their opinion was "that trace elements may well be the critical constituents" (p. 387). In his monograph on North American elk, Murie (1951:237, 309–13) reviewed and discounted the importance of common salt both as a constituent of lick earths and as a nutrient to be supplied to free ranging elk. Heimer (1973:63) speculated that Dall sheep counterveiled high phosphorus loads from winter and spring alpine vegetation by geophagy of calcium- and magnesium-rich earths that provided cations for excretion of the phosphorus. Notwithstanding occasional published skepticism, the widespread belief remains that licks are sought out for the benefit of their sodium content.

Licks of prehistoric mammals

Paleontological studies of several late Pleistocene sites have revealed that some springs have been used by large herbivores more or less continuously over thousands of years, suggesting that these springs were mineralized licks. One lick that received considerable scientific attention during settlement of the Ohio valley was Big Bone Lick in Boone County, Kentucky. Located about 4.8 km from the Ohio River and about 32 km downriver from Cincinnati, it was accessible to river traffic and frequently visited by early travelers. Jillson (1936) wrote a historical chronicle of events associated with the lick. The name derives from the great quantity of bones that were dug from the bog or "jelly ground" that surrounds the springs. This lick was the focus of several well-developed game trails and was frequented by native Americans to make salt. Naturalists and military personnel exploring the region often visited the lick and made bone collections, one of which went to Benjamin Franklin. Another shipment went to Thomas Jefferson, who showed particular interest in the skeletal remains found in the lick. The list of large vertebrates recorded from this lick, which includes extinct Pleistocene species, comprises a remarkable assemblage of mammals—the ground sloth, horse, tapir, elk, musk-ox, moose, bison, bear, mastodon, and mammoth (Jillson 1936:117). The spring waters of this lick arise from the Saint Peter sandstone that lies beneath a section of younger rocks, the uppermost of which is the Eden group consisting of calcareous blue shales (p. 111) that outcrop in the surrounding hills. Salinity of the waters was rather low; 500–600 gal of water were needed to make a bushel of salt, a ratio that suggests the water had about 8000 to 10,000 ppm sodium chloride. The literature also contains numerous references to the sulfurous nature of the spring water; however, the most interesting observations made by an early investigator regarding our initial belief that the role of sulfur in ecosystems had long been underrated was that of the English geologist Charles Lyell, who visited the lick very early in the nineteenth century. In describing the deposits (Jillson 1936:107) Lyell noted: "The greatest depth of the black mud has not been ascertained; it is composed chiefly of clay, with a mixture of calcareous matter and sand, and contains 5 parts in 100 of sulphate of lime with some animal matter." At that time, the lick evidently occupied about an acre.

An accumulation of bones less spectacular than that at Big Bone Lick but equally important is Boney Spring in Benton County, west central Missouri. Excavation and study of this site was carried out by Saunders (1977). Remains of 31 mastodons were unearthed and osteological evidence indicated 4 ground sloths had died there. Numerous other mammalian, reptilian, and piscine fossils also were recorded. Boney Spring is situated in an intermediate terrace of the Pomme de Terre River and the water issuing from the spring, now dry, probably derived from Ordovician dolomites that outcrop in the valley wall. Saunders painstakingly mapped the areal and vertical distribution of the bones. The stratigraphically highest deposit of bones was largest in diameter and the lowest stratum was the least, indicating that the spring grew in size with time. His sequential maps indicated that the number of animals visiting the spring increased as it increased in size. The deposits are remarkably circular (largest and topmost is 10 m in diameter) in plan view. In cross section, the spring contains a central, granular tufa that has been dated at 16,190 ± 400 years B.P. The bone bed is a lenticle, 0.78 m at its thickest, that directly overlies the tufa; the sediment in the bed is a gray clay and its accumulation is believed to be contemporaneous with animal use of Boney Spring and the thanatacoenosis that it bears (Saunders 1977:71). Saunders concluded that the spring was used as a source of sodium by the mastodons, which (he speculated) were under nutritional stress (p. 103). This stress was assumed to be traceable to a climate change toward a drier regime, evidence for which is found in deciduous pollen mixed with spruce at the level of the top of the bone bed (a stratum that dated back about 13,500 years). Saunders cited modern studies that describe concentrations of elephants assumed seeking sodium at pools and dry licks during drought periods. From these observations, Saunders concluded that the mastodon and the associated mammals sought out Boney Spring for a similar reason. Whether this spring did bear sodium-rich water is a moot point; Saunders (p. 98) cited data for another modern spring, 3.5 km to the north-northwest, that had concentrations of only 4.1 mg of sodium and 20 mg of sulfate per liter. It seems unlikely that such a low concentration of sodium would be effective in alleviating sodium stress (see App. 11).

Another accumulation of fossils at a probable lick has been reported from Saltville in western Virginia. Here thick mucks overlie the Maccrady formation (Mississippian), which is a sequence of shales and dolomites. Ray et al. (1967) recorded bones of musk-ox, caribou, mastodon, mammoth, and bison from the muck sediments. Saline waters here evidently bore both sodium chloride and calcium sulfate, but no analyses of the mucks are available. Probably not unfittingly, a creek to the north of the springs was given the name Lick Skillet.

Licks and the expanding frontier

Jakle (1968, 1969) made a study of place names, commerce in salt, and the economic development of the salines or springs that yielded strong brines in the Ohio River basin. His study had its origins in the late nineteenth-century speculation of F. J. Turner (Jakle 1969:687) that the availability of salt had a marked constraint on the settlement of the Ohio valley. Salt was needed not so much for a condiment as for a preservative. Large quantities of salt were used for preserving the meat of livestock raised on the frontier; game harvested there; and green hides, most of which were shipped east in the trade established between frontier and eastern population centers.

The advantage of the presence of natural licks in harvesting game is evident in Bryant's description (1918) of the lick complex along the North Fork of the Trinity River in the Klamath Mountains of California. According to a resident of the area, about 10,000 deer had been killed at the licks; and Bryant personally noted the lack of fear shown by deer using the licks when he visited the locality.

Jakle (1969) concluded that the paths developed by buffalo between licks, particularly in Kentucky, served as open tracts along which fortified stations were built and around which the settlers first developed agriculture. Evidently the eastern native Americans did not use the salines such as the Scioto saline in Jackson County, Ohio, until they adopted salt as a condiment in emulation of the settlers. Regional relocations and subsequent resettlements of native Americans in the eighteenth century were to licks farther west (Jakle 1969:705); and grants of land at salines (such as those at Shawneetown, Illinois) were made by the federal government to bands and tribes.

CHAPTER TWO

Animal licks in perspective

Physical settings

Descriptions of licks in the journals and observations of travelers and naturalists of the late eighteenth and early nineteenth centuries are vivid. For example, Faux (1823) gave the following description of a lick, of which the exact location on the Illinois frontier is unknown: "I saw a lick of singular size extending over nearly half an acre of land, all excavated three feet, that is to say, licked away, and eaten, by buffaloes, deer, and other wild animals. It has the appearance of a large pond dried. The earth is soft, salt, and *sulphurous,* and they will resort to it" (Jakle 1969:689, our italics).

The famous licks of Kentucky, such as the Blue Licks along Licking River in Nicholas County, were described in the late eighteenth century as being "pressed down . . . to a depth of three or four feet . . . where all the old trees have their roots bare of soil to that depth" (Roe 1970:842–43). This phenomenon of eating soil away from roots can be seen in a photograph of a lick in Pennsylvania used by white-tailed deer (Fig. 2.1). Transpiration creates a flow of soil water—capillary flow, in large part—toward the roots, and solutes tend to collect in greater concentration in the soil around the roots.

Licks often occur at the base of hills or bluffs bordering streams. The description by McBride (1946:20) of licks in southern Ohio (Scioto County) is probably typical of most sites in the Appalachian Plateau: "[Licks] . . . emerge from between layers of sandstone and shale and usually occur at or near stream level, although in some instances they are found as much as one hundred and fifty feet above stream level." McBride found 58 licks in the area and proceeded to "improve" 29 of them by cleaning them—evidently by deepening wet licks and removing debris. He felt that use, measured by track incidence, increased in 12 licks.

Fig. 2.1. White-tailed deer lick (sample 186) at base of oak tree, Centre County, Pa. (Courtesy S. Liscinsky)

The licks of mountain sheep and goats are often quite different. Cowan and Brink (1949) studied many licks in national parks of the Canadian Rockies, concluding that these species "come almost exclusively to dry earth exposures," many of which are "white earth slopes" (p. 381). Mountain goats at such a lick in British Columbia are shown in Fig. 2.2. Here dry, friable rock, which may be essentially unweathered, is eaten. In other habitats in the parks of the Canadian Rockies, Cowan and Brink note that moose and mountain caribou preferred "wet, muck licks and mineral springs," white-tailed and mule deer came more frequently to wet licks, and "elk visit both types with almost equal freedom" (p. 381). A moist, deep, organic soil site used by woodland caribou in British Columbia is depicted in Fig. 2.3 and a lick in organic-rich materials used by moose on the Kenai Peninsula of Alaska is depicted in Fig. 2.4.

Fig. 2.2. Mountain goats at a dry lick near Radium Hot Springs, southeast British Columbia. (Courtesy D. Downing)

Fig. 2.3. Lick (sample 234) used by woodland caribou and moose in east central British Columbia. (Courtesy K. Fujino)

Fig. 2.4. Moose lick (sample 175) in the Kenai Peninsula, Alaska. (Courtesy P. A. LeRoux)

There are occurrences of salts that are licked. Among the samples received in this study were salt-encrusted shales and coal. Undoubtedly animals avidly lick rock surfaces at licks where travertine deposits are being formed, such as Elk Lick Falls in Kentucky. It is possible that ruminants observed licking the roofs and walls of caves and occasionally stalactites are obtaining alkaline-earth carbonates and not sodium chloride. In most such reports, it is difficult to determine if observers made a correct distinction between salinity and alkalinity.

Location, geology, and physiography

Casual inspection of a moderate-scale map of northeastern United States reveals the common occurrence of *lick* in names of features (Fig. 2.5), particularly those in association with small tributary streams known regionally as *runs*. Jakle (1969) published a map of such names found on topographic maps of the Ohio River basin (see Fig. 2.6). The distribution of names suggests that licks occur in several places in the thick sequence of Paleozoic rocks of this region. The age of rocks in which licks are situated ranges from Recent alluvium to Precambrian. Topographically, licks in eastern and western United States tend to be situated (1) in streambeds, (2) at the base of an upland where natural seepage is likely to occur, or (3) at the contact of a porous, waterbearing formation (such as sandstone) with an underlying impervious rock (such as shale). Faults,

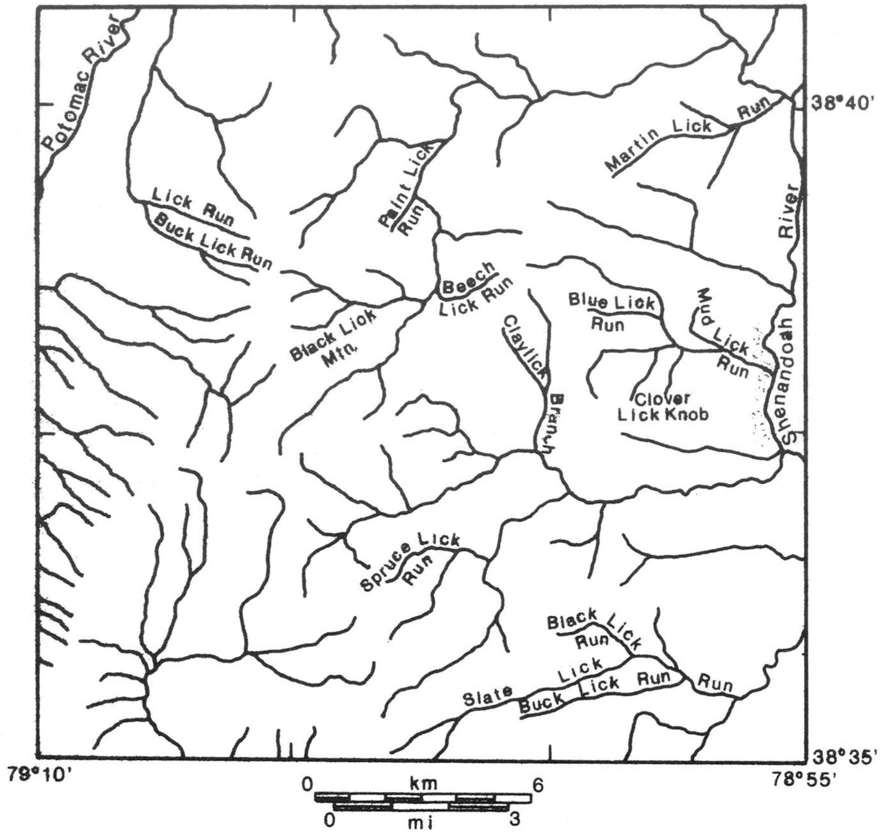

Fig. 2.5. Sketch map of a portion of Rockingham County, Va., and Pendleton County, W.Va., depicting features with *lick* in their names. (Adapted from Dry River District Map, George Washington National Forest [Va. Comm. Game Inland Fish. USDA For. Serv. 1972])

which often control groundwater conditions and create seeps or springs, are also the loci of licks. In many places where the landscape is dominated by exposed rock, the reason for siting of the lick is not apparent. In some cases a site is apparently selected because weathering of exposed rock has made the material friable and, therefore, palatable.

A large diversity of lithologies is represented; however, the licks of northeastern United States are in sedimentary rocks, particularly in the cyclothymic sequences of Pennsylvanian-age rocks of the upper Ohio River basin. The types of rocks associated with the lick earths studied are given for each sample in App. 1.

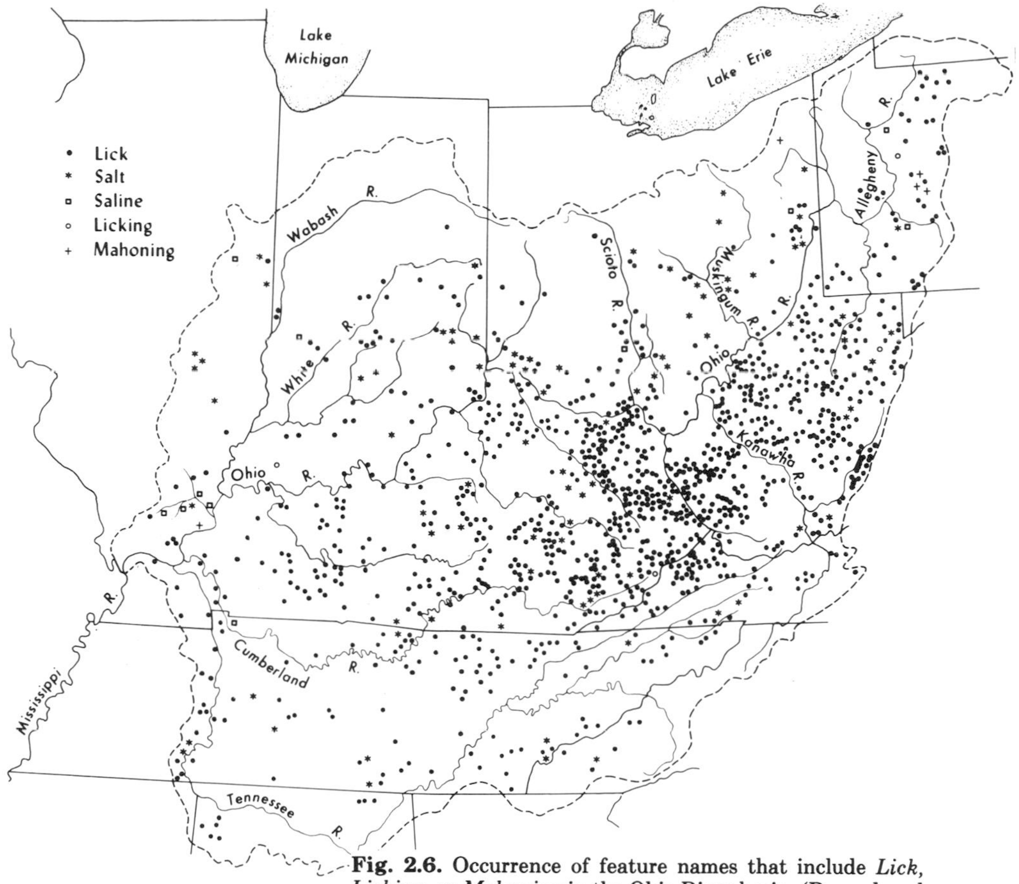

Fig. 2.6. Occurrence of feature names that include *Lick, Licking,* or *Mahoning* in the Ohio River basin. (Reproduced by permission of J. J. Jakle and American Names Society)

Native ruminant ranges without licks

In many respects, ranges without licks are as important in understanding the significance of geophagy in the life history of ruminants as ranges that possess licks. In our circularization of biologists, representatives in Massachusetts and Nevada responded that they knew of no licks in their states. Hungerford (1970:858) did not find licks used by the mule deer on the Kaibab Plateau in northern Arizona; however, this population did use salt blocks and the salt-affected soil occurring beneath the blocks. Russo (1956) noted that he knew of no mineral licks used by desert bighorn sheep in Arizona.

A substantial difference in the behavior of bison with respect to lick use appears when the immense herds of the West are compared with the comparatively small populations of the East. Descriptions of these differences and the apparent lack of lick use by the pronghorn antelope follow.

BISON. The first descriptions of encounters with bison were made by Spanish explorers in Texas. Later, scattered groups of bison were found by the Virginia colonists. Bison were first recorded in the interior by Sieur de LaSalle and Louis Hennepin, with sightings in the upper Mississippi valley (Haines 1970; Roe 1970).

In the Appalachians, bison could find sufficient forage only in broad valleys and in situations such as the bluegrass region around Lexington, Kentucky. Bands or herds were nomadic to the extent that well-developed trails or roads, as the explorers called them, were persistently followed. These trails connected the grazing lands with the famous licks of the dissected Appalachian Plateau. Numerous place names identify the former range of bison in the folded Appalachian Mountains and the Piedmont.

The best known lick used by bison in the South was known as the Great Buffalo Lick in Oglethorpe County, northeast Georgia. Evidently this lick was a kaolin deposit that William Bartram described as a white, sweetish, not salty, "fattish clay" (Goff 1975:385). Shay (1978) reviewed the distribution of bison in archeological materials in the eastern part of the Prairie Peninsula and concluded that the bison was in the process of expanding its eastern range when first encountered by Europeans. Shay concluded that a substantial population was not established on the prairies east of the Mississippi River until after A.D. 1000.

The presence of the distinct trails described in the Northeast contrasts markedly with the habits of bison on the High Plains and evidently with those on the tall grass prairie. On the broad rolling expanses of grassland, the bison clearly prospered and formed the immense herds that are so often referred to. The herds moved in broad fronts, perhaps 80 km (50 mi) wide (Haines 1970:14), moving to the south with the approach of winter, to the north with the approach of spring, over rather short distances to find shelter from winter winds, or to new pasture. The only reference to salt licks or geophagia with respect to bison on the Great Plains is in the trade literature of the American Colloid Company regarding the promotion of VolClay®, a sodium-saturated montmorillonite. Extensive deposits occur in the Black Hills, and the product literature relates that "the buffalo led miners to the rich sodium bentonite deposits of the Black Hills." This clay is promoted as a feed supplement. Bison do eat soil in preference to salt blocks in the National Bison Range of Montana (Babe May 1976, pers. commun.).

Certainly the Plains tribes would have developed strategies for hunting bison at licks if licks were present; but the evidence suggests that herds wandered almost aimlessly on the Plains. Prior to the introduction of the horse, native Americans were confined to following the bison on foot. Where terrain and herd behavior conditions permitted, the famous "buffalo jumps" were used for collecting large numbers of animals (Frison 1978).

PRONGHORN ANTELOPE. The pronghorn shared much if not all of the bison's range, occupying a food resource niche provided by browse—particularly several of the sagebrushes—and forbs. Except for the spring flush and flushes associated with late summer rains, grasses are not readily taken by antelope and are not a major category of their diet. Browse and forbs are particularly rich in minerals when compared with grasses, which might account for the few observations of pronghorn use of licks. Audubon and Bachman (1851) noted that "antelopes are remarkably fond of saline water or salt and know where all the 'salt licks' are found. They return to them daily, if near their grazing grounds, and lay down by them, after licking the salty earth or drinking the salt water" (in Yoakum 1967). Skinner (1922) observed pronghorns using "soda licks" and salt placed for bison in the Yellowstone valley, but Kitchen (1974) did not mention such behavior in his study of antelope on the National Bison Range. Prenzlow et al. (1968) studied antelope in Larimer County, Colorado, and observed animals eating soil in areas of salt blocks placed for cattle. The salt-impregnated soil was eaten in preference to direct lick use of the blocks. Frison (1977:422) noted that antelope, particularly the males, exhume shallow Indian burials on the Northern Plains and scatter the bones. Presumably, these excavations were associated with osteophagia of the small bones or geophagia of the soil surrounding the bones, which had been calcium- and phosphate-enriched by their slow dissolution.

SOIL AND PLANT CHEMISTRIES OF THE GREAT PLAINS. How did the soils and forages of the native range of the Plains bison differ from the soils of the humid, eastern states? To answer this, data have been taken directly or calculated from original sources on the levels of calcium, magnesium, sodium, and potassium in soils of New York, Georgia, and Kentucky for comparison with the same elements in soils of the Dakotas, Wyoming, and Montana (Table 2.1) The data are total contents of each element and do not strictly represent the amount available to plant roots, which is one of the most important research goals in soil chemistry. From the standpoint of ingestion of soil incidental to feeding (important in the review of studies laid out later), the differences in base contents become important.

The problem of relating soil levels of any given element to that in areal plant parts is well expressed in the study of Severson et al. (1977), who collected western wheatgrass, silver sagebrush, and all the biomass at 21 sites in the northern Great Plains (some of their plant data are in

Table 2.1 Geometric mean (GM) and geometric deviation (GD) of total calcium, magnesium, sodium, and potassium contents in surface soils of central and eastern United States.

State	Ca				Mg				Na				K				Source
	n	GM, %	GD	error	*n*	GM, %	GD	error	*n*	GM, %	GD	error	*n*	GM, %	GD	error	
New York	94:94	0.47	1.84	...	94:94	0.50	1.47	...	...	...	...	...	94:94	1.73	1.30	...	Bizzell 1930
Georgia	23:30	0.12	1.80	...	30:30	0.026	2.48	...	6:30	<0.02	...	...	30:30	0.073	2.85	...	Connor and Shacklette 1975
	29:30	0.31	1.44	...	30:30	0.25	2.19	...	29:30	0.21	2.82	...	30:30	1.2	1.98	...	
Kentucky	91:96	0.14	3.23	1.56	96:96	0.19	1.57	1.08	92:96	0.25	3.07	1.35	96:96	1.0	1.55	1.14	Connor and Shacklette 1975
	105-108	0.26	2.60	1.83	108:108	0.18	1.37	1.22	...	...	...	...	108:108	1.0	1.42	1.14	
North Dakota	29:29	1.77	2.11	...	29:29	0.68	...	1.90	29:29	1.07	1.26	...	29:29	1.39	1.28	...	Hopper et al. 1931
Colorado	167:168	0.9	2.37	1.04	168:168	0.40	2.21	1.22	168:168	0.98	1.70	1.05	168:168	2.9	1.40	1.01	Connor and Shacklette 1975
Wyoming and Montana	48:48	0.62	3.18	1.20	48:48	0.50	1.71	1.17	...	...	...	...	48:48	2.6	1.25	1.16	Connor and Shacklette 1975
Montana, Wyoming, and Dakotas	21:21	1.1[a]	0.52[b]	...	21:21	1.1[a]	0.45[b]	...	21:21	1.1[a]	0.58[b]	...	21:21	2.3[a]	0.26[b]	...	Severson et al. 1978

Note: Entries under *n* are proportions of noncensored determinations. For analyses below the detectable limit, appropriate adjustments for censoring were made for the mean (Connor and Shacklette 1975). Error refers to analytical variation. Counties from which samples were obtained are listed in App. 5.

[a]Arithmetic mean.

[b]Standard deviation.

Table 2.2. Selected nutritional properties of three grasses from Yellowstone Park, Wyo., and buffalo grass from Edwards Plateau, Tex.

Grass	Locality	Stage of growth	*n*	N, %	S, %	P, %	Ca, %	Mg, %	Na, %	K, %	Source
Bromus inermis	Yellowstone Park	bloom	2	2.08	.17	.32	.31	.13	.08	2.22	Hamilton and Gilbert 1972
Phleum pratense	Yellowstone Park	bloom	2	1.49	.22	.15	.23	.15	.11	2.00	Hamilton and Gilbert 1972
Festuca idahoensis	Yellowstone Park	bloom	2	1.98	.15	.32	.34	.17	.11	2.15	Hamilton and Gilbert 1972
Buchloe dactyloides	Edwards Plateau	early growth	2	1.91 ± 0.01	n.d.	.25 ± .02	.44 ± .03	.13 ± .01	n.d.	0.87 ± 0.10	Fraps and Cory 1940
Buchloe dactyloides	Edwards Plateau	mature	3	1.16 ± 0.05	n.d.	.16 ± .01	.46 ± .09	.11 ± .01	n.d.	0.42 ± 0.05	Fraps and Cory 1940

Tables 5.10–5.13). Samples of the A and C horizons were also taken at each site. Despite use of a variety of extracts, Severson et al. were unable to relate unequivocally the elemental concentrations in the plants with total concentration, extractable contents, and certain physical and mineralogical characteristics of associated soils. Later, multiple regression showed that about 67% of the magnesium content of wheatgrass was explained by negative correlations with total soil calcium and potassium and by positive correlations with total soil magnesium and (interestingly) calcite (Severson et al. 1977: Table 10).

Sulfur has been omitted in almost all these soil and plant analyses; but note that, of the soil samples, 8000 ppm is the probable upper limit (only 1 sample in 20 is expected to exceed this level) in the Fort Union shale of the northern Great Plains and 2100 ppm (dry weight) is the probable upper limit in sagebrush in the Green River basin (U.S. Geol. Surv. 1977: Apps. 2 and 3, respectively). In comparison, the probable upper limit of sulfur in surface soils of New York calculated from the 94 samples of Bizzel (1930) is 1220 ppm.

Buffalo grass is the common grass over the western High Plains and evidently was the most important food of the bison herds using this range. Wenger (1943:5) stated, "Buffalo grass has no superior as a productive source of livestock feed on at least a part of the rolling, erosive and unproductive areas in the Central Plains." He extols its palatability and nutritional qualities, particularly in regard to its winter utilization. Regrettably, Wenger does not give any nutritional data. Beetle (1950) states only one analysis for a sample taken in August 1906 at Wheatland, Wyoming. Following is that proximate analysis (as percent dry weight):

ash	11.60
ether extract	2.42
crude fiber	26.81
crude protein	8.34
nitrogen-free extract	50.83

Fraps and Cory (1940) give some analyses for buffalo grass taken on the Edwards Plateau of Texas. These data are presented in Table 2.2, together with analyses by Hamilton and Gilbert (1972) of three grasses from pastures used by bison in Yellowstone National Park. Comparison of these data with those for the bases and phosphorus in the tables of Chap. 5 shows that these foods contain satisfactory nutrient levels. Yellowstone Park grasses are particularly high in sodium; and their high potassium content suggests that the plants were in early stages of blooming or that, in general, unusually high potassium contents are encountered there.

The data for grasses in Table 2.2 and for grasses of the Great Plains in Tables 5.10–5.14 indicate that mineral nutrition of grazers over this broad region is high. However, sodium levels (Table 5.12) are low by standards set for domestic livestock in much of the Northern Plains. Perhaps any physiological distress caused by these levels (if indeed they are low) is counterveiled by ingestion of the sodium-rich soil adhering to the often sparse short grasses. If so, the interplay of these factors combined to create a favorable mineral balance for the huge herds of Plains bison and obviated the necessity of finding and using licks.

Phenology of lick use

Animals in temperate regions invariably seek out salt licks during late spring and early summer. Thus, Dalke et al. (1965) found that elk and deer in north central Idaho sought out licks two to three weeks after calving in the first week of June. In his study of Dall sheep at Dry Creek lick in Alaska, Heimer (1973) found that, in a three-year period, use of this lick peaked between June 6 and June 27. He speculated that the warming spring temperatures and subsequent snow melt were dominant factors affecting the time of first lick use. Weeks and Kirkpatrick (1976) reported that peak use of licks by white-tailed deer in southwestern Indiana occurred in April and May, decreased by 50% in June, and remained steady thereafter through December. No use of salt licks by these deer was observed from January through March. Calef and Lortie (1975) noted that barren ground caribou using a lick on the Firth River, Yukon Territory, remained in that area during the June calving period. Best et al. (1977), studying moose in the Swan Hills of Alberta, found that the time of lick use there extended from April 9 to June 9. Singer (1975) observed mountain goats using the Walton lick in Glacier National Park from April 15 to September 15, 1975, with peak use in early July.

Soil series, rock formation names, and ungulate ecology

In this study it became apparent that names of many natural features on maps relate to the biogeochemistry of some wildlife species and that knowledge of the bedrock geology, structural geology, and the soils of the area further substantiate conclusions drawn from the chemical study of the lick earths. For example, at least four soil series mapped in northeastern United States contain the work *lick* or *mahoning,* the Delaware tribe's name for lick. One of these series is Lickdale, first described from Lebanon County, Pennsylvania (National Cooperative Soil Survey 1970). Lickdale is a very poorly drained soil with a black, very dark gray, or very dark grayish-brown A1 horizon and a gray or grayish-brown B horizon. The common textures are loams and silt loams; but silty clay loams, clay loams, and sandy clay loams are also included in the series. The solum is characteristically strongly or very strongly acid. The Ap horizon in the type description, which was silt loam in texture, exhibited a slightly sticky to slightly plastic consistency, suggesting unusual colloidal chemistry. A number of the lick samples submitted to us displayed these properties (see descriptions in App. 1). Lickdale soils occupy upland flats and depressions with slopes up to 5% and are developed from acid sandstones and shales. At the time of this writing, there were no chemical data for the series (R. L. Shields, pers. commun.). A visual survey of county soil survey maps for eastern Pennsylvania reveals that early in this century Lickdale was mapped in many small tributaries (runs) in areas of sandstone and shale outcrops, which are the natural settings for the numerous licks that are named on maps of West Virginia and Kentucky (less so for Ohio, western Pennsylvania, and Virginia).

Stonelick series contrasts markedly with Lickdale series. The type locality for Stonelick soils is along East Fork of Little Miami River (near Elklick Road) in Clermont County, Ohio. The typical profile consists of a brown sandy-loam surface horizon overlying alternating strata of alluvium of loamy-sand or sandy-loam texture (National Cooperative Soil Survey 1979*b*). The surface horizon is mildly alkaline and subjacent layers are calcareous and are of Recent or Wisconsinan age. This soil type occurs in south central and southwestern Ohio and Indiana. Analyses of one profile in Cass County, Indiana (Purdue Univ. Agric. Exp. Stn. USDA Soil Conserv. Serv. 1977), yielded a pH of 7.2 for the Ap (0–15 cm) and C1 (15–41 cm) horizons; the calcium carbonate equivalent for these horizons was 6.1 and 6.8%, respectively.

Licking series was first described in Stark County, Ohio (National Cooperative Soil Survey 1979*a*). The soil occupies high terraces and slack-water valley fills in southeastern Ohio, northwestern West Virginia, and adjacent parts of Kentucky. The soil typically consists of a dark brown silt loam surface horizon overlying a subsoil that is yellowish brown silty clay loam, in turn overlying olive-brown silty clay horizons. The pH of the surface horizon and upper subsoil ranges from 4.5 to 6.0; but under the surface the pH rises to 7.3, and in the stratified sediments comprising the parent materials the pH may be as high as 7.8. From the characteristics that we can learn from the series description, this soil, like the Lickdale series above, embodies many of the properties that we anticipate finding in a lick earth. Licking soils occur on steep slopes of up to 25%; therefore in some places erosion would expose the neutral, clay-rich subsoil and provide a lick site.

Unlike the three soils described above, Mahoning soils occur on drift plains and moraines of northeastern Ohio (National Cooperative Soil Survey 1971). The soil takes its name from Mahoning County, Ohio, through which the Mahoning River flows. Because of its physiographic setting, the extensive Mahoning series cannot be interpreted relative to ungulate biogeochemistry as can other soils bearing lick names, which (in contrast to the Mahoning soils) have a patchy distribution or are confined to valleys. Nevertheless, note that both Mahoning and Licking series may be found in the same landscape.

From the descriptions and data available, it is evident that the properties of these soil series with lick-associated names are consistent with the physiochemical properties of lick-earth materials from northeastern United States that we studied. Each soil in its setting is unique. Lickdale probably represents a source of bases because the occurrence of this soil in nearly flat or depressional areas favors accumulation of bases from surface drainage and from groundwater sources. The surrounding landscape consists of upland soils that are acid and relatively impoverished of bases. The natural setting of the Stonelick soils in valleys also favors localized concentrations of calcium and magnesium. Presumably deer and elk used a lick near the type locality of this soil series where, considering the name, we can assume travertine or tufa was accumulating. A valley setting for a lick-associated soil could be explained by anoxic acid waters rising through the calcareous alluvial strata and dissolving some of the cabonate mineral, which subsequently precipitated upon the loss of carbon dioxide. This geologic scenario is reminiscent of Boney Spring, Missouri. Presumably the Licking series occupies the region of the country where licks are most numerous. The texture and chemistry of lick earths combined with their physiographic setting set the stage for the exposure of alkaline-rich earths by erosion and by pawing ungulates.

Another, presumably ungulate-associated, place name in an alkaline-rich environment is the hamlet of Deer Plain, located along the lower Illinois River on a terrace that lies about 12 m above the floodplain (Flock 1979). The surficial 3 m of the terrace is composed of clay-rich sediments that evidently were laid down in a slack-water environment in the lower Illinois valley when high waters in the Mississippi River caused ponding or even reversal of flow in the lower reaches of the Illinois River. The extractable alkaline earths of these upper sediments are unusually high—41.7 me (milliequivalents) calcium/100 g and 13.0 me magnesium/100 g (Flock 1979), averages of two Booker clay and silty clay profiles. Undoubtedly the proportion of magnesium increases with depth. It can be assumed that in primitive times the terrace supported a "floodplain prairie"—the Booker soil is a product of grassland vegetation and floodplain prairies occur in the Illinois valley (Zawacki and

Hansfater 1969). The name Deer Plain suggests that deer frequented the broad terrace, probably descending from the surrounding, strongly dissected, forested uplands to feed on the especially nutritious, wet prairie grasses and sedges. Perhaps against the bluff were lick-seeps where the wet, calcium- and magnesium-rich clays were available to deer. This lower Illinois River terrace was undoubtedly a prime hunting ground of prehistoric native Americans, as testified by numerous archeological sites in the lower valley—notably the now famous Koster site near the north end of the terrace in Jersey County that dates back to as far as 7000 years B.P. (Struever and Holton 1979).

In eastern United States at least five rock formations or units bear the word lick. Butts (1910) described a sandstone and conglomerate of wide extent in central Alabama as the Lick Creek sandstone member of the Pottsville formation, taking its name from a stream in Jefferson County, Alabama. McFarlan (1943:106), in his description of the geology of Kentucky, described a member of the Pottsville formation of eastern Kentucky that Morse named Salt Lick Beds. These beds, which are up to 0.6 m thick, consist of flattened silty or sandy limestone concretions.

Three Lick Bed is one of five shale units identified in the Ohio shale (Fulton 1979) of Kentucky. Perhaps the numerous features with lick names associated with the outcrop of the Ohio shale in eastern and central Kentucky (see maps of Potter 1978; Fulton 1979) are related to outcrop of this unit.

Stout (1943) lists the Elk Lick limestone and marly shale and Elk Lick coal (usually missing) members in the upper part of the Conemaugh series of southeastern Ohio. The Birmingham shale underlies the limestone, creating conditions for seep formation. A substantial thickness of sandstone and lesser clay beds, coal, and local limestone in the lower Conemaugh is referred to as the Mahoning member (Stout 1943; Condit 1912).

The Licking Creek limestone (Heldenberg group, Devonian) is named for outcrops along the creek of that name near Warren Point, Franklin County, Pennsylvania (Swartz 1939). At the type locality, the limestone is underlain by shale, the Mandata, which suggests that conditions are right for the development of calcium-bearing seeps and springs. Near Clifton Forge in northern Virginia, Edmundson (1958) reports that this limestone is 92–95% $CaCO_3$.

Two geologic formations not associated with the word lick that provide base-rich habitats should be noted. The Caribou member of the Slave Point formation (Devonian) occurs over a wide area in northern Alberta (Green 1970) and is composed of limestones (in part dolomitic), minor shales, and gypsum. Also, in northern and northwestern Alberta

(Green 1970) there is outcrop of Wapiti formation (Cretaceous), which includes feldspathic, clayey sandstones; bentonitic mudstone and bentonite; and scattered coals. From knowledge of lick earths and bedrock substrates associated with ungulate populations, it seems that these names appropriately associate cervid populations with the lithologic substrates of their regional ecosystems.

Zoology of geophagy

Although all ruminant species in North America have been reported to use licks, substantial differences in the degree of lick use exist over the range of some species, such as the desert bighorn sheep and the pronghorn antelope. (The apparent lack of use by the once great herds of bison on the Great Plains has been discussed.)

Among the notations accompanying the samples received by us, S. Liscinsky observed woodchucks and squirrels using a lick in Centre County, Pennsylvania (sample 186). T. E. Ramsey noted raccoons, small rodents, and birds at a lick in Tehama County, California (sample 155). An interesting observation by J. Huffer of honeybees using a deer lick in Augusta County, Virginia (sample 160) recalls the fact that pioneers in the Northeast found licks by "lining bees" (J. A. Jakle, pers. commun.). Weeks (1974) observed that squirrels and woodchucks in Martin County, Indiana, frequented roadsides during certain seasons and that the animals licked the road surface. He noted that woodchucks, at times, seemingly ate considerable quantities of gravel from the roadsides.

Cattle shared licks with deer in Ohio (samples 182, 183) and Missouri (sample 377). Sample 399, from a lick used by cattle in Washtenaw County, Michigan, is from an area of saliferous waters (near Saline River in Saline Township) and may never have been used by native ruminants.

In addition to the great variety of ungulates using licks in Africa (Chap. 8), chimpanzees have been observed eating earth materials.

Butterflies are often observed to alight on the moist soil surrounding puddles or on bars in streams. Arms et al. (1973) laid out trays of sand moistened with solutions of various salts in such a situation. They measured number of visits and minutes of time spent by tiger swallowtails (*Papilio glaucus*) on the trays and found that these butterflies selected sodium salts. They did not seem to select among nitrate, phosphate, or chloride anions and the level of minimum detection was thought to be about 10^{-3} M. The water of natural soil (23.7% moisture) used by swallowtails in this experiment contained 28 mg Na/l or about 1.2 me Na/l.

GEOPHAGY IN CARNIVORA. Brown and black bears evidently

consume soil in late spring and summer in the Kenai Peninsula of Alaska. Chatelain (1950) found "gray, clayish dirt" in bear feces in an area of moose licks. He made no attempt to relate the geophagia to nutrition, but it is noteworthy that grasses and sedges occur in 71% of the scats collected during this period.

W. J. Allred (in Honess and Frost 1942), in describing licks used by bighorn sheep in the Gros Ventre Mountains in Wyoming, noted one lick that was heavily utilized by bears. The analyses of this lick were CaO, strong; P_2O_5, 0.201%; Cl, 0.008%; SO_4, trace.

Dixon (1939) recounts observations of a lick above Big Meadows in Yosemite National Park where bears, mule deer, and porcupines eat soil. Analytical work on the sample yielded discrepant results: One laboratory found ferric sulfate to be the major constituent; another laboratory found calcium chloride to constitute about one-third and sodium chloride about one-half the sample.

GEOPHAGY IN BIRDS. During his trip down the Ohio River in 1810, the early American ornithologist Alexander Wilson made the following observation of Carolina parakeets: "At Big-Bone Lick, thirty miles above the mouth of Kentucky River, I saw them in great numbers. They came screaming through the woods in the morning, about an hour after sunrise, to drink the salt water, of which they, as well as the pigeons, are remarkably fond" (Wilson and Bonaparte 1878:110). Another indication that the parrot family eats soil or salt is found in the photograph in Nicolai (1975: Plate 24) of red-and-yellow and red-and-green macaws "pecking at salty clay" in a vertical bank along Rio Manu in Peru. Perhaps the old saw about Polly wanting a cracker is relevant to this apparent proclivity to sodium chloride. Mourning doves observed near Kansas, Illinois, exhibited a special affinity for the saline ponds and flats that resulted from drillings for natural gas (W. B. Taber 1954, pers. commun.). Russell (1974) also noted that doves frequent these kinds of sites and called attention to the fact that shooting at salt licks has at times been outlawed (p. 76). The sample from T. E. Ramsey, Tehama County, California (sample 155), was from a lick used by band-tailed pigeons and other birds (not identified). March and Sadleir (1972), in a study of band-tailed pigeon use of springs and sites for grit-gathering in southwestern British Columbia, concluded that calcium was sought after. These investigators collected waters from 11 springs and pools and analyses indicated high levels of sodium and calcium. Although grit was not analyzed, after calcium chloride was spread to control road dust in one area, pigeons would gather grit coated with the salt. March and Sadleir concluded that pigeons were seeking calcium to satisfy needs created by reproduction. Jeffery et el. (1980:235) photographed the trapping of band-tailed pigeons at a mineral spring in western

Washington. They report that "coastal pigeons typically touch the ground only at mineral springs and are on the ground only a few seconds each day." Wiley and Wiley (1979:49–50) noted that white-crowned pigeons sought salt-laden sandy earth on the coast of Puerto Rico. The use of natural salts by the parrot and pigeon families deserves study. Geophagy in these strict granivores is perhaps related to the high potassium in their diets (see App. 12).

Osteophagy in free-ranging wild ruminants

Chewing and eating bones has been commonly observed among rodents; however, osteophagy has seldom been observed in ruminants. We have cited bone eating by giraffes in Tanganyika. Dagg and Foster (1976) also mentioned osteophagy by giraffes but did not discuss this behavior in detail. In 1000 observations of mule deer in Big Bend National Park, Texas, Krausman and Bissonette (1977) noted only three cases of osteophagy; these involved one each by an adult male, an adult female, and a yearling male in the months of September, November, and December. Lovas (1958) found an antler fragment in the rumen of a mule deer from the Little Belt Mountains, Montana. In his table of diets summarized by season, antler material was found only in the January–March period. It is significant that from the extensive surveys of rumen contents of white-tailed deer in the eastern portion of its range, evidence of bone consumption was not found. The rare discoveries of madstones or bezoars reported by Mosby and Cushwa (1969) suggest that bone fragments would not have been missed by other investigators.

CHAPTER THREE

Geophagy in livestock nutrition

FRENCH concluded that for centuries native Africans had driven their cattle herds to licks. This aspect of animal husbandry presumably reaches back into antiquity and was an integral part of the domestication of *Bos*. In east and west Africa, cattle were taken to an area where soil or friable rock was eaten *in situ* or the herders mined the earth and provided it as a supplement, sometimes feeding it in troughs (French 1945:104).

The earliest reference in the serial scientific literature to a specific nutritional advantage gained from geophagy appears to be that of Rigg and Askew (1934), who found that a twice weekly drench of an alluvial soil was effective in avoiding anemia in sheep. Each drench consisted of 7–9 g soil that had 6% iron oxide extractable with one-tenth normal oxalic acid. In a second experiment, using limonite (71.6% iron oxide), Rigg and Askew demonstrated that it was not the iron content of the soil that improved health of the sheep. They concluded that the soil was supplying some other essential element that was important in hemoglobin synthesis; it subsequently was found to be cobalt.

In 1962 Cooley and Burroughs reported that silica sand added at 2% by weight to high-concentrate beef rations increased weight gain and feed conversion by about 5%. Higher and lower levels of sand, a silicious variety used in cleaning masonry, were less effective. This sand did not increase serum or urine silica. Sand added to higher roughage diets was without merit, so they suggested that sand imparted characteristics found in the high roughage diets to concentrate rations. Measurements to evaluate rate of digesta passage were inconclusive.

Healy (1973) reviewed his findings on the relation of ingestion to the grazing behavior of sheep and dairy cattle. He found that the amount of soil eaten increased with increased stocking rates. Grazing reduced the height of the forage plants, and trodding and puddling of the soil contaminated the plants with soil more readily. Healy believed soil contamination of pasture forages was generally on the order of 2% fresh weight of the forage (15–20% d.w.). Soils having strongly developed structure were associated with lower contamination levels. Annual loads of soil ingested ranged from 190 to 630 kg for cows in the dairy herds studied and from several to 75 kg for sheep. Soil intakes were occasionally higher during wet, cloudy periods and during slow forage growth. In arid regions, air-borne soil collects on vegetation. Study of soil ingestion by twin calves indicated substantial individual variation and that variation of soil ingestion is about threefold within a herd. Healy suggests that this relatively great difference in ingestion rates in marginal nutritional situations may account for the better condition of those animals taking larger amounts of soil. He also believed that the abraded, much shortened incisors in sheep could be attributed to soil ingestion.

Healy's research (1973) on the nutritional contribution of ingested soil included both *in vitro* and *in vivo* studies. *In vitro,* he worked with two soils of contrasting character, equilibrating the soil (1 g/ml at 5°C for 15 hr) with rumen (pH 7.4), duodenal (pH 3.0), and ileal (pH 8.2) liquors. Calcium, magnesium, and zinc declined in concentration in the presence of soils in rumen liquor, whereas aluminum, manganese, selenium, and zinc increased. Iron and phosphorus either increased or decreased. In the same soils equilibrated with duodenum liquor, calcium, magnesium, and manganese increased slightly, zinc increased from 20 to 30%, phosphorus decreased somewhat, iron decreased by 65–80%, and copper decreased dramatically by 50% in one soil and increased by about 50% in the other soil. Aluminum increased by 3 times and selenium increased about 2 to 5 times. Results for ileal liquor mixed with the two soils gave uniform but variable declines for calcium (about 30% in one soil), magnesium (about 10% in one soil), and zinc (about 10 and 20%). Iron and phosphorus behaved nonuniformly: An increase and a decrease of about 10% occurred for iron, and an increase of about 5% in one case and a decrease of almost 10% in the other occurred with phosphorus. Nonuniform increases also occurred with aluminum (much less than 1 and 8 times), manganese (much less than 1 and about 4.5 times), selenium (about 2 and 5.5 times), and copper (about 12 and 30 times). Certain of the changes seem related to known soil properties. For example, soil with high phosphorus-fixing capacity is associated with decreases in phosphorus in all three liquors. The amount of selenium in the liquor was directly related to that in the soil.

In his *in vivo* studies, Healy (1973) administered isotopes with 100 g soil as a single drench to each of two sheep. As a proportion of the isotope, he found absorbed levels to be ^{75}Se, 34%; ^{65}Zn, 14%; ^{60}Co, 1%; ^{54}Mn, 0.4%. Healy cautions that the values are minimal because no tissues were analyzed. Among the data the absorption of selenium is substantial; and although the level of cobalt uptake is low, Healy recalled the work of Rigg and Askew (1934) who successfully cured cobalt deficiency with 20 g soil per week.

Working with G. F. Wilson, Healy (1973:585) found that after 6 weeks on an experimental diet of about 11% soil, the amount of selenium in the blood of newly weaned lambs was raised substantially over that of the control group. Selenium was higher in the blood of those lambs eating the soil with higher native selenium content. In collaboration with N. D. Grace, Healy (1973:586) studied the absorption of calcium and magnesium when 100 g of Egmont brown loam or 100 g of Waikiwi silt loam was fed with 600 g of herbage to groups of wethers. After about 6 weeks of acclimatization, urine and feces were monitored for calcium and magnesium over a 6-day period. Substantial retentions were apparent (Table 3.1). The investigators offered no explanations but pointed out that mineral contributions of the soils themselves did not seem large enough to account for the retentions. In his studies of flocks grazing on limed or unlimed pastures, Healy (1973:587) found that those grazing on limed fields had fecal calcium levels 25% higher than those grazing on unlimed fields. As differences in the calcium content of ryegrass and clover on the two areas could not account for the differences in calcium turnover rates, Healy concluded that the sheep on the limed field ingested limestone particles directly. Healy (p. 574) observed that soil ingestion was higher on pastures where earthworms were abundant and concluded that casts were eaten. According to Edwards and Lofty (1977), earthworm casts are notably rich in calcium and magnesium. Healy et al. (1972:782) also concluded that "ingested soil was a source of iodine and prevented the development of goitre in lambs from ewes on high stocked areas." Ewes on pastures stocked at high rates ingested more of the soil, which was found to contain about 1.5 ppm iodine. Soil content of feces of sheep on the high rate of stocking averaged 33% compared with 4% soil in feces of sheep at lower density rates.

Working with hypocupremic ewes, Suttle (1975) found that a diet of 10% soil was associated with a decline in serum copper over a 5-week period. An increase of urinary zinc and molybdenum occurred during this

Table 3.1. Calcium and magnesium in wethers fed herbage amended with 100 g/day of Egmont and Waikiwi soils (Healy and Grace, in Healy 1973).

	Ca, g/day		Mg, g/day	
Treatment	intake	retention	intake	retention
Control	3.120	−0.396 ± 0.17	0.924	−0.088 ± 0.07
Egmont brown loam	3.440	−1.449 ± 0.22	1.029	+0.324 ± 0.19
Waikiwi silt loam	3.290	−1.299 ± 0.32	1.109	+0.298 ± 0.16

period, suggesting that soil sources of these elements were available, but Suttle did not believe that these metals had adversely affected copper metabolism.

In Tennessee, Miller et al. (1977) fed 450 and 900 g/day of subsoil from Tellico clay loam to Holstein cows (nonpregnant). Compared with the control animals, decreases of 5 and 9% in dry-matter digestibility and 2 and 6% in protein digestibility were found. There were no changes in apparent digestibility of organic matter, crude fiber, nitrogen-free extract, and combustible energy. Apparent absorption of potassium was decreased by 12 and 25% with increased clay supplementation; no changes were noted for calcium, magnesium, and phosphorus; fecal loss of magnesium was significantly higher with 900 g soil fed. Increases of 20 and 19% in apparent digestibility of the ether extract also were noted. This finding is particularly significant in view of the findings of Rindsig and Schultz (1970), who noted almost twice the fat content in cow's milk with 5% bentonite added to a concentrate ration. Their explanation was that bentonite affected digesta transit rate; increased rumen pH; and by virtue of its ion exchange properties, influenced gastrointestinal chemistry. Fecal nitrogen tended to increase and overall balances of phosphorus and magnesium were negative with bentonite addition. The balance for calcium was lowered but was not significant.

The nature of the soils contaminating forages has diverse effects. Mayland et al. (1977) reported that heifers grazing crested wheatgrass growing on calcareous loamy-sand soils of southern Idaho ingested substantial amounts of adventitious soil. Using titania contents of the feces and of the soil, they calculated that during mid-June and early August, 0.72 and 0.99 kg soil/animal/day, respectively, were ingested. These computed intakes of soil were equivalent to 14–20% of dry fecal weight. In support of these relatively high levels of soil intake, Mayland et al. cited a personal communication of a veterinarian who concluded that 5 years of tooth wear in New Mexico cattle was equivalent to 12 years in Idaho cattle.

Torii (1978) summarized the extensive use of zeolites, particularly the species clinoptilolite, used in feeds for swine and poultry in Japan. Zeolites, unlike the planar phyllosilicates, are three-dimensional aluminosilicates that bear alkalis and alkaline earths and possess substantial cation-exchange capacity. Feeding 6% zeolite to swine in a commercial operation reduced gastric ulcers and pneumonia; and the overall death rate fell from 4.0 to 2.6%. In another study, zeolite in the rations of dams reduced the incidence of diarrhea in neonates; and their weight at the end of 5 weeks was 60 to 80% greater than that of controls, presumably a result of zeolite supplementation. In other experiments, zeolite fed to swine at 0.5% level was found to improve feed efficiency by at least 5%.

Availability of elements in earth materials

Smith and Halsted (1970), using rats, investigated the availability of zinc in a clay traditionally eaten by certain Iranian populations in whom zinc deficiency is common. They fed rats a 20% clay diet, pretreated with three-normal hydrochloric acid and therefore devoid of calcium carbonate and native exchangeable cations. Measured by weight gain, this diet performed intermediate to diets supplemented with zinc carbonate at levels of 4.4 and 7.1 ppm zinc—the latter level being considered sufficient for maximal rat growth. The authors concluded that 13% of the zinc in the clay was available—a surprisingly high level, as it must be presumed that the acid pretreatment removed substantial amounts of readily available zinc. An alternate explanation would be that additional zinc was made available as a result of the acid destruction of the structure of the clay materials. In any event, the calculated level of zinc available to the rats is nearly identical to the 14% ^{65}Zn that Healy (1973) found was absorbed by sheep.

The radioisotope studies of Healy (1973) with sheep in which low absorption levels of transition metals were found have been cited. Iodine derived from soil ingestion apparently contributed to thyroid health of ewes on heavily stocked pastures (Healy et al. 1972). Miller et al. (1978) emphasized the importance of soil type when evaluating balances of minerals derived from soils. Although Healy (1973) found that sheep derived both calcium and magnesium from soils, Miller et al. (1977) were unable to confirm a similar relationship to their studies of dry cows and pigs. They nonetheless found a significant linear fecal excretion of potassium with soil treatment. Because comprehensive chemical and mineralogic characterizations were not carried out on any of these soils, Miller et al. (1978) were correct in cautioning against comparisons among studies. From observations of animal health and weight gains reported in these various studies, it appears that ingestion of as much as 10%, but usually about 3%, soil may be beneficial to livestock. Miller et al. (1978) also came to this conclusion in their review of livestock geophagy.

The work of Miller et al. (1978) concerning content of soil found in the different parts of the gastrointestinal tract is important in consideration of availability. These authors fed 0.9 kg soil/day to steers averaging 482 kg in weight. Calculated in relation to the total amount found in the gastrointestinal tract, most of the soil was recovered from the stomach (Table 3.2). As particle size decreased (brown clay), the amount of soil recovered from the stomach was reduced. The partitioning of soil by weight in any one of the four stomach compartments appeared to be in proportion to the volume of a given compartment, although in the case of brown clay a substantially larger amount was found in the omasum. The concentration of earth material in the rumen was also

Table 3.2. The proportion (%, calculated from Miller et al. 1978) and total weight (kg) of earth materials recovered from stomach compartments of steers fed 0.9 kg/day of earth material and a control treatment of hay plus concentrates.

Material or proportion	Stomach[a]	Compartment[b] rumen	omasum	abomasum	Total weight
Sandy loam	88.8	81	12	7	4.85
Red clay	77.4	82	10	8	2.82
Control	94.4	84	14	2	0.34
Brown clay[c]	67.0	56	37	7	0.87
Compartment proportion[d]		88	6	6	

[a]Percentage of soil in stomach as proportion of total soil in gastrointestinal tract.
[b]Weight in each compartment as proportion of total in stomach.
[c]Clay-size fraction from sandy loam.
[d]Volumetric proportions of adult bovine stomach (Warner and Flatt 1965).

consistent with the occurrence of stratification by density layering of the liquid within the rumen (Hungate 1966:173–74). The lowest layer, on the floor of the rumen, was referred to as the high-density layer consisting of "high-density particles," minerals, and protozoa. The fact that substantial earth material collects on the floor of the rumen does not mean that dissolution or exchange phenomena making mineral elements available for absorption in the rumen would not also provide for absorption of these "free" elements lower in the gastrointestinal tract. However, there is reason to believe that substantial mineral absorption occurs in the rumen. For example, Grace (1970) found that 95% of absorption of magnesium in sheep occurred in the stomachs, probably in the abomasum where calcium was also absorbed (Hungate 1966:201). It also was estimated that sodium absorption in the rumen and abomasum in sheep, coupled to absorption of ionized volatile free acids, accounted for a flux of 850 me Na/day, equivalent to about 20 to 25% of the acid flux (Hungate 1966).

Geophagy may substantially change the color of the rumen. Healy and Wilson (1971) found that lambs fed two soils at 14% (later, 11%) of the diet of lucerne and corn developed dark gray or black rumina in contrast to the cream-colored rumina of control animals. The omasal surface and the abomasum were only slightly colored. The black deposit on the rumen papillae dissolved easily in six-normal hydrochloric acid; and analyses of the acid extract and of this tissue revealed that one of the soils produced significantly higher levels of calcium, magnesium, iron, manganese, aluminum, titanium, copper, nickel, and phosphorus. Healy and Wilson speculated that the deposit on the rumen papillae was a phosphate, although they did not rule out the possibility that it could be a pigmented organic compound or that, if it was a mineral deposit, it might adversely affect absorption of nutrients. Nockels et al. (1966), feeding wethers, concluded that several conditions were necessary for darkened ruminal walls to develop. They suggested that several factors probably explained the development of a pigmented rumen: a substrate of keratinized tissue, rapid growth of the stomach tissue together with limited abrasion, a supply of iron (level not defined), and an acid pH. In these circumstances, iron was deposited in a thickened stratum corneum and adhering vesiculated cells of the rumen. Hamada et al. (1970) found that darkening occurred with a purified diet containing 0.03% iron and that the intensity of darkening increased as the ratio of sodium to potassium decreased.

Judging from these studies where soil supplementation was not a part of the experimental design, the soils likely contributed significant amounts of iron—Healy and Wilson (1971) reported from 2 to 5% iron content in the several clay-size fractions that they separated from the soils. Also, the soils evidently were not effective in neutralizing any acidity produced by the pelleted diets and they were ineffective in stimulating motility in the rumen papillae.

Alkali supplements in livestock production

Considerable evidence, some of which has been considered above, indicates that many of the earth materials sought by wild animals are alkaline salts or basic in reaction; hence it is worthwhile to consider effects of feeding alkali supplements to domesticated animals. Livestock on low-roughage diets experience a decline in rumen pH, which is detrimental to the ruminal microflora. In such diets, feeding of alkaline salts has been found to be beneficial. For example, Hinders et al. (1961) found rumen pH to decline from 6.9 to 6.0 in dairy cows on pelleted alfalfa, but pH rose to 6.65 and titratable acidity was lowered in cows fed pellets supplemented with 25 g Na_2HPO_4 (dibasic sodium phosphate). Cows on unsupplemented alfalfa consumed 15.6 kg sodium chloride and 7.6 kg bone meal, whereas the supplemented group took 1.8 kg and 2.8 kg, respectively.

Emery et al. (1964) fed sodium bicarbonate and calcium carbonate to cows fed grain ad libitum with 0.9 kg roughage. Sodium bicarbonate was fed to one test group at a level of 0.45 kg/day and calcium carbonate to another test group at a level of 0.27 kg/day. Milk fat increased to 0.86% (absolute) in cows given sodium bicarbonate, but no effect was found in cows given calcium carbonate. Neither salt affected milk production, nonfat solids, or protein and no differences were found in blood ketones, glucose, pH, sodium, lipids, or lactic acid. Grain intake declined 10 to 20% in the sodium bicarbonate trial. Volatile fatty acids and lactic acid increased ($P \leq .05$) in the rumens of the calcium carbonate group over those levels in the control group.

Bhattacharya and Warner (1968) fed cattle, wethers, and rabbits pelleted diets supplemented with $NaHCO_3$ (sodium bicarbonate), Na_2CO_3 (sodium carbonate), and $Ca(OH)_2$ (calcium hydroxide). Assays of rumen physiology and blood chemistry for cattle and wethers were similar to the results for dairy cows described above. Calcium hydroxide in rabbits' diet (at 2% level) significantly ($P \leq .01$) increased intake of a complete rabbit diet. They grew faster and cecal pH and free fatty acids in plasma increased. In feeding wethers, lime was found to be superior to the sodium salts for feed consumption. The authors speculated that high levels of volatile free acids in the rumen were absorbed more slowly over a longer period, and that a satiety signal related to blood volatile-free acids was reached at a higher level of food intake.

To summarize, results of feeding clay, sand, soils, and carbonate salts are almost uniformly efficacious. The underlying reason, or reasons, for better health and improved weight gains is not clear, although the buffering and alkaline pH-producing effects of these supplements appear to be important.

CHAPTER FOUR

Some pathologies of major mineral imbalance

WILD RUMINANTS in North America resort to licks mainly in spring and early summer. This specific use pattern suggests a cyclical need for some lick-borne factor. Evaluation of ruminant livestock illnesses in relation to lick-use pattern provides insight into nutritional stresses experienced by wild animals and how they cope with them by geophagy. Two conditions in cattle, tetany and diarrhea, are closely associated with the spring flush of grass. Hypocalcemia is associated with the period of late gestation and early lactation. Calculosis also is considered here because of its relationship to alkaline-earth nutriture. Important interspecific differences in susceptibility to these conditions likely exist.

Tetany

Tetany in ruminants is characterized by convulsions and paralysis, which if untreated usually end in death. The condition is most common in cattle, but it also is observed in goats and sheep and often occurs after the first growth of grass in the spring or after a flush of autumn growth. The cause is attributed to lowered serum magnesium levels; and striking recoveries of animals, seemingly on the verge of death, are obtained by intravenous treatment with magnesium salts. In their review, Grunes et al. (1970:335–36) cited incidences in West Virginia, Maryland, Kentucky, Georgia, Texas, Nevada, Idaho, Utah, and California. The disease is known worldwide where conditions are suitable, and much that we know about tetany is traceable to research done in the Netherlands.

In the conterminous United States, tetany has been associated with grazing pastures containing the grass genera *Lolium, Agropyron, Dactylis, Bromus,* and *Hordeum* (Grunes et al. 1970:342). Small grains used as pasture also have been found to result in tetany. The vigorous early growth of grass is associated with high nitrogen and potassium levels. Potassium concentrations, expressed as milliequivalents of potassium per milliequivalent of calcium plus magnesium, have been used as indexes in evaluating the potential of a grass to produce tetany (pp. 345–46). Forages having values greater than two were associated with tetany in about 1% of the herd; and as the ratio increased, incidence also increased.

Because tetany has been produced in cattle by feeding potassium chloride with either citric acid or transaconitic acid (Grunes et al. 1970:339), it was suggested that these acids in some way cause tetany, perhaps by complexing magnesium. Increases in transaconitic acid have been found in nitrate-amended grasses. Histamine, which occurs at variable levels in plants, also has been implicated through interaction with potassium on muscular excitability. In spite of these alternate hypotheses, most animal research has focused on the cation balance factor. It is significant that grass forage from pasture fertilized with chicken manure causes tetany; chicken manure is rich in both potassium and nitrogen and can be spectacular in its fertilizing benefits.

Wilcox and Hoff (1974) set forth a hypothesis to explain the occurrence of tetany on the basis of ammonium levels in soil during the spring. They concluded that two factors prevailed: (1) After fall and winter leaching, soil nitrogen available to vascular plants was largely in the ammonium form; (2) at the same time, lowered soil temperatures adversely affected nitrifying bacteria, thereby preserving the ammonium form. The absorption of ammonium ions by plants is associated with decreased absorption of calcium and magnesium; the amounts of these alkaline earth ions taken up can be particularly low if their absolute levels in exchangeable forms are low in the soil. After absorption by the plant, ammonium ions in the cytoplasm draw upon labile carbohydrate sources for detoxification of amino acids and amides. The result is low levels of readily available energy to ruminants eating spring forages. Forage high in nitrogen yields relatively substantial amounts of ammonium in the rumen and causes pH to rise, which may cause phosphate salts of calcium and magnesium to precipitate. In addition, ammonium also may affect the transport of the cations across the gut wall. An elevated serum ammonium draws on glucose reserves for detoxification, thereby setting the stage for hypoglycemia, the onset of which may have been initiated earlier because of reduced carbohydrate levels in the rumen. Wilcox and Hoff (1974) concluded that the antecedents of tetany in grazing mammals relate to several factors in forage plants: high nitrogen and low carbohydrate levels combined with lowered magnesium and calcium levels as a result of poor absorption of these ions in spring.

Another set of factors in the soil system favors high potassium to low calcium plus magnesium contents in plant tissue. As soil moisture content increases, the activity of monovalent ions relative to polyvalent ions in the outer layers of the soil's colloidal system increases; this phenomenon is called the *dilution effect* (Wiklander 1964). Empirical evidence (Thomas and Hipp 1968) shows that the ratio of potassium to calcium plus magnesium in plant tissue declines with decreasing soil moisture content. There also is evidence (Thomas and Hipp) from experiments that the rate of diffusion of ^{86}Rb decreases with decreasing soil moisture. Content of rubidium in corn plants was found to be proportional to the rate of diffusion. Because rubidium behaves chemically and physically in many ways like potassium, it is felt that potassium also would display similar physicochemical properties. The importance of soil moisture level in the availability of ions to roots is apparent from these studies. The importance of winter precipitation in temperate and boreal regions in raising levels of water stored for subsequent plant production is widely appreciated. This pattern of moisture accumulation, including the incidence of spring precipitation, influences the relative abundance of mineral nutrients absorbed by vascular flora.

Stout et al. (1976) published preliminary results of a continuing study of blood composition of high-producing cows on 15 farms in Pennsylvania that are located in several geologic provinces. The experiment was designed to measure farm, animal-within-farm, stage-of-lactation, seasonal, and yearly effects. When two years of data were analyzed without regard to season because of unavoidable data losses, highly significant ($P \leq .01$) farm effects were found for serum calcium, magnesium, and potassium. Because higher level interaction effects of year, season, and farm for potassium ($P \leq .05$) and of year and farm for calcium, magnesium, and potassium ($P \leq .01$) were found, it is difficult to assess the relative importance of values found between farms (and between bedrock substrate) to the serum analytes.

In a study of mineral nutrient contents of several kinds of forages in five regions in Pennsylvania, Stout et al. (1977) found significant differences for 11 elements. They analyzed a total of 7908 forage samples, one of the most comprehensive surveys reported in the literature. If the National Research Council level of recommended nutrients for dairy cows and Pennsylvania State University standards are used as bases for comparison, none of the forage samples equaled the 0.35% phosphorus recommended by the council or the 0.40% recommended by the university. Calcium found in corn silages was below the 0.47% recommended for dairy cows, and calcium in grass hay was only marginally sufficient in all regions except the Piedmont. Magnesium in corn silage was below recommended levels except in the Piedmont and the glaciated portions of the Allegheny Plateau, and magnesium levels in grass hay was deficient in all regions. Magnesium was also deficient in mixed hays in which grass was dominant except in the Piedmont region. Overall, magnesium tended to be higher in forages from the Plateau and Piedmont regions.

Sodium levels in corn silage and in grass hay were 1/6 to 1/10 as much as the 0.06% level recommended by the Council for beef cattle and far below the level of 0.18% suggested for dairy cattle. Copper in corn silage was about 8 ppm in all regions—somewhat below the Council recommendation of 10 ppm. Zinc except for grass hay from the Allegheny Plateau of western Pennsylvania was consistently below the Council recommendation of 40 ppm.

These results for animal blood chemistry and for forages are important for free ranging native ruminants in that they identify widespread deficiency or marginal levels of minerals in forages over extensive geographic areas. The relationship of adequate (or in some cases luxuriant) potassium nutrition relative to deficient or marginal levels of calcium and magnesium in forages creates conditions for the development of tetany in native ruminants. In addition, low sodium content of forages consumed by livestock (discussed in Chap. 9) could exacerbate the onset of tetany due to high potassium content.

Hypocalcemia

Lowered serum calcium levels—as low as 2 mg/100 ml (usually about 6 mg/100 ml in cows)—are sometimes seen in parturient cows, goats, sheep, and swine (Blood et al. 1979). The condition is known as hypocalcemia, parturient paresis, or milk fever and is characterized by general muscular weakness, circulatory collapse, and/or drowsiness. These conditions are traceable to lowered extracellular calcium that markedly reduces acetylcholine release at the neuromuscular junction during invasion of the action potential (Iggo 1977). Several factors are believed to be important in creating the symptoms associated with hypocalcemia: (1) malabsorption of calcium in the intestine; (2) loss of calcium in colostrum beyond amounts that can be replaced from depots or intestine; and (3) reduced skeletal mobilization (Blood et al. 1979:827). Susceptibility to hypocalcemia varies markedly in a herd and is found in animals in which the serum drop in calcium occurring at parturition is steeper than normal. Conflicting evidence exists for involvement of the parathyroid or adrenal glands. Westerhuis (1974) hypothesized that a low calcium diet immediately prepartum will stimulate compensatory bone resorption; by this physiological strategy, plasma calcium levels are held high and calcium requirements for milk are satisfied. Westerhuis's dietary trials largely substantiated this hypothesis and led to the recommendation that prepartum feed contain at least 0.5% calcium. After calving, 250 g calcium carbonate is given orally and calcium in the diet is increased to 1.0%. Magnesium is provided at 30 g/day before and after calving to maintain plasma levels and prevent tetany. Westerhuis found that on such a calcium regimen, cows not prone to milk fever possessed substantial capability of intestinal absorption. The oral dose of 250 g calcium carbonate at calving created high calcium and phosphorus levels in plasma immediately postpartum, whereas a cow on the same feed without the oral supplement required three days to achieve such plasma levels.

Diarrhea

The loose, semifluid consistency of feces from ungulates using lush spring pasture has been attributed to increased water intake. Rook and Balch (1959) studied cows transferred from a silage and concentrate diet to a freshly cut cocksfoot herbage. Total water intake increased from 69.3 to 76.3 kg/day. Dry matter intake decreased from 14.5 to 12.3 kg/day and fecal dry matter dropped from 11 to 7 kg/day. Sodium intake increased from 14 to 32 g/day and potassium from 226 to 368 g/day with the cocksfoot regimen. Fecal excretion of sodium increased from 5.8 to 9.7 g/day and potassium decreased from 71.2 to 64.0 g/day; these values represent declines of intakes from 41 to 30% and 32 to 17%, respectively. The authors concluded that the fluid consistency was "due to a decreased output of fecal dry matter with virtual absence of mature plant structural components, and results from physical rather than physiological causes."

Some evidence supports the relationship inferred between cell-wall components and fecal consistency. In a study of dry cows, Paquay et al. (1970*a*) found correlation coefficients of .82 ($P \leq .01$) between fecal water and digestible crude fiber and pentosans. Fecal water was also highly correlated ($r = .89$, $P \leq .01$) with dry-matter intake and with fecal potassium ($r = .80$, $P \leq .01$), which showed about the same association as urinary potassium with urinary water ($r = .76$, $P \leq .01$). In a parallel study with lactating cows given a free choice of water, Paquay et al. (1970*b*) found similar correlations (between fecal potassium and fecal water, $r = .49$, $P \leq .01$, and between urinary potassium and urinary water, $r = .74$, $P \leq .01$).

Hebert and Cowan (1971) describe mountain goats coming to Toby Creek lick in southeastern British Columbia as being diarrhetic. They further describe the onset of lick use as contemporaneous with the change from the winter diet of browse to the spring diet of forage and the "change in feces from hard dry pellets to soft amorphous masses or to diarrhoea" (p. 609). Similar morphological changes were noted by Tener (1954) in musk-ox on the Fosheim Peninsula, Ellesmere Island, Northwest Territories, who described the "round, hard caribou-like dung" associated with the "coarse diet" of winter and a softer dung, like that of the domestic cow, which was associated with "green, often succulent, vegetation" (p. 16). Weeks and Kirkpatrick (1976:621) similarly described feces of white-tailed deer in south central Indiana; fresh stools collected in early March before lick-use began were significantly lower in sodium than those collected in April through June (50 ppm versus 124 ppm). Potassium contents followed the same pattern (4973 ppm versus 9616 ppm, $P \leq .01$); and by dividing the April-through-June sample into groups of less than and greater than 50% ash, Weeks and Kirkpatrick (p. 621) found more than twice as much potassium in the high organic matter group (11,521 versus 4284 ppm, $P \leq .01$). However, if the inorganic fraction is considered potassium-free (and therefore a diluent) and the potassium content of the organic fraction is calculated, the levels are nearly equal—13,475 ppm in the less than 50% organic matter group and 14,875 ppm in the greater than 50% organic matter group. A similar calculation for sodium shows an almost fourfold increase, from 137 to 500 ppm with geophagy. In a somewhat parallel study, Hebert and Cowan (1971) held two mountain goats in captivity for 10 days, first providing them with "field-dried forage (about 10% moisture)" and then with a succulent forage. Following the change in diet, the fecal sodium values increased from 138 to 299 ppm (evidently dry weight).

Urinary calculosis

The occurrence of urinary calculosis in livestock—especially males—is related to the gross mineral nutrition of these animals; this relationship in turn offers insight into geophagy of free-ranging, native ruminants. The inorganic calculi are composed of phosphatic or silicious accumulations. Often the stones are spheroidal and are made up of concentric layers of varying thickness and contrasting color (Jensen and Mackey 1979). An organic matrix makes up as much as 5% by weight of the stone and may form in discrete layers. Evidently this material is largely mucopolysaccharide and is similar to colloidal material characteristic of urine produced by healthy, functioning kidneys. Some authors have speculated that this material represents nucleating centers for subsequent precipitation and growth of the inorganic portion of the stone. Calcium, magnesium, and ammonium phosphates are characteristic of one class of calculi; the specific mineral form is dependent on urine pH and the ratio of components. Silicious calculi, which are invariably noncrystalline, evidently form in response to polymerization of silicic acid in urine; and their development would be favored in the acid range.

So-called calculogenic diets—diets high in phosphorus relative to calcium (ratio about one) that can produce phosphatic calculi over very short periods—have been compounded (Jensen and Mackey 1979). Grain sorghum is a major component of one of the most commonly employed diets, all of which are characterized by low calcium and magnesium contents and low ratios of calcium to phosphorus. Silicious uroliths are formed by feeding "prairie grass hay" (Bailey 1977:241) and are seen in animals on poor pastures of semiarid regions during fall and winter.

Urinary calculosis occurs throughout the ungulates and among the carnivora. Recent livestock production techniques involving confined feeding and rations with large amounts of concentrates have resulted in urinary calculosis of epidemic proportions — 70% of a herd might be found to be affected by phosphatic calculi. Beef cattle losses nationwide have been estimated at 0.6% annually (Jensen and Mackey 1979:262), equivalent to 600,000 animals in a herd of 100,000,000. The incidence of this disease and the ease of its development emphasize the importance of a proper balance of calcium and magnesium in the diet of these animals and the "fine tuning" effect of these elements on their metabolism. Observation of pica in confined animals is traceable, at least in part, to malnutrition in respect to these elements. Evidently only Langman (1978) has found kidney stones in wild ruminants—specifically a giraffe population that ate soil around or in termite mounds.

Renal calcification is produced easily in the rat by a magnesium-deficient diet (Seelig and Bunce 1972:73), and magnesium was beneficial in reducing formation of calcium oxalate uroliths in rats made deficient in vitamin B_6. It is significant that even at magnesium levels associated with normal weight gains, microscopic uroliths occurred in as many as 30% of the rats. Seelig and Bunce (p. 75) observed that the human "stone belt" in conterminous United States coincides closely with low-magnesium soils.

The common occurrence of silicious uroliths in free-ranging cattle in the far northern Great Plains is of interest. The condition in its advanced, pernicious state—usually seen in males—is commonly known as water belly and is due to rupture of the bladder or a ureter and collection of urine in the abdominal cavity. Bison flourished on these ranges; it is doubtful if they often were similarly affected, as it would have been the subject of comment in the literature. Were plasma silica levels of bison equal to those of cattle on the same range? If we assume similar or even higher levels, it must be concluded that the physiology and biochemistry of the bison kidney is substantially different from that of cattle and better adapted to the mineral environment of the shortgrass plains.

CHAPTER FIVE

Food habits and nutritional requirements

Selective behavior in feeding

Wild animals exercise substantial selectivity in their choice of foods. Such selectivity is evident even among domesticates on tame pasture. It can be assumed that certain foods are most common in the diet because of their contribution of energy, protein, vitamins, and/or minerals. In most studies the relative importance assigned to a given plant species reflects its weight or volume fraction in the rumen, its occurrence in fecal material, or choices made by observing foraging animals. However, from close physical proximity to or contact with an animal—as in the study by Healy, who tamed deer and walked with them through the successional growth of a cutting in the Allegheny Forest—we appreciate the extraordinary selective processes that are operating. In Healy's words (1971:722): "The deer were always selective when feeding. They sniffed about and ate individual leaves or twigs ... appeared to detect differences among individual leaves on the same plant ... [and] preferences varied for individual plants of most species." Through these observations we gain a perspective of our sampling design for native foods that makes it suspect.

Harper et al. (1967:32) earlier used almost the same description of feeding by free-ranging elk in California: "animals would walk through an area with their noses to the ground as though smelling the vegetation. They stopped occasionally to *feed on a selected plant* [our italics]."

Biologists have observed that deer utilize certain widely disseminated plants when growing on certain soils. For example, Hundley (1959), working in Montgomery and Pulaski counties in the Ridge and Valley Province of western Virginia, attempted to measure nutritional differences among dogwood, black locust, and red maple growing on four different geologic formations. His data for these plants, sampled in alternate months during the year, reflected substantial effect of formation on moisture and protein contents. Hundley also analyzed phosphorus, calcium, cobalt, and manganese; he found levels of manganese and cobalt generally greater than 0.07 ppm (a level considered sufficient for cattle) and some greater than 150 ppm (a level thought toxic). In retrospect, it is regrettable that Hundley's choice of sites did not take advantage of a greater range in rock type—Muskingum soil (one of the two soils he studied) occurred on three of the four rock formations, although the soil chemical data that he gives for the sites are very similar. His analysis of variance of the data does not indicate where among the formations the differences occur.

Fertilizing soils in mule deer range in Lincoln County, south central New Mexico, resulted in increased production of plant biomass and utilization by deer (Anderson et al. 1974). Protein in leaves of wavyleaf oak increased from 19 to 24% at 38 days after fertilization with 101 kg N/ha as urea. Judged by pellet group counts, deer used the fertilized plots in preference to the surrounding vegetation during summer, fall, and winter. Lack of significance in the spring may have been due to a shift to succulent forbs and grasses. Fertilization with ammonium sulfate did not have the effect that urea did on nitrogen level. As measured by leader growth, only mountain mahogany production was significantly increased; wavyleaf oak and fourwing saltbush were not affected, but unusually dry years may have affected the utilization of nitrogen by these species.

Thomas et al. (1964:250) fertilized crested wheatgrass and bluegrass plots in the Black Hills with nitrogen and hydrated lime. As measured by observation and pellet count, deer grazed the fertilized plots seven times more than the control plots and preferred the crested wheatgrass, perhaps because of its greater succulence (75 versus 69% moisture). Nitrogen fertilization increased nitrogen, phosphorus, calcium, and potassium in the grasses. Hydrated lime increased calcium content. The fertilized plots were used one week earlier and more than two times as long as the control plots.

Swift (1948) noted that deer in Centre County, Pennsylvania, were using wheat and clover that grew on small elevations in preference to forage produced on lower sites. Samples taken on May 18 from the selected areas were submitted to proximate analyses and determination of several elements. The data in Table 5.1 are extracted from Swift's tabulation. The most striking aspect of these analyses is the ability of the deer to select forage with a lower ratio of potassium to calcium plus magnesium, thereby reducing the hazard of tetany. Jung et al. (1975) published serum data for the tetany-prone beef cattle in Centre County. The average magnesium content was 1.09 mg/100 ml in the herd suffering tetany, whereas it was 2.29 mg/100 ml in the disease-free cattle. These lines of evidence—the observation of deer exercising considerable choice

Table 5.1. Composition of forage selected by white-tailed deer in Centre County, Pa. (Swift 1948). The ratio K/(Ca + Mg) is calculated on the basis of equivalents.

Crop and use	N[a], %	K, %	Ca, %	Mg, %	Ash, %	$\frac{K}{Ca + Mg}$
Wheat						
grazed	2.69	2.63	0.43	.16	8.8	1.94
not grazed	2.68	3.15	0.29	.14	8.6	3.09
Clover						
grazed	4.87	2.05	1.73	.27	8.9	0.48
not grazed	4.02	2.63	1.36	.37	9.3	0.68

[a]Calculated from crude protein divided by 6.25.

of forage and the confirmation of hypomagnesia in cattle—point to a local stress situation in native herbivores.

Another perspective can be gained by determining the incidence of place names containing the word *lick*. Thirteen contiguous quadrangles (7.5 minute series; 1:24,000 scale) were chosen from Centre County for comparison with the seven contiguous quadrangles (same scale) that cover Adams County to the south in the Piedmont region. No names containing *lick* occur in Adams County but five are found in the Centre County sample. The difference may be explained by magnesium contents of cultivated and native herbaceous growth in the two counties. Corn silage grown in Adams, Montgomery, Bucks, and Pike counties fell in the range of 0.21 to 0.24% magnesium, whereas silage from Centre County fell in the range of 0.15 to 0.18% (Stout et al. 1977). This difference reflects a magnesium deficiency that native ruminants have attempted to alleviate through geophagy. Note that Elk Gap, an appellation suggesting ancient use of the lick by cervids other than deer, provides one access to Lick Hollow.

TASTE. Taste obviously affects food selection, appetite for specific foods, and learning. The sense has two components: one innate and the other learned. Humans detect four primary tastes: sweet, sour, bitter, and salty. An alkaline taste evidently is produced by sensation in several kinds of receptors; this taste has been categorized as a complex sensation and not a primary taste (Goatcher and Church 1970:974). It may be, as Kare and Beauchamp (1977:717) have expressed it, that "each animal lives in a separate taste world" and, instead of the four-way categorization applicable to humans, responses by other mammals are based on much more sophisticated levels of discrimination. For classification the categories pleasant, unpleasant, and indifferent are used.

Sweet and saline solutions have been the objects of most studies and the rat most often has been the test animal. Sodium chloride solutions over a wide range of concentrations are tolerated by some domesticates; there are periodic reports of sodium intoxication in pigs (Kare and Beauchamp 1977:719). Indifference to salt at low concentrations is displayed by the cat and the chicken.

Arnold (1964) reported results for a group of sheep, some of which were sodium stressed by circumventing the salivary ducts by parotid fistula, others by giving access to several forages ranging from 0.02 to 0.85% sodium content. At any one sampling time the fistulated group tended to eat more sodium-rich forage, and this forage made up more of their diet as the effects of the fistula became more pronounced with time. Recently, Arnold and Dudzinski (1978) estimated that sheep cannot detect levels of sodium in herbage below 0.10% and consequently do not select below this level.

By a classic experiment, Richter (1937) demonstrated that adrenalectomized rats possess the ability to drink sufficient 3% saline to compensate for sodium losses. Blair-West et al. (1965) reported that sheep with parotid fistula showed preference to sodium bicarbonate solution. Chickens exhibit similar compensatory behavior. Wood-Gush and Kare (1966) placed a group of growing broilers on a calcium-deficient diet. These birds, subsequently fed diets containing either 1 or 4% calcium as calcium carbonate, ate significantly more of the enriched diets than the control group. In another experiment, birds from the calcium-deprived group made significantly more attempts to peck bizarre objects like buttons—behavior that in essence could be called pica. These examples suggest that compensatory sensory-feeding mechanisms operate among vertebrates in environments with mineral deficiencies.

From paired experiments with jungle fowl and domestic chickens and with Norway and laboratory rats, Kare and Beauchamp (1977:719) made the significant conclusion that the wild animals, unlike the domestic animals, were "more responsive to nutritional and physiological consequences of their diet than to the sensory qualities." This conclusion is especially relevant to the act of geophagy in wild populations and provides a basis for the belief that the act is of fundamental physiological significance to the well-being of the animal.

Evidently the only work on taste in native North American ungulates is that of Rice and Church (1974), who found that male black-tailed deer (in contrast to females) are attracted to a sour taste. They further pointed out that high levels of organic acids of low molecular weight were present in the spring flush of grass. This response also may be associated with the preference of deer for fruits of *Prunus* and *Malus*. Earlier, Crawford and Church (1971) found that black-tailed deer rejected sodium chloride solutions of greater than 0.89% (165 me/l)

Table 5.2. Taste preferences for solutions among black-tailed deer and domesticated ruminants (Crawford and Church 1971).

	Taste group			
	sweet	sour	bitter	salty
Deer	*x*	*x*	*x*	a
Sheep	*x*			
Goats	*x*	*x*	*x*	*x*
Cattle	*x*	*x*		

[a]Sodium acetate is taken.

and that neither sex showed preference below this concentration. However, sodium acetate tastes salty, and deer showed a preference over water for a 0.12 to 4.25% solution (15 to 52 me/l) of this salt. Note also that acetic acid was taken over the range 0.007–0.45% (1–75 me/l), and from these results the authors inferred that the acetate radical was being selected. From other research done by Church's group, taste preferences of species are shown in Table 5.2. The absence of a preference for sodium chloride is of prime interest in this study.

Food habits

It is apparent that geophagy can only be understood in the perspective of nutrition. The major features of the food habits of all ungulate species resorting regularly to licks are summarized here. In most instances reviews are utilized. The annual diet pattern is presented in a seasonal framework wherever possible. The format conforms to the following grouping of months: spring—April, May, June; summer—July, August, September; fall—October, November, December; winter—January, February, March.

MOOSE. Peek (1974) reviewed food habit studies of moose in North America. Data are presented in Table 5.3. The moose is the only persistent browser of the North American ruminants, with forbs, grasses, and sedges relegated to minor importance except during spring and summer. Aquatic macrophytes are taken in early summer, with preference shown for yellow pond lily, pondweeds, and horsetails. However, Peek (p. 209) points out that significant populations occur in the Matanuska Valley of Alaska, the Gallatin Mountains of Montana, and the Cobequid Hills of Nova Scotia where bog habitats do not constitute a significant portion of the landscape.

Among browse species, willows are important in the diets of Alaskan and Shiras moose. Balsam fir, quaking aspen, and paper birch assume importance in the diet of moose in central and eastern Canada. Other important species are red-osier dogwood, pineberry, mountain maple,

Table 5.3. Food habits of moose (from Peek 1974).

Range	Season	Classes of food
Alaska (Kenai Peninsula)	summer	66% birch leaves; 25% forbs; 10% grasses, sedges and aquatics, mushrooms when found
	winter (Feb.–May)	72% birch stems, 26% low-bush cranberry, 6% willows and alder
Montana (Ruby River)	early winter	67% willow
	late winter	59% willow
Montana (Red Rock Lake)	winter	99.8% browse (mostly willows)
	summer	86% browse (mostly willows)
Montana (Gallatin Mountains)	winter	82% browse (25% willows)
Montana (Yellowstone National Park)	summer	88% willow, 9% aquatics, 2% grasses and forbs
Montana (Gravelly Mountains)	summer	71% forbs (mostly sticky geranium), 29% browse, 1% grasses
British Columbia (Wells Gray Park)	winter	75% willow and false box
Alberta (Cypress Hills Park)	winter (Feb.)	56% serviceberry, 21% quaking aspen, 12% *Prunus pennsylvanica*
Newfoundland	winter (heavily used range)	47% balsam fir, 20% white birch, 13% raspberry
	winter (lightly used range)	44% spruce and fir, 22% salices, 11% alder
	winter (cutover range)	29% fire cherry, 25% white birch, 15% balsam fir, 10% quaking aspen
Minnesota (Baxter State Park)	winter (yard)	54% balsam fir, 23% mountain maple (99% browse)
Quebec (Laurentide Park)	late winter (yard)	86% balsam fir, 14% white birch
Ontario (St. Ignace Island)	winter	27% balsam fir, 12% white birch, 9% mountain maple, 9% red-osier dogwood
Michigan (Isle Royale)	winter	80% mix of quaking aspen, white birch, balsam fir, mountain-ash, red-osier dogwood, yew, willows (pattern among years depends on browsing intensity and its effect on production)

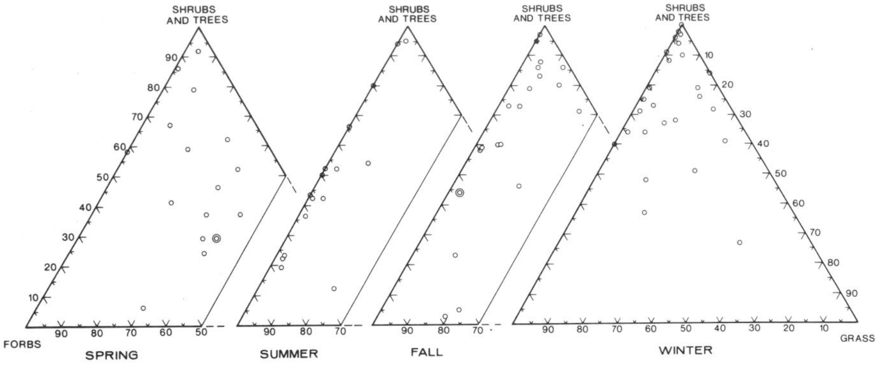

Fig. 5.1. Proportions of shrubs and trees, grass, and forbs in mule-deer diets of the four seasons (from Kufeld et al. 1973). Encircled dots represent two diets of the same proportions.

and beaked hazel. Peek (pp. 209–10) concluded that it was not fruitful to generalize about moose forage requirements other than to note that many preferred species are common in early successional stages; consequently, fire and logging are important agents in the maintenance of prime moose habitat. In winter, choice of browse is limited by snow depth.

MULE DEER. Kufeld et al. (1973) has summarized food habits information reported in 99 studies of the mule deer of the Rocky Mountains. The relative importance of various plant groups by seasons is presented in Fig 5.1. Substantial seasonal shifts occur in the consumption of grasses and forbs—grasses are important in spring when they are succulent and rapidly growing; forbs, for similar reasons, are important during summer and, to a lesser extent, in fall. As might be expected for a species having an extensive range, variation in their food habits is substantial. This variation is apparent in Fig. 5.2, which indicates that the seasonal lines do not depart substantially from those expected (bearing in mind that the sample sizes are small) in normal distributions. The spring and summer lines are the most similar. Mule deer consumed the greatest amount of browse in winter, and its prevalence in the winter diet most nearly followed a normal distribution for the sample. Availability (in view of snow cover) may explain in part this shift in diet.

WHITE-TAILED DEER. Throughout its range the white-tail is more likely to be a grazer of grasses and forbs than a browser of woody foliage and twigs. The studies of Korschgen (1954) in Missouri, Harlow (1961) in Florida, Cushwa et al. (1970) in much of the southeastern United States,

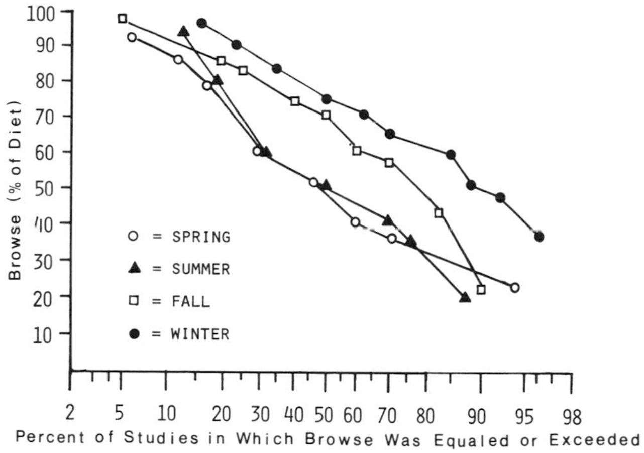

Fig. 5.2. Seasonal incidence of browse among studies of the diet of mule deer, plotted on a probability scale (from Kufeld et al. 1973).

Sotala and Kirkpatrick (1973) in Indiana, and Stiteler and Shaw (1966) in northeastern United States suggest that deer seek out succulent, green foods. For example, greenbrier and Japanese honeysuckle are particularly important in winter over much of the northeast. The results of many of these studies are summarized in Table 5.4. Cushwa et al. sought to evaluate the importance of woody or lignified items in the diets of deer from four physiographic regions in the Southeast. Food that could be classified as woody was less than 0.5% in the fall and winter diet of 416 deer. Twigs that were taken were succulent and, the authors suggested, akin to being herbaceous in texture. Mast and fruit are significant in the diet when available, especially in late summer and fall. The ubiquitous occurrence of Ascomycetes and Basidiomycetes in the diet suggests that these fungi impart important nutritional factors.

Few data are available for the extensive area of mixed forest in northern Michigan, Wisconsin, and Minnesota that is the habitat for a large, important deer herd. Rogers et al. (1981) have reviewed the available data. In spring, the new growth of grasses and forbs comprise 90% of the diet, and deer especially seek openings supporting this vegetation. In summer, leaves of nonevergreen plants, mushrooms, and fruit are important; the leaves of young aspen, especially leaves of suckers, are sought. Like moose, deer will forage on certain aquatic macrophytes and filamentous algae. Sedges, grasses, and evergreen forbs gain in importance in autumn with the decline in availability of nonevergreen plant leaves. Woody browse, cedar, and spruce leaves are taken in winter.

Table 5.4. Food habits of white-tailed deer in its southern and eastern range.

Range	Season	Classes of food, %					Remarks	Source
		browse	forbs	grass	fruit	fungi		
Arizona (south central)	spring	76	10	tr	9		also 5% *Yucca baccata* flower and stalk; fruit includes 33% *Quercus turbinella*	McCulloch 1973
	summer	14	16	tr	70			
	fall (22 Nov.–1 Dec.)	73	20	6	tr			
	fall (10–28 Oct.)	48	35		17		fruit of *Garrya flavescens*	
	fall (28–31 Dec.)	39	28	5	16		fruit (dried pods) of *Acacia gregii*	
	winter (10–16 Jan.)	68	27	5				
	winter (26 Feb.–24 Mar.)	50	40	10				
Texas (Rolling Plains)	spring	70	27				*Phoradendron* sp. important	Quinton and Horejsi 1977
	summer	45	53	2				
	fall	24	73				*Opuntia* sp. main food	
	winter	73	25				*Phoradendron* sp. important	
Texas (east, near Karnack)	spring (May)	28	50	12	7	3		Short 1971
	summer (July)	45	7	7	7	34		
	fall (Sept.)	27	3	3	66	1	fruit includes 64% acorns	
	fall (Nov.)	20		1	78	1	fruit includes 75% acorns	
	winter (Feb.)[a]	32, 54	1, 0	17, 10	22, 32	10, 1	fruit includes 0 and 25% acorns, respectively	
South Dakota and Wyoming (Black Hills)	May–Sept.	56	26	14		4	fungi includes lichens	Hill and Harris 1943
	Oct.–Dec.	61	25	11		3	fungi includes lichens	
	Jan.–Apr.	75	9	13		3	fungi includes lichens	
Missouri	spring	20	28	21	21	0.1		Korschgen 1962
	summer	14	55	0.4	5	0.8	forbs include 16% *Lespedeza stipulacea*	
	fall	15	13	2	65	0.2	fruit includes 49% acorns	
	winter	22	13	6	57	0.5	fruit includes 56% acorns	
Indiana (southwest, Martin County)	spring	35	26	18		3	also 8% unidentified leaves	Sotala and Kirkpatrick 1973
	summer	40	30		5	5	also 9% unidentified leaves	
	fall	21	14	2	65	2	also 2% unidentified leaves; fruit includes 31% acorns, 19% persimmons, and 14% wild crab apples	
	winter	17	34	28	12	2	also 2% unidentified leaves	

Table 5.4 *(continued)*

Range	Season	Classes of food, % browse	forbs	grass	fruit	fungi	Remarks	Source
Ohio	spring	29	16	1	44		fruit includes 40% grain of *Zea mays*	Nixon et al. 1970
	summer	64	6	2	1	7		
	fall	43	8		38	2		
	winter	33	7		51		fruit includes 35% grain of *Zea mays*	
Southern Appalachians	spring	92			8		3% of browse and forbs was stems and twigs[b]	Cushwa et al. 1970
	summer	70			0.5	28	0.7% of browse and forbs was stems and twigs[b]	
	fall	17		3	76	2		
	winter	84			14	1		
Coastal Plain	spring	89			6	1	25% of browse and forbs was stems and twigs[b]	Cushwa et al. 1970
	summer	80			2	12	14% of browse and forbs was stems and twigs[b]	
	fall	15		1	77	2	4% of browse and forbs was stems and twigs[b]	
	winter	96			2	2	6% of browse and forbs was stems and twigs[b]	
Piedmont	spring	100					5% of browse and forbs was stems and twigs[b]	Cushwa et al. 1970
	summer	36			21	37	6% of browse and forbs was stems and twigs[b]	
	fall	59			34	4	15% of browse and forbs was stems and twigs[b]	
	winter	22		75	1	1	0.2% of browse and forbs was stems and twigs[b]	

[a]Data for Feb. 1967 and 1968, respectively.
[b]Stems and twigs were mainly succulent rather than hardened or lignified.

Table 5.5. Food habits of elk (Kufeld 1973).

Season and Range	Classes of food, % browse	forbs	grass	Remarks
Spring				
Montana			87	average of 8 studies
Summer				
Montana	6	64	30	average of 7 studies
Colorado	1	41	58	
Colorado	10	12	78	
Idaho	55	20	25	
Manitoba	52	26	22	
Fall				
Montana			73	average of 9 studies
Colorado			92	
Idaho	40	20	40	
New Mexico	77	2	21	
Manitoba	55	8	37	
Winter				
Montana	9	8	84	average of 6 studies
Montana, Idaho, and Washington	15	2	65	5% moss and lichens
Colorado	57			} average 22% grass, 10% forbs
Idaho	82			
New Mexico	62			
Manitoba	62			
Alberta			97	Jasper National Park

ELK. Kufeld (1973) reviewed food habit studies of elk. These data (see Table 5.5) indicate that elk are primarily grazers in spring and fall. Forbs assume importance during summer and browse can be important in any season. These generalizations suggest a pattern of opportunism whereby elk exploit the nutrient flushes in vegetation as they sequence during the growing season—grasses in spring and fall and forbs as they mature in summer. In winter, browse, as for moose and mule deer, is their primary food.

CARIBOU. In a study of the Arctic herd at Prudhoe Bay, Alaska, White et al. (1975) found that caribou in late June sought out *Dryas* heath communities that were typified by calciphilic vegetation. Near the middle of July, wet meadows dominated by *Dupontia fisheri* with a "rich variety of herbs, salices, and grass-like species" were utilized. The results of analysis of 20 esophageal fistula specimens taken during July yielded 55% Cyperaceae, 29% Salicaceae, 2% Gramineae, 8% herbs, and 2% lichens. The sedge *Eriophorum augustifolium* was the most common food plant.

In the Arctic Archipelago, Parker (1978) found that willow (*Salix arctica*) constituted about 80% of the summer diet of caribou but declined to 5% in the winter when mosses made up about 85% of the total. Sedges and grasses made up only 3% of the summer intake and forbs from 5 to 7%. Parker's description indicates that substantial woody material is taken.

Freddy and Erikson (1975) found that woodland caribou in the Selkirk Mountains of northern Idaho and northeastern Washington eat species of the arboreal lichen *Alectoria* almost exclusively during winter and spring even though green forage was available in June. Earlier observations by Edwards and Ritccy (1960) in Wells Gray Park, British Columbia, indicated that caribou eat chiefly *Alectoria* and grasses with subordinate amounts of mosses, sedges, and herbs during May through August. From October through December, leaves of evergreen shrubs and *Alectoria* are eaten. Unlike Parker's observations of Arctic caribou noted above, no evidence was found that woody plant parts are used by the woodland caribou of British Columbia.

MUSK-OX. Spencer and Lensink (1970) inferred from the distribution of musk-ox in different habitats on Nunivak Island, Alaska, that *Carex* spp. are the main items in the diet and that other foods such as beach ryegrass, crowberry, and Labrador tea are of subordinate importance. Parker (1978) reported that musk-ox in the Arctic Archipelago are largely sedge grazers. Barnett et al. (1977) found that in their classification of terrain on Melville Island, musk-ox showed marked preference for grazing the sedge meadows. The distribution of these meadows in the Arctic largely conforms with the area of marine submergence associated with the disappearance of the last ice sheet. The close association of this ecotype with fine-textured sediments is a good example of geologic influence on the distribution of an herbivore population. Tener (1965) found that soils of summer ranges in the Thelon and Lake Hazen areas of the Northwest Territories were neutral or alkaline whereas soils of the winter range of the Thelon herd were strongly acidic (pH 3.8–4.0). Tener (p. 20) speculated that willow was probably the most important summer food of musk-ox in these areas.

BISON. The bison appears to be a strict grazer. Of course, virtual extinction in the late nineteenth century did not leave the immense herds on the Great Plains for field study in the twentieth century. Meagher (1973) observed the Yellowstone herd over a number of years; her analyses of rumen contents showed that grass and sedges constituted 91% of the diet throughout the year and 99% of the diet during the fall and winter. Browse species ranged from only a trace to 2%. Meagher pointed out that bison are uniquely adapted to foraging in deep snow with their large heads and strong necks that can throw the snow aside and uncover the meadow grasses.

Peden et al. (1974) compared the foraging and efficiency of bison with sheep and cattle on the shortgrass plains. They found that bison have a greater preference for warm season grasses than sheep and cattle; are less selective than cattle; and can better digest low-protein, poor quality forage. Sheep tend to consume fewer grasses than either cattle or bison. The overall interspecific strategy of these ruminants is to select forage most easily digested.

PRONGHORN ANTELOPE. The pronghorn is predominantly a browser (Table 5.6). Perhaps its feeding habits evolved in a way that complements the habits of the American bison, with which it shared much of its range. Their partitioning of the food resources of the plains paralleled the contrasting feeding habits of ungulates on the Serengeti Plains of Africa (Bell 1971); there the diet of the diminutive Thomson's gazelle is composed of almost 40% dicotyledons, many of which are fruit, whereas the topi, wildebeest, and zebra are exclusively graminivorous. Here, these noncompetitive food habits coincide with the onset of the dry season—the larger species graze the tall grasses, thus making the low-growing or prostrate protein-rich forbs available to the gazelle. Bell (pp. 118–19) used the endogenous nitrogen excretion rate equation ($N = K + W^{0.74}$) to explain the apparent inverse selectivity for protein with size of ruminant.

Buechner (1950) described a similar but somewhat more competitive relationship among pronghorn, elk, white-tailed deer, and bison in the Wichita Mountains of southwestern Oklahoma. In late spring and summer, forbs make up 99% of the pronghorn diet in contrast with the bison diet, which consists of 99% grasses. Deer and (to a lesser extent) elk compete with pronghorn, taking 99% and 24% forbs, respectively. The association of pronghorn with bison on the plains may be analogous to that of the Thomson's gazelle with graminivorous species. In the drier habitat of southwestern Utah, browse dominated by black sagebrush assumes importance in the annual diet of antelope (Beale and Smith 1970). However, spring rains bring on flushes of succulent forbs and grasses that are eaten; and during June and July of wet years, forbs with browse make up about 90% of the diet.

The significance of browse and forbs in the pronghorn diet is apparent from observations of mortality in a herd using grassland along the Milk River in northeastern Montana (Martinka 1967). Winter starvation occurred when antelope were confined to grassland in contrast with antelope occupying ranges consisting of sagewort, creeping juniper, and mixed forbs.

MOUNTAIN GOAT. Hibbs (1966) compiled literature on the mountain

Table 5.6. Food habits of pronghorn antelope. All data from rumen-content determinations.

Range	Season	Classes of food, %			Remarks	Reference
		browse	forbs	grass		
Oregon	year (?)	69	21	7		Yoakum 1958[a]
Oregon–Nevada	n.d.	93	6	1	61% sagebrush	Mason 1952
California (northeastern)	spring	64	35	1		Ferrel and Leach 1950
California (northeastern)	summer	47	50	1		Ferrel and Leach 1950
Utah (western)	summer/fall	58	42	0.3	late summer, 59% *Artemisia nova*	Beale and Smith 1970
Idaho	winter	85	10	5		Fichter and Nelson 1962[a]
Montana	year	64	20	16		Cole and Wilkins 1958[a]
Montana	year	63	34	3		Cole 1956
Montana	fall (mostly)	85	7	7	*Artemisia tridentata* most common	Buck 1947[a]
Wyoming	summer/fall	86	10	4		Baker 1953[a]
Colorado	year	—90[b]—		0.1-9	0.1–3% grass in winter, 2–4% in May, Sept.	Douglas 1953[a]
Colorado	year	47	20	7		Tileston and Yeager 1962
Colorado (northern)	fall	73	24	0.8	sampled in Oct., 2.3% cactus	Scarvie and Arney 1957[a]
South Dakota	year	66	23	11	north of Black Hills	Terwilliger 1946[a]
South Dakota	winter	83	—17—		83% sagebrush	Bever 1951[a]
Nebraska	n.d.		82	10	also 1.6% corn and barley	Sather and Schildman 1955[a]
Oklahoma	year		99	1	Wichita Mountains	Buechner 1950

[a]Includes cactus.
[b]Original reference not seen, data from Yoakum 1967.

Table 5.7. Food habits of mountain goat (from Hibbs 1966).

Range	Season	Classes of food
South Dakota (Black Hills)	winter, spring	60% lichens and mosses; 20% bearberry; 10% ponderosa pine; 10% miscellaneous including ferns, grasses, currant, rose, willow, daisy[a]
Montana (Red Butte)	summer	96% shrubs, 3% grasses, 1% forbs (96% gooseberry)
	winter	63% grasses, 35% shrubs
Idaho	winter	96% grasses and mountain mahogany
Western Canada	summer	63% grasses, 23% willow, 14% herbs

[a] *Erigeron.*

goat; data are generalized in Table 5.7. He concluded that the goat is principally a grazer of grasses and forbs; however, several studies indicate that browse also is important when it is available.

BIGHORN SHEEP. Todd (1972) reviewed the food habits of bighorn sheep (Table 5.8). Grasses and sedges are staple foods of the Rocky Mountain and California subspecies, whereas browse assumes dominance in the diet of the desert bighorn. Like the mountain goat, some bighorn populations subordinate their consumption of grasses over other forages because of availability. The growth of forbs is closely tied to seasonal rainfall; they are much sought after and, when preference indexes are available, may be found to rank above grasses and browse (Todd, p. 7). Forb use tends to be highest during spring and early summer. Where they are a component of the flora, cacti are taken. Bighorn sheep commonly paw and dig while feeding—perhaps looking for tender shoots—and may consume much soil in the process. Fecal soil levels may reflect this behavior. In their study of sheep in Death Valley, California, Welles and Welles (1961) concluded that browse and forbs constitute a much greater proportion of diet than grasses; however, annual variation in diet is related closely to variation in rainfall.

Major mineral elements

Licks are common in certain ranges and almost nonexistent in others, and because of the importance of geophagy in spring we can hypothesize in a given case whether some populations are predisposed to geophagy. For this reason, for the geographic breadth of this study, and for a look at the variability of forages (both seasonal and spatial), data for the bases—calcium, magnesium, sodium, potassium—and phosphorus in native forage are summarized. Data for these cations were particularly

Table 5.8. Food habits of bighorn sheep (from Todd 1972).

Range	Season	Classes of food, %			Remarks
		browse	forbs	grass	
Rocky Mountain bighorn sheep					
Wyoming	winter-spring	19	30	48	2% sedges
Colorado	September	19	6	75	
Idaho	winter	9	—86—		5% moss and lichens
	spring	22		77	1% moss and lichens
	summer	14		86	
	fall	25	—66—		9% moss and lichens
Montana	not identified	3	7	90	
California bighorn sheep					
California (Sierra Nevada)	early winter			25–35	forbs and browse about equal
	late winter			40–50	*Eriogonum* greater than browse
British Columbia	fall-winter-spring	24	72	72	
Desert bighorn sheep					
Nevada (Silver Peak Range)	n.d.	8	32	60	
Nevada (Desert Game Range)	winter	17	2	81	
	spring	8	25	67	
	summer	3	3	94	
	fall	44	8	48	

scrutinized to detect seasonal patterns and range differences that would support data derived from analyses of earth materials. Although we did not analyze the lick earths for phosphorus, we have included it in the survey because so much attention has been attached to calcium nutrition vis-à-vis phosphorus levels in the diet. There also is good reason to suspect that phosphorus is the limiting factor, or one of the limiting factors, in phenotypic development and fertility of deer in southeastern United States. Much later discussion centers on white-tailed deer populations east of the 100th meridian; therefore the survey of native forage compositions has been confined largely to this region. With certain notable exceptions it is the best-known area of the continent in respect to deer nutrition, although current knowledge of the nutritional value of native forages is at best sketchy.

MINERAL REQUIREMENTS. Very few studies of mineral nutrition of native ruminants have been made. When Halls (1970) reviewed the subject, he could cite only the work of Magruder et al. (1957) for calcium and phosphorus in white-tailed deer. When analyses have been carried out on native forages, investigators usually compare element levels found with those recommended for domestic sheep and cattle. Even for domesticates, few studies exist for minerals other than calcium and phosphorus. Data for calcium, magnesium, sodium, potassium, and phosphorus recommended by subcommittees of the National Research Council for beef, dairy cattle, and sheep, together with the scant information available for deer, are summarized in Table 5.9. The levels of calcium and phosphorus found sufficient for deer generally are consistent with recommended levels for sheep and cattle, although the 0.56% phosphorus that Magruder et al. suggested would provide optimum antler development in white-tailed deer is substantially higher than phosphorus levels recommended for domesticates, especially levels required by lactating animals. Thus these recommended levels provide valuable bases for comparison with the mineral contents of native forages.

The energy and protein nutritional needs of native ruminants are not discussed here in detail; rather, we assume that soils having a large supply of readily available calcium, magnesium, and potassium will support vegetation correspondingly rich in minerals and protein and containing optimum fiber and carbohydrate. Ecosystems thus endowed have the highest carrying capacity and generally support the largest phenotypes of the species present (see Chap. 8).

CALCIUM. Calcium, like sodium, is under significant homeostatic control. Serum levels of the several circulating fractions are maintained by withdrawal from bone when dietary sources are inadequate. Growth, pregnancy, and lactation exert specific elevated needs for calcium (Table 5.9). Ullrey et al. (1973) concluded that a diet containing 0.40% calcium was needed by recently weaned white-tailed deer. Deer on a diet of 0.18% calcium produced antlers of low specific gravity, and ash content and a basal section of the antler exposed substantial cortical porosity. In contrast, deer on a 0.62% calcium (and 0.26% phosphorus) diet developed significant histopathology in the coccygeal vertebrae. This pathology is characterized by disarray or lack of uniform alignment of cartilage cells in the distal end. Earlier, Magruder et al. (1957:9) recorded best antler growth in pen-reared deer on a ration including 0.64% calcium and 0.56% phosphorus. Other pen-reared deer survived but had stunted growth on diets containing 0.25 to 0.30% calcium (with phosphorus held constant at 0.56%).

In light of these findings on captive deer, calcium levels in native forages become more meaningful. Calcium data from several widely separated studies are presented in Table 5.10. Clearly, the 0.25% calcium levels reported by Magruder et al. (1957) must represent the lower limit for forage in the conterminous states. The 0.4% level recommended by Ullrey et al. (1973) is met or exceeded by many food plants over most of

Table 5.9. Macroelement requirements of beef and dairy cattle, domestic sheep, and white-tailed deer.

Species, sex, and stage of development	Animal weight, kg	Nutrient content of diet, % Ca	Mg	Na	K	P	Source
Beef cattle							National Research Council 1963
wintering pregnant heifers	318	.16			a	.15	
	455	.16				.15	
lactating cow	409–55	.24		c	a	.18	
normal growth, heifers and	182	.29	b	≈.06	a	.21	
steers	364	.18				.15	
	455	.15				.15	
Dairy cattle							National Research Council 1978
lactating cow	≥ 500	.54	.20[e]	.18	.80	.38	
growing heifers and bulls		.40	.16	.10	.80	.26	
Domestic sheep							National Research Council 1975
lactating ewe	60	.50	.06	.04–.10[f]	.50[f]	.36	
lamb, finishing	40	.31		.16		.19	
White-tailed deer							
fawn, weaned		.40	.14–.15[g]			.25–.27	Ullrey et al. 1973
fawn, weaned		.46–.51[h]				.26[h]	Ullrey et al. 1975
male, optimum antler growth		.64				.56	Magruder et al. 1957

[a] "It seems unlikely that a deficiency would occur under most practical conditions" (p. 9).
[b] Requirement undetermined, calf requires 0.6g mg/45.4 kg body weight/day.
[c] Lactating beef cow requires 11 g Na/day.
[d] Yielding 26 to 35 kg milk/day.
[e] Mg is suggested to be 0.25% under conditions conducive to grass tetany.
[f] General macromineral requirements (Table 10 in source)).
[g] Not a variable in experiment; this level required in feed for satisfactory growth.
[h] P level is optimized to Ca at 0.46–0.51%.

Table 5.10. Calcium contents of plants used as foods. Contents are given as the mean and its standard error (geometric deviation, GD, in three cases). Data are mostly for the area east of the 102d meridian. Under plant part: L means leaf; S means stem; I means inflorescence; F means fruit.

Locality	Forage	Plant part	Season	*n*	Mean, %	Standard error, %	Source
Southern Alberta and Saskatchewan	*Agropyron smithii*		Apr.	5	0.46		Clark and Tisdale 1945
			May	9	0.38		
			July	14	0.26		
			Oct.	6	0.38		
	Bouteloua gracilis		Apr.	4	0.41		
			June	8	0.46		
			Aug.	5	0.36		
			Oct.	8	0.35		
	Atriplex Nuttalli		May	6	1.02		
			July	13	1.10		
			Oct.	6	1.46		
	Eurotia lanata		May	7	0.89		
			June	7	0.92		
			July	4	0.84		
			Oct.	5	1.49		
	Artemisia cana		May	3	0.71		
			Oct.	2	0.97		
	Prunus melanocarpa		June	2	1.89		
	Amelanchier alnifolia		July	2	0.99		
Wyoming, Montana, and Dakotas (northern Great Plains)	*Agropyron smithii*	L, S	summer	21	0.23	1.29 (GD)	Severson et al. 1977
	Artemisia cana	L, S, I	summer	19	0.57	1.33 (GD)	
Wyoming and Montana (Powder River Basin)	*Bouteloua gracilis*	L, S,	fall	46	0.25	1.96 (GD)	Erdman and Gough 1975
Wyoming (Big Horn Mountains)	forbs		July	29	1.25	0.20	Beetle 1956
	grasses, sedge		July	3	0.45	0.04	
Wyoming (Laramie Plains)	browse		July	3	0.95	0.18	McCreary 1939
	forbs		June	4	1.52	0.30	
	grasses		summer	5	0.36	0.07	
Wyoming (Ferris Mountains)	browse		June	4	0.72	0.13	McCreary 1939
			Aug.	3	1.31	0.45	
Wyoming (Red Desert)	browse		Sept.	8	1.38	0.31	
	browse	L, L+S	Nov.	25	1.24	0.08	Hamilton 1958
		L, L+S	Apr.	25	1.12	0.08	
	browse	S	Nov.	25	0.73	0.06	
		S	Apr.	25	0.69	0.04	

Table 5.10 *(continued)*

Locality	Forage	Plant part	Season	*n*	Mean, %	Standard error, %	Source
Wyoming (Red Desert) *(cont.)*							
	forbs	L	Nov.	4	1.14	0.16	
		L	Apr.	4	1.07	0.20	
	forbs	S	Nov.	4	0.60	0.03	
		S	Apr.	4	0.56	0.03	
	grasses	L, S	Nov.	2	0.49	0.05	
		L, S	Apr.	2	0.30	0.06	
Wyoming (Laramie Range)	browse	L	Apr.	8	1.03	0.15	Hamilton 1958
		L	Aug.	34	1.41	0.09	
	browse	S	Apr.	8	1.21	0.20	
		S	Aug.	34	1.01	0.07	
Wyoming (Salt Range)	browse	L	Aug.	40	1.47	0.10	Hamilton 1958
		S	Aug.	40	0.90	0.06	
Wyoming	browse[a]	L, S	summer	12	2.30	0.23	Hamilton and Gilbert 1972
	Agropyron spp.	L, S	summer	15	0.33	0.01	
North Dakota	*Agropyron smithii*		summer	1	0.23		Hopper and Nesbitt 1930
North Dakota	*Agropyron smithii*		spring	12	0.53	0.03	S. D. Fairaizl, pers. commun. 1978
			summer	12	0.57	0.05	
			fall	12	0.51	0.02	
			winter	12	0.41	0.04	
South Dakota (Black Hills)	browse (deer)	L	June, July	5	1.00	0.12	Gastler et al. 1951
		S	Jan.	17	0.18	0.02	
Oklahoma (Payne County)	grasses		spring	3	0.30	0.003	Waller et al. 1972
			summer	3	0.31	0.02	
			fall	3	0.29	0.02	
			winter	3	0.28	0.04	
Texas (Edwards Plateau)	*Quercus virginiana*	L	spring	2	0.67	0.06	Fraps and Cory 1940
		L	summer	4	0.84	0.22	
		L	fall	8	0.67	0.27	
		L	winter	3	1.68	0.35	
		F	winter	3	0.14	0.02	
	forbs	L, S	spring	3	1.85	0.92	
	Opuntia spp.	F	spring	1	3.90		
		L, S[b]	summer	1	2.03		
		L, S[b]	fall	3	10.61	0.50	
		L, S[b]	winter	3	8.56	1.19	
	Bromus catharticus	L, S, I	spring	1	0.24		

Table 5.10 *(continued)*

Locality	Forage	Plant part	Season	*n*	Mean, %	Standard error, %	Source
Louisiana (Rapides Parish)	grasses		spring	3	0.36	0.02	Duncan and Epps 1958
			summer	3	0.35	0.09	
			winter	2	0.24	0.04	
	Helianthus augustifolius		June	1	1.64		
			Aug.	1	2.34		
			Mar.	1	1.32		
	Pinus palustris		June	1	0.20		
			Aug.	1	0.24		
			Sept.	1	0.33		
Louisiana (Winn Parish)	browse	L	spring	5	0.71	0.11	Blair and Epps 1969
		L	summer	5	1.00	0.16	
		L	fall	5	1.06	0.16	
		L	winter	2	0.80	0.18	
Minnesota (phosphorus-deficient area)	prairie hay			51	0.44	0.02	Eckles and Gullickson 1932
Missouri	deer foods	L, S	summer	17	1.95	0.17	Torgerson and Pfander 1971
		L, S, F	winter	14	0.82	0.24	
Wisconsin	browse	L, S	summer	3	0.89	0.08	Gerloff et al. 1964
	Aster spp.	L, S	summer	8	0.94	0.14	
	Fragaria virginiana	L, S	summer	2	1.12	0.10	
	grasses	L, S	summer	10	0.26	0.04	
	Liliaceae	L, S	summer	8	1.19	0.19	
West Virginia (Hampshire County)	deer foods		winter	18	0.78	0.12	Towry et al. 1974
New York (Adirondack Region)	deer foods		winter	20	0.96	0.09	Bailey 1967
North Carolina (Edgefield County)	*Lonicera japonica*	L, S	spring	3	0.63	0.16	Thorsland 1966
		L, S	summer	3	0.72	0.14	
		L, S	fall	3	1.83	0.08	
		L, S	winter	3	0.95	0.22	
	Smilax spp.	L, S	spring	3	0.57	0.05	
		L, S	summer	3	0.83	0.04	
		L, S	fall	3	1.14	0.10	
		L, S	winter	3	1.18	0.14	
Florida (Everglades)	deer foods		spring	10	1.50	0.26	Loveless 1959
			summer	9	1.56	0.29	
			fall	10	1.70	0.28	
			winter	10	1.55	0.30	

[a]Includes 6 *Atriplex Nuttalli* samples.
[b]Mostly joints or branchlets.

the range of the white-tailed deer. Notable concentrations of calcium in plants growing on limestone terranes have been reported for Edwards Plateau in Texas and the Everglades, Florida. Data for deer forages in Missouri indicate that deer foods growing on the soils overlying certain Paleozoic rocks of the interior plateaus probably contain similarly enriched levels of calcium.

Forage calcium tends to increase with maturity (see Table 5.10). This seasonal increase accompanies growth and concomitant skeletal development in the ruminants and antler development in cervids, and it provides for the increased needs of lactation in females. Bighorn sheep in Rocky Mountain National Park have been observed eating calcium-rich mud at the edge of Sheep Lake; based on this observation, Packard (1946:14) speculated that these animals suffered a calcium deficiency.

MAGNESIUM. No studies are known that identify a magnesium requirement for white-tailed deer or other free-ranging ungulates. For dairy cattle, the National Research Council (1978) has set 0.16% as the desirable dietary level of magnesium for the dry, pregnant cow; the growing animal; and maintenance of the mature bull. For lactating animals, the suggested minimum level is raised to 0.20%. From the few data available (Table 5.11), it appears that most native forages contain at least 0.15% magnesium. Western wheatgrass and grama grass from the northern Great Plains are notable in respect to their low magnesium content.

SODIUM. Like potassium, few data exist for sodium in native forages. The values found in the literature range almost three orders of magnitude (Table 5.12)—the lowest levels occur in plants of the northern Great Plains and certain moose forage of Isle Royale and the highest levels in deer foods of the Everglades. With respect to season, the data suggest that sodium increases in the winter diet of the white-tailed deer of southwest Indiana and the Everglades. The high levels of sodium in Everglades plants is reflected in half the plant genera reported by Loveless (1959)—in hydrophytes, browse, forbs, and the fern *Osmunda regalis.* The National Research Council (1978) has set the sodium requirement in feed of the lactating cow at 0.18% and of the growing calf and mature bull at 0.10%. Levels in all the native foods except the Everglades samples and aquatics on Isle Royale fall substantially below the cattle recommendations. (The relevance of this comparison is open to argument.) Botkin et al. (1973) concluded that sodium content of moose forage on Isle Royale was low—even to the extent of limiting population size—except for aquatic vegetation, which was sought out in summer. But as these authors observed (p. 2745), "The animals are productive and show no gross symptoms of sodium deficiency."

Table 5.11. Magnesium contents of plants used as foods. Contents are given as the mean and its standard error (geometric deviation, GD, in three cases). Data are mostly for the area east of the 102d meridian. Under plant part: L means leaf; S means stem; I means inflorescence; F means fruit.

Locality	Forage	Plant part	Season	n	Mean, %	Standard error, %	Source
Wyoming, Montana, and Dakota (northern Great Plains)	*Agropyron smithii*	L, S	summer	21	0.07	1.36 (GD)	Severson et al. 1977
	Artemisia cana	L, S, I	summer	19	0.22	1.29 (GD)	
Wyoming and Montana (Powder River Basin)	*Bouteloua gracilis*	L, S	fall	46	0.08	1.63 (GD)	Erdman and Gough 1975
Wyoming	browse[a]	L, S	summer	12	0.68	0.12	Hamilton and Gilbert 1972
	Agropyron spp.	L, S	summer	15	0.39	0.14	
Wyoming (Red Desert)	browse	L, L+S	Nov.	25	0.21	0.01	Hamilton 1958
		L, L+S	Apr.	25	0.47	0.05	
		S	Nov.	25	0.31	0.02	
		S	Apr.	25	0.34	0.02	
	forbs	L	Nov.	4	0.32	0.03	
		L	Apr.	4	0.33	0.05	
		S	Nov.	4	0.30	0.03	
		S	Apr.	4	0.28	0.02	
	grasses	L, S	Nov.	2	0.33	0.01	
		L, S	Apr.	2	0.30	0.02	
Wyoming (Laramie Range)	browse	L	Apr.	8	0.24	0.01	Hamilton 1958
		L	Aug.	34	0.33	0.02	
	browse	S	Apr.	8	0.20	0.02	
		S	Aug.	34	0.17	0.01	
Wyoming (Salt Range)	browse	L	Aug.	40	0.39	0.02	Hamilton 1958
		S	Aug.	40	0.22	0.01	
North Dakota	*Agropyron smithii*		summer	1	0.14		Hopper and Nesbitt 1930
North Dakota	*Agropyron smithii*		spring	12	0.14	0.02	S. D. Fairaizl, pers. commun. 1978
			summer	12	0.14	0.02	
			fall	12	0.10	0.01	
			winter	12	0.08	0.01	
Texas (Edwards Plateau)	*Quercus virginiana*	L	spring	2	0.18	0.01	Fraps and Cory 1940
		L	summer	4	0.18	0.09	
		L	fall	8	0.17	0.06	
		L	winter	3	0.16	0.02	
		F	winter	3	0.06	0.01	
	forbs	L, S	spring	3	0.20	0.04	

Table 5.11 *(continued)*

Locality	Forage	Plant part	Season	*n*	Mean, %	Standard error, %	Source
Texas (Edwards Plateau) *(cont.)*	*Opuntia* spp.	F	spring	1	0.71		
		L, S[b]	summer	1	0.26		
		L, S[b]	fall	3	0.85	0.08	
		L, S[b]	winter	3	0.97	0.10	
	Bromus catharticus	L, S, I	spring	1	0.09		
Louisiana (Rapides Parish)	grasses		spring	4	0.56	0.10	Duncan and Epps 1958
			summer	3	0.52	0.03	
			winter	2	0.41	0.01	
	Helianthus augustifolius		June	1	1.18		
			Aug.	1	1.26		
			Mar.	1	1.28		
	Pinus palustris		June	1	0.33		
			Aug.	1	0.33		
			Oct.	1	0.39		
Minnesota (phosphorus-deficient area)	prairie hay			39	0.23	0.02	Eckles and Gullickson 1932
Wisconsin	browse	L, S	summer	3	0.31	0.02	Gerloff et al. 1964
	Aster spp.	L, S	summer	8	0.38	0.03	
	Fragaria virginiana	L, S	summer	2	0.41	0.09	
	grasses	L, S	summer	10	0.23	0.03	
	Liliaceae	L, S	summer	6	0.23	0.02	
West Virginia (Hampshire County)	deer foods (mostly browse)		winter	18	0.09	0.02	Towry et al. 1974
New York (Adirondack Region)	deer foods		winter	20	0.14	0.01	Bailey 1967
North Carolina (Edgefield County)	*Smilax* spp.	L, S	spring	3	0.20	0.04	Thorsland 1966
			summer	3	0.22	0.01	
			fall	3	0.26	0.05	
			winter	3	0.28	0.01	
	Lonicera japonica	L, S	spring	3	0.33	0.04	
			summer	3	0.29	0.07	
			fall	3	0.53	0.06	
			winter	3	0.42	0.07	
Florida (Everglades)	deer foods		spring	10	0.48	0.16	Loveless 1959
			summer	9	0.45	0.17	
			fall	10	0.60	0.23	
			winter	10	0.60	0.18	

[a]Includes 6 *Atriplex Nuttalli*
[b]Mostly joints or branchlets.

Table 5.12. Sodium contents of plants used as foods. Contents are given as the mean and its standard error (geometric deviation, GD, in three cases). Data are mostly for the area east of the 102d meridian. Under plant part: L means leaf; S means stem; I means inflorescence. Michigan samples evidently were taken in summer and/or fall.

Locality	Forage	Plant part	Season	*n*	Mean, %	Standard error, %	Source
Wyoming, Montana, and Dakotas (northern Great Plains)	*Agropyron smithii*	L, S	summer	21	0.0024	1.79 (GD)	Severson et al. 1977
	Artemisia cana	L, S, I	summer	19	0.0067	2.42 (GD)	
Wyoming and Montana (Powder River Basin)	*Bouteloua gracilis*	L, S	fall	46	0.0063	1.51(GD)	Erdman and Gough 1975
Wyoming	Browse[a]	L, S	summer	12	1.29	0.50	Hamilton and Gilbert 1972
	Agropyron spp.	L, S	summer	15	0.14	0.01	
North Dakota	*Agropyron smithii*		summer	1	0.43		Hopper and Nesbitt 1930
North Dakota	*Agropyron smithii*		spring	16	0.0058	0.0014	S. D. Fairaizl pers. commun. 1978
			summer	16	0.0062	0.0023	
			fall	16	0.0030	0.0002	
			winter	16	0.0050	0.0008	
Michigan (Isle Royale)	*Abies balsamea*	L, S		26	0.00028	0.00006	Botkin et al. 1973
	Acer spicatum	L		22	0.00093	0.00013	
		S		31	0.00087	0.00019	
	Betula alleghaniensis	L		5	0.00195	0.00020	
		S		5	0.0024	0.0009	
	Betula papyrifera	L		31	0.0016	0.0003	
		S		45	0.00107	0.00015	
	Corylus cornuta	L		8	0.00044	0.00022	
		S		7	0.00091	0.00015	
	Populus tremuloides	L		14	0.00070	0.00013	
		S		20	0.00064	0.00008	
	Serbus americana	L		35	0.00075	0.0005	
		S		49	0.00056	0.00007	
	Carex aquatilis			2	0.0380	0.0324	
	Carex rostrata			2	0.0250	0.0145	
	Chara sp.			5	0.1020	0.0320	
	Equisetum fluviatile			1	0.157	. . .	
	Potamogeton gramineus			6	0.620	0.103	
	Potamogeton richardsonii			1	0.723	. . .	
	Utricularia vulgaris			3	0.805	0.045	

Table 5.12 *(continued)*

Locality	Forage	Plant part	Season	n	Mean, %	Standard error, %	Source
Indiana (Martin County)	deer foods		spring	30	0.0048	0.0006	Weeks and Kirkpatrick 1976
			summer	28	0.0049	0.0004	
			fall	37	0.0041	0.0005	
			winter	33	0.0081	0.0005	
New York (Adirondack Region)	deer foods		winter	20	0.055	0.005	Bailey 1967
Florida (Everglades)	deer foods		spring	10	1.02	0.32	Loveless 1959
			summer	10	0.70	0.25	
			fall	10	0.56	0.20	
			winter	10	1.19	0.35	

[a]Includes 6 *Atriplex nutalli* samples.

POTASSIUM. The level of potassium recommended by the National Research Council (1978) for feed for pregnant and lactating cows, growing calves, and mature bulls is 0.80%. We know of no recommendations for captive native ruminants or minimal levels associated with free-ranging ungulates as in the case of calcium for white-tailed deer. Concentration of potassium in forages tends to decrease with maturation, particularly in grasses. This characteristic is especially apparent in the data for deer foods from Missouri and Indiana (Table 5.13). For native food plants that have been studied, only western wheatgrass from the Northern Plains and leaves of live oak from the Edwards Plateau contain less than 0.80% potassium, and forbs and browse in these areas contain substantially greater amounts. Potassium deficiency in the diet of any wild ruminant is very unlikely in the geographic areas considered in our study.

PHOSPHORUS. Although phosphorus probably is not sought in geophageous activity, we wish to consider on a regional basis the levels of phosphorus in native forages and the nutritional well-being of native ruminants. For weanling white-tailed deer, Ullrey et al. (1975) concluded that 0.28% phosphorus was desirable when calcium ranged 0.46–0.51%. Earlier, Magruder et al. (1957) noted stunted survival when phosphorus and calcium were in the 0.25–0.30% range. Judging from phosphorus data reviewed in Table 5.14, a level of 0.25% is not often attained over much of United States east of the 100th meridian. The early flush of grasses in the northern Great Plains, selected forbs, and deer foods of the Everglades are notable for their relatively high concentrations. In contrast, the relatively low concentrations in Louisiana plants support Lay's contention (1957) that phosphorus is significant in the control of deer populations in the uplands of the South. Judging from livestock nutrition standards and the results of the few studies done to date with native ruminants, phosphorus levels in native forage seem low; yet these ranges may support large populations and produce large phenotypes. Cowan and Brink (1949) found 60–100 ppm phosphorus in Spray River (Banff National Park) and Athabasca Falls (Jasper National Park) licks, respectively. This phosphorus was extractable by one-normal hydrochloric acid. The authors concluded from these low levels that phosphorus was not sought after in these licks.

Table 5.13. Potassium contents of plants used as foods. Contents are given as the mean and its standard error (geometric deviation, GD, in two cases). Data are for the area east of the 102d meridian. Under plant part: L means leaf; S means stem; I means inflorescence; and F means fruit.

Locality	Forage	Plant part	Season	*n*	Mean, %	Standard error, %	Source
Wyoming, Montana, and Dakotas (northern Great Plains)	*Agropyron smithii*	L, S	summer	21	0.31	1.46 (GD)	Severson et al. 1977
	Artemisia cana	L, C, I	summer	19	1.1	1.16 (GD)	
North Dakota	*Agropyron smithii*		spring	12	0.93	0.12	S. D. Fairaizl, pers. commun. 1978
			summer	12	0.75	0.09	
			fall	12	0.31	0.03	
			winter	12	0.28	0.04	
Texas (Edwards Plateau)	*Quercus virginiana*	L	spring	2	0.85	0.02	Fraps and Cory 1940
		L	summer	4	0.59	0.05	
		L	fall	8	1.08	0.18	
		L	winter	3	0.72	0.11	
		F	winter	3	0.59	0.07	
	forbs	L, S	spring	3	2.14	0.45	
	Opuntia spp.	F	spring	1	3.01		
		L, S[a]	summer	1	1.46		
		L, S[a]	fall	3	1.88	0.25	
		L, S[a]	winter	3	2.46	0.28	
	Bromus catharticus	L, S, I	spring	1	1.06		
Louisiana (Rapides Parish)	grasses		spring	2	2.40	0.46	Duncan and Epps 1958
			summer	4	1.34	0.16	
			winter	3	0.86	0.06	
	Helianthus augustifolius		June	1	5.00		
			Aug.	1	3.49		
			Mar.	1	2.83		
	Pinus palustris		June	1	0.86		
			Aug.	1	0.70		
			Oct.	1	0.60		
Wisconsin	browse	L, S	summer	3	1.44	0.64	Gerloff et al. 1964
	Aster spp.	L, S	summer	8	2.62	0.74	
	Fragaria virginiana	L, S	summer	2	1.50	0.24	
	grasses	L, S	summer	10	0.89	0.08	
	Liliaceae	L, S	summer	8	3.69	0.46	

Table 5.13 *(continued)*

Locality	Forage	Plant part	Season	n	Mean, %	Standard error, %	Source
Missouri	deer foods		summer	17	2.41	0.36	Torgerson and Pfander 1971
			winter	14	0.79	0.14	
Indiana (Martin County)	deer foods		spring	30	2.02	0.08	Weeks and Kirkpatrick 1976
			summer	28	1.50	0.06	
			fall	37	1.24	0.06	
			winter	33	1.11	0.04	
West Virginia (Hampshire County)	deer foods (mostly browse)		winter	18	0.59	0.09	Towry et al. 1974
New York (Adirondack Region)	browse		winter	20	0.60	0.06	Bailey 1967
North Carolina (Edgefield County)	*Smilax* spp.	L, S	spring	3	2.43	0.21	Thorsland 1966
			summer	3	2.19	0.33	
			fall	3	0.71	0.14	
			winter	3	0.95	0.11	
	Lonicera japonica	L, S	spring	3	2.33	0.10	
			summer	3	2.26	0.20	
			fall	3	1.80	0.07	
			winter	3	2.11	0.36	
Florida (Everglades)	deer foods		spring	10	1.27	0.19	Loveless 1959
			summer	10	1.10	0.22	
			fall	9	1.28	0.33	
			winter	10	1.43	0.34	

[a]Mostly joints and branchlets.

Table 5.14. Phosphorus contents of plants used as foods. Contents are given as the mean and its standard error (geometric deviation, GD, in one case). Data are mostly for the area east of the 102d meridian. Under plant part: L means leaf; S means stem; I means inflorescence; and F means fruit.

Locality	Forage	Plant part	Season	*n*	Mean, %	Standard error, %	Source
Southern Alberta and Saskatchewan	*Agropyron smithii*		Apr.	5	.05		Clark and Tisdale 1945
			May	9	.22		
			July	14	.15		
			Oct.	6	.06		
	Bouteloua gracilis		Apr.	4	.07		
			June	8	.19		
			Aug.	5	.15		
			Oct.	8	.12		
	Atriplex Nuttalli		May	6	.48		
			July	13	.20		
			Aug.	6	.10		
	Eurotia lanata		May	7	.30		
			June	7	.23		
			July	4	.22		
			Oct.	5	.09		
	Artemisia cana		May	3	.45		
			Oct.	2	.22		
	Prunus melanocarpa		June	2	.20		
	Amelanchier alnifolia		July	2	.35		
Wyoming and Montana (Powder River Basin)	*Bouteloua gracilis*		fall	46	.06	1.5 (GD)	Erdman and Gough 1975
Wyoming (Big Horn Mountains)	forbs		July	29	.38	0.02	Beetle 1956
	grasses, sedge		July	3	.33	0.03	
Wyoming (Laramie Plains)	browse		July	3	.23	0.03	McCreary 1939
	forbs		June	4	.35	0.08	
	grasses		summer	5	.14	0.02	
Wyoming (Ferris Mountains)	browse		June	4	.37	0.05	McCreary 1939
			Aug.	3	.23	0.06	
Wyoming (Red Desert)	browse		Sept.	8	.18	0.05	McCreary 1939
	browse	L. L+S	Nov.	25	.44	0.05	Hamilton 1958
			Apr.	25	.21	0.01	
	browse	S	Nov.	25	.20	0.01	
			Apr.	25	.19	0.01	
	forbs	L	Nov.	4	.22	0.02	
			Apr.	4	.19	0.01	

Table 5.14 *(continued)*

Locality	Forage	Plant part	Season	*n*	Mean, %	Standard error, %	Source
Wyoming (Red Desert) *(cont.)*							
	forbs	S	Nov.	4	.18	0.01	
			Apr.	4	.19	0.01	
	grasses	L, S	Nov.	2	.10	0.00	
			Apr.	2	.08	0.01	
Wyoming (Laramie Range)	browse	L	Apr.	8	.17	0.01	Hamilton 1958
			Aug.	34	.25	0.01	
		S	Apr.	8	.16	0.01	
			Aug.	34	.18	0.01	
Wyoming (Salt Range)	browse	L	Aug.	40	.32	0.02	Hamilton 1958
		S	Aug.	40	.20	0.01	
Wyoming	browse[a]	L, S	summer	12	.28	0.01	Hamilton and Gilbert 1972
	Agropyron spp.	L, S	summer	15	.27	0.01	
North Dakota	*Agropyron smithii*		July		.21		Hopper and Nesbitt 1930
North Dakota	*Agropyron smithii*		spring	3	.32	0.02	Whitman et al. 1951
			summer	3	.18	0.03	
	Stipa comata		spring	3	.25	0.01	
			summer	3	.15	0.02	
	Bouteloua gracilis		spring	2	.26	0.04	
			summer	3	.18	0.03	
	Andropogon gerardi		summer	3	.24	0.04	
	Andropogon scoparius		summer	3	.14	0.04	
South Dakota (Black Hills)	browse, deer	L	June/July	5	.34	0.05	
		S	Jan.	17	.18	0.02	
Oklahoma (Payne County)	grasses		spring	3	.086	0.025	Waller et al. 1972
			summer	3	.081	0.005	
			fall	3	.045	0.009	
			winter	3	.052	0.020	
Texas (Edwards Plateau)	*Quercus virginiana*	L	spring	2	.21	0.05	
		L	summer	4	.09	0.004	
		L	fall	8	.10	0.01	
		L	winter	3	.10	0.01	
		F	winter	3	.07	0.01	
	forbs	L, S	spring	3	.20	0.03	

Table 5.14 *(continued)*

Locality	Forage	Plant part	Season	*n*	Mean, %	Standard error, %	Source
Texas (Edwards Plateau) *(cont.)*							
	Opuntia spp.	F	spring	1	.23		
		L, S[b]	summer	1	.10		
		L, S[b]	fall	3	.09	0.02	
		L, S[b]	winter	3	.12	0.03	
	Bromus catharticus	L, S, I	spring	1	.25		
Texas (near Nacogdoches)	browse, deer	L	spring	20	.42	0.04	Short et al. 1975
		L	summer	20	.12	0.01	
		L	fall	20	.12	0.01	
		L	winter	20	.10	0.01	
		S	spring	20	.37	0.04	
		S	summer	20	.10	0.005	
		S	fall	20	.14	0.05	
		S	winter	20	.10	0.005	
Louisiana (Rapides Parish)	grasses		spring	3	.08	0.02	Duncan and Epps 1958
			summer	3	.05	0.01	
			winter	2	.12	0.005	
	Helianthus augustifolium		June	1	.12		
			Aug.	1	.09		
			Mar.	1	.18		
	Pinus palustris		June	1	.10		
			Aug.	1	.09		
			Sept.	1	.08		
Louisiana (Winn Parish)	browse, deer	L	spring	5	.18	0.03	Blair and Epps 1969
		L	summer	5	.16	0.01	
		L	fall	5	.11	0.01	
		L	winter	2	.10	0.005	
Minnesota (phosphorus-deficient area)	prairie hay			51	.10	0.01	Eckles and Gullickson 1932
Missouri	deer foods		summer	17	.24	0.03	Torgerson and Pfander 1971
			winter	14	.16	0.06	
Wisconsin	browse	L, S	summer	3	.17	0.05	Gerloff et al. 1964
	Aster spp.	L, S	summer	8	.20	0.02	
	Fragaria virginiana	L, S	summer	2	.34	0.01	
	grasses	L, S	summer	10	.16	0.02	
	Liliaceae	L, S	summer	8	.28	0.01	

Table 5.14 *(continued)*

Locality	Forage	Plant part	Season	*n*	Mean, %	Standard error, %	Source
West Virginia (Hampshire County)	deer foods (mostly browse)		winter	18	.16	0.005	Towry et al. 1974
New York (Adirondack Region)	deer foods		winter	20	.16	0.01	Bailey 1967
Florida (Everglades)	deer foods		spring	9	.21	0.05	Loveless 1959
			summer	9	.19	0.05	
			fall	9	.19	0.05	
			winter	9	.26	0.05	

[a] Includes 6 *Atriplex Nuttalli* samples.
[b] Mostly joints or branchlets.

CHAPTER SIX

Samples, analytical procedures, and data analyses

Samples and general chemical protocol

Samples from licks were obtained by the generous cooperation of many biologists. A circularization of federal, state, and provincial conservation agencies early in 1975 asked that samples be taken when convenient during routine field operations. Sampling instructions, polyethylene sample bags, and cards for site descriptions were included with the cover letter. The instructions suggested obtaining subsamples—each to a depth of 15 cm (6 in.)—at the rate of 6/46.5 m^2 (500 ft^2) of lick area. These subsamples were composited and placed in the sample bag.

A total of 276 samples of salt lick earths representing different licks were studied; 103 samples represented subsamples taken in the vicinity of 12 licks. The samples that were analyzed for all physical and chemical components were selected to represent as wide a geographical range as possible (Figs. 6.1, 6.2, 6.3, 6.4) as well as a balanced representation with respect to the species of ungulates that had been reported frequenting the respective lick sites. (See App. 1 for description of lick earths and settings.) The hypothesis that sulfur was one of the principal elements ruminants were seeking at licks stimulated this study; accordingly, determinations were made for total sulfur content of all lick samples. Water extractions of the lick earths for the readily soluble portions of the major elements present were made, and another series of extractions using normal, neutral ammonium acetate was carried out. This latter treatment displaced calcium, magnesium, sodium, and potassium from exchange sites on aluminosilicate minerals and organic matter. It was assumed that these extractable minerals were also extractable in the rumens of ungulates.

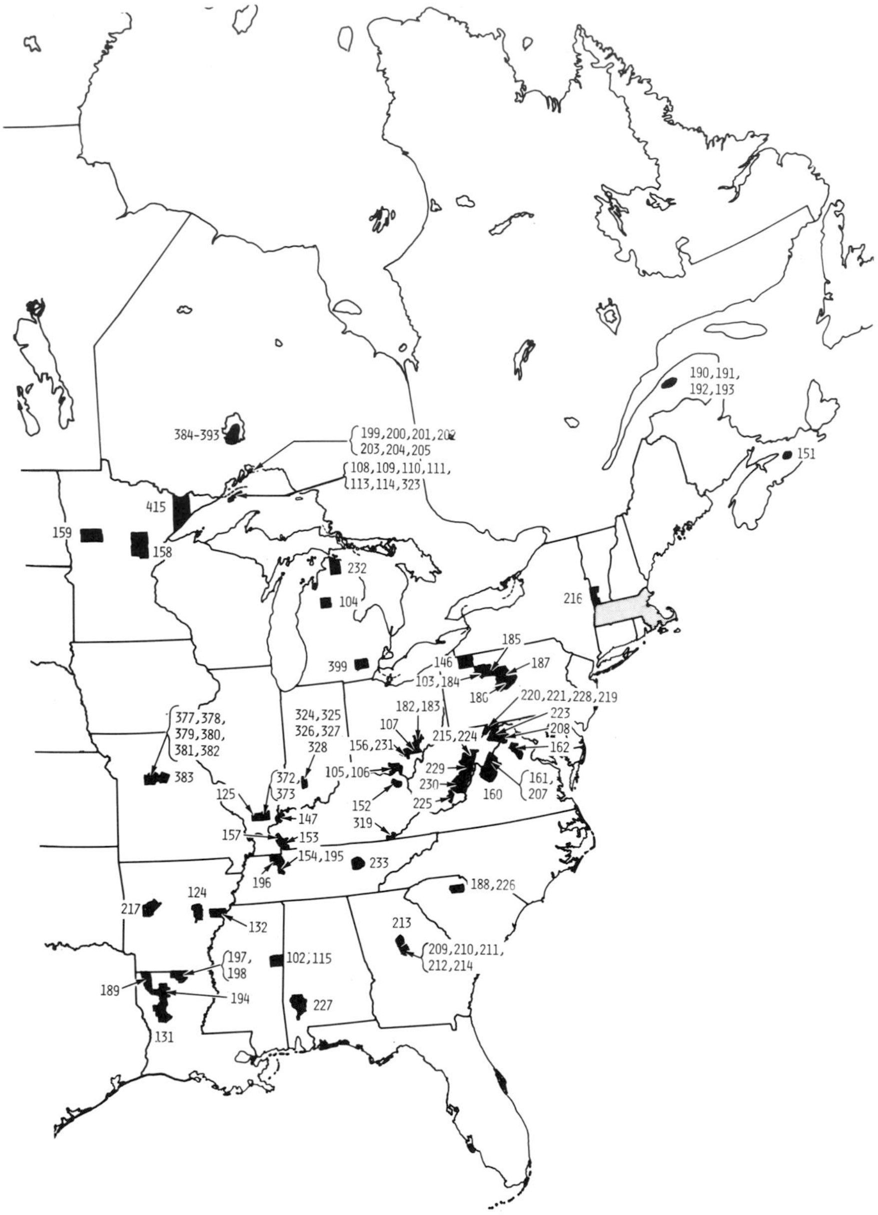

Fig. 6.1. Locations of lick samples from eastern conterminous United States and Canada.

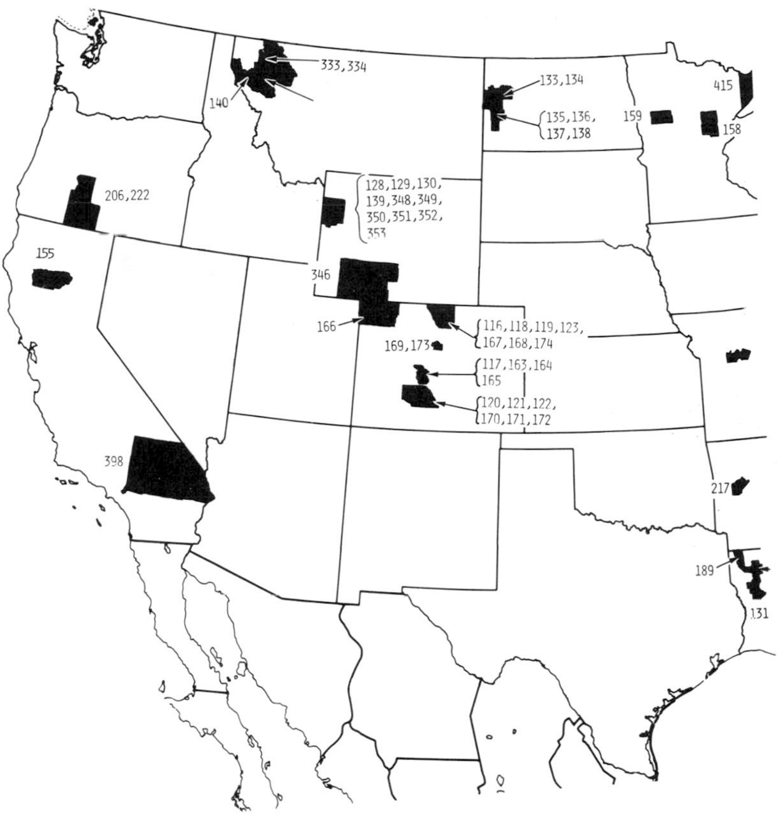

Fig. 6.2. Locations of lick samples from western conterminous United States.

Preparation of lick materials

The samples usually were received in the plastic bags that had been supplied to potential cooperators. Upon receipt, all samples were dried at 65°C and then crushed to pass a 2-mm stainless steel screen. Fragments remaining on the screen (usually roots, bark or twigs, and root fragments) were stored for subsequent identification (see descriptions in App. 1). The material less than 2-mm diameter was then passed through a riffle splitter. From one of the final aliquots, about 20 g were ground to pass a 0.25-mm screen. The grinding was done either mechanically between rotating tungsten carbide faces or manually with an agate mortar and pestle. A Wiley mill was used to cut the fibrous plant material of peats. After drying at 110°C, the finely ground material was stored in a tightly closed, screw-capped vial that had been cleaned with acid.

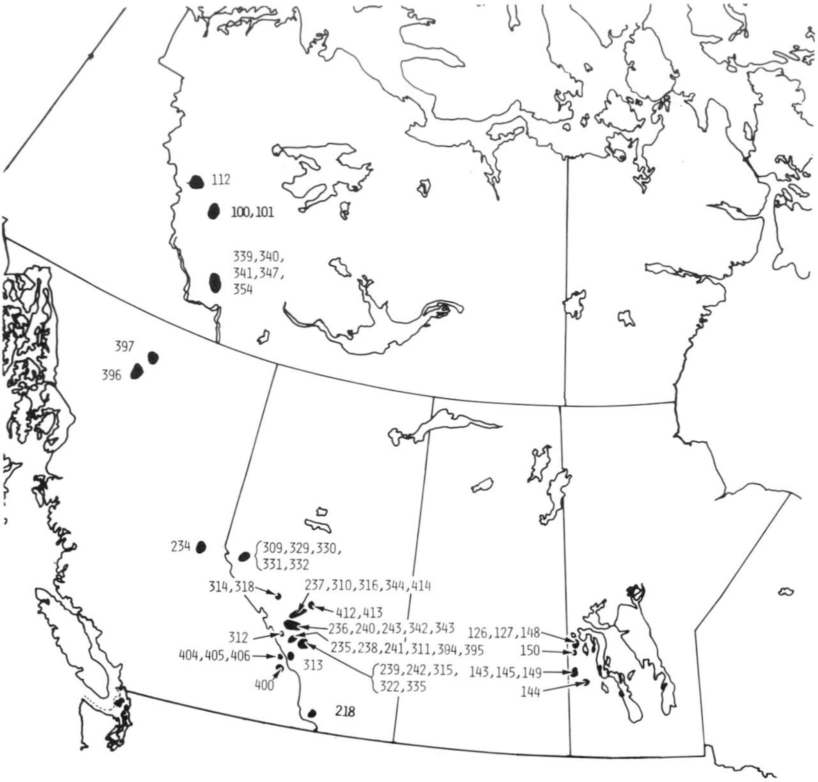

Fig. 6.3. Locations of lick samples from western Canada.

Description of samples

During the final stage of sample splitting, one-half the last split was set aside for description. Description of the physical characteristics of material generally followed the protocol given in the *Soil Survey Manual* (Soil Survey Staff 1951). An aliquot of the sample was wetted with water and the color determined in daylight, using the Munsell system. Texture was determined on the sample by rubbing between the fingers. A number of texture reference specimens were used for comparison. The presence of carbonates was determined by observing effervescence in the presence of six-normal hydrochloric acid. A hand lens was used for the examination of specimens that exhibited very little or no reaction. Odor was noted in some samples and an attempt was made to compare the odors among samples, keeping descriptive adjectives to a minimum. (Odor description is, of course, subjective.)

The fragments that remained on the 2-mm screen were weighed and

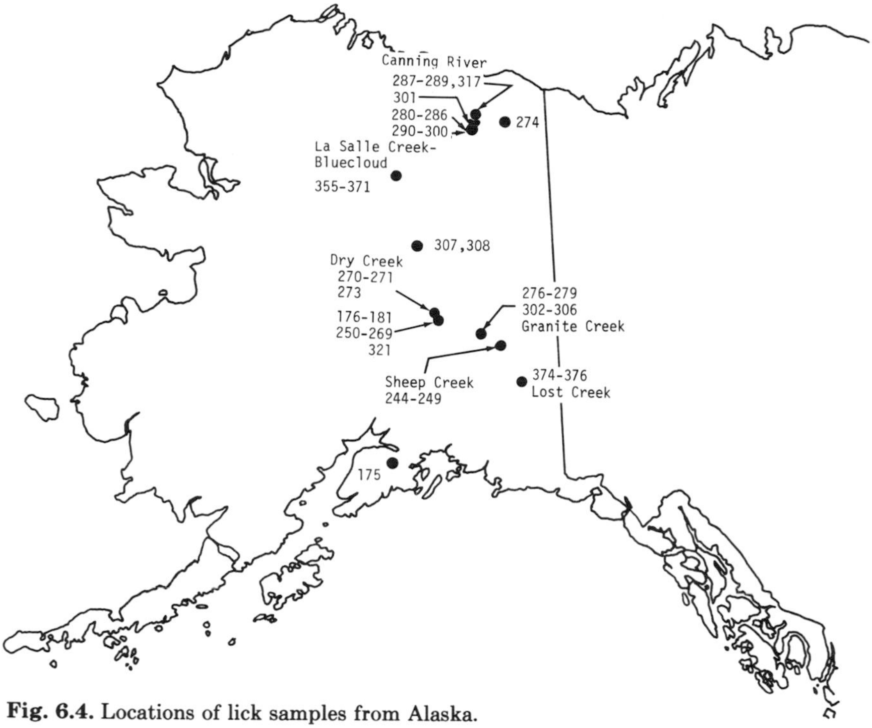

Fig. 6.4. Locations of lick samples from Alaska.

their percentage of the total sample weight determined. It was apparent that some of the fragments were aggregates, which were washed with a stream of tap water on the screen. The remaining residue was dried and weighed. Finally, the indurated fragments were identified by megascopic methods commonly used in rock and mineral identification.

WATER EXTRACTS. The composition of soil water always has been of interest to soil scientists, and one of the greatest challenges to soil science has been the relation of soil-water solutes to crop production. Early in this century, agricultural experiment stations examined methods to extract the water that filled pores, channels, and interstices of soil as it moved to or was intercepted by roots. From the resultant body of research, substantial emphasis was placed on a test called the saturation paste extract. This extract is prepared by slowly adding deionized water to a known volume of soil and carefully mixing until at some moisture content the soil "glistens as it reflects light, flows slightly when the container is tipped, and the paste slides freely and cleanly off the spatula for all soils but those with a high clay content" (Richards 1954:84). The paste is allowed to stand from 4 to 24 hours; then the water is extracted and collected under vacuum. The conductivity, pH, and ionic content of this water are determined, and on another paste aliquot, the water content of the paste before extraction also is determined. These data, particularly conductivity, are of great value in assessing the salinity of soil, and a substantial body of literature has accumulated relating the salt content of the paste extract to crop production. The method is almost universally used in evaluating suitability of soils in arid-region agriculture.

This water extraction method was applied to about half of the salt lick samples. Because NaCl is very soluble in water (about 36 g/100 cm^3), this procedure was deemed sufficient to extract any salt that may have been present in the samples. Ions found in this extract would be readily available to an animal.

The water extract was analyzed for conductivity, pH, and Cl immediately after separation from the soil; the extract remaining was acidified with HCl and stored at 4°C for later analyses. Sulfur was determined by reducing SO_4 in a mixture of HI, H_3PO_2, and HCOOH to H_2S, which was subsequently collected in NaOH (Johnson and Nishita 1952). The H_2S was determined potentiometrically with $Cd(NO_3)_2$ or by titration with $HgCl_2$ in the presence of dithizone (Acker 1956). Chloride was determined by the Mohr titration method. Flame emission photometry was used in the analysis of sodium and potassium; atomic absorption spectrophotometry was used in determination of calcium and magnesium with lanthanum added for suppression of matrix effects.

AMMONIUM ACETATE EXTRACTS. If the results of water extraction of the lick materials demonstrated that sodium chloride was not the factor sought, what was the important element or elements? To us, the course of inquiry pointed to finding a commonality among the samples from licks used by the same species. Because licks occupy rather small portions of the landscape and the range of the animals using them, they must contain an enhanced level (or levels) of the element (or elements) sought. How is this enhanced concentration determined? Perhaps extracting the mineral elements in lick materials with liquors from the stomachs and intestines of the ungulates using the lick and determining the elements extracted would be the most logical methodology (from the standpoint of physiological realities) short of performing a complex factorial experiment with penned animals of each species. Either of these approaches were beyond our resources.

The mineral extraction of soils using neutral or slightly acidified salt has been used by soil scientists for many years. The complement of cations or (less frequently) anions removed has been related to many attributes of the soil and to biomass production. The use of ammonium acetate adjusted to pH 7 has gained substantial current support (Schollenberger and Simon 1945). This method extracts the soluble salt fraction and the cations adsorbed to the colloids present. We chose to analyze extractable levels instead of total content because of the ambiguity that must arise when data for total content are interpreted with respect to a biological system. Several previous studies (Table 7.6) have reported ammonium-acetate extract results for licks. However, it was often difficult to determine the nature of the analytical technique used in some studies.

TOTAL SULFUR. An aliquot of lick material was digested with sodium hypobromite and the hydrogen sulfide generated (as in the case of sulfur in the water extract) was collected in sodium hydroxide. The hydrogen sulfide thus collected was determined either potentiometrically or by titration.

CALCIUM CARBONATE. Carbonate minerals were not specifically identified but were calculated as calcium carbonate from the weight of carbon dioxide collected in soda lime. The gas was generated by treating an aliquot of lick material with 50% phosphoric acid and the gas was scrubbed of water and sulfides. The method and apparatus are described by Jeffery (1975).

IDENTIFICATION OF MINERALS. Ultimately, the explanation of effects of soil ingestion must confront the question of equilibriums among a mixture of crystalline and noncrystalline solids, an organic phase, and solutes in a liquid phase. Even given abiotic circumstances, this situation is very complex, so when the effects of microbes and flux in the gastrointestinal tract are added, the complications of understanding the effects of changing even one variable are considerable. Nevertheless, much progress has been made in our understanding of nutrient requirements of ruminants and of certain interactions among the subsystems.

The fact that wild ruminants will eschew salt blocks to eat the soil surrounding or underneath the blocks suggests a significant interaction between the mineral and the sodium ions in the soil solution. We discount the importance of chloride ions, because (compared with cation exchange capacity) very little anion exchange capacity exists in typical soil or earth materials and chloride is leached rapidly. The most reactive portion of soil and lick earth is the clay-size ($< 2\ \mu m$) fraction; therefore we might anticipate that lick earth is enriched in (1) clay-size material and in (2) some particularly reactive species. Consequently we selected samples with a broad geographic and species representation for identification of clay minerals.

Clay was separated from the earth by siphoning off a suspension or by centrifugation (Jackson 1973). In a few cases organic matter was removed with sodium hypochlorite. Identification was based on X-ray diffraction analyses of two aliquots of clay, one saturated with magnesium and another with potassium. Both aliquots were solvated with glycerol.

Diffraction patterns were obtained with filtered copper radiation over the reflection range of 2–32° two-theta, which includes diagnostic reflections for the common minerals anticipated to be present in the earths. Sensitivity of this procedure varies substantially with the mineral species sought and the instrument configuration used. Based on our experience, the limit of detection is about 5% by weight of constituent except in the case of quartz, for which detection limits are lower.

Three phyllosilicates were estimated semiquantitatively. Abundances of illite, kaolin, and smectite were estimated by multiplying the net intensities of their first order reflections by 2, 3, and 1; summing these data; and proportioning them. Smectite is defined by a reflection at 17.7 Å in the Mg-saturated sample, vermiculite by the increase in intensity of the 10 Å reflection together with decreases in 7 Å and 14 Å lines of K-saturation, and chlorite by a 14 Å line obtained from the K-saturated sample. Illite and kaolin were identified in the Mg-saturated sample. Quartz was identified by its (101) reflection and calcite and dolomite by their (104) reflections.

In several of the samples apparent salt efflorescences occurred. A small amount of these salts was mounted on a microscope slide with Canada balsam, and an attempt was made to identify the mineral(s) with the petrographic microscope.

Data analyses

At the outset of this study we anticipated the problem of ascertaining the importance of specific elements in lick earths in comparison with concentrations in regional soils. For example, few data for total sulfur in soils exist, particularly in the conterminous United States. Ideally, elemental levels of random samples of earth materials adjacent to a lick can be compared with those of the lick samples. Comparisons of this type in the literature are cited later; however, in this study we did not feel we could justifiably impose further on cooperators by asking them to obtain control samples as well. Among other problems in taking control samples is the substantial variability of many properties in native soils, even in soil bodies classified at the most refined levels (Beckett and Webster 1971; Tidball 1976).

In many instances the mean and variances for collections of soils appear to follow a log-normal distribution (Miesch 1967), which best describes the population variability. The Kolmogorov-Smirnov statistic was used to determine the applicable population distribution for the lick earths. This test is most robust when used with sample sizes of 60 or more, a number that was not met in many of our applications. However, in data sets that were near or greater than 60 the Kolmogorov-Smirnov statistic indicated that the distributions of most analytes were log normal; therefore we calculated geometric means and deviations for much smaller sample sizes that were assumed to be taken from a log-normal population. In these cases the geometric mean (GM) and the geometric deviation (GD) are given. About 67% of the samples lie between the product (GM) (GD) and the quotient (GM) $(GD)^{-1}$ and 95% lie between (GM) $(GD)^2$ and (GM) $(GD)^{-2}$.

In several instances we compared lick earths with nonlick earths, and we chose stepwise discriminant analysis to determine the linear combination of variables that best distinguished the sample populations from one another. Licks used by different species were similarly tested. In all instances the statistical programs (SPSS) of Nie et al. (1975) were used.

CHAPTER SEVEN

Ungulate licks in North America

WE HAD the unique advantage of assessing a continentwide collection of licks—earths that represent most of the major climatic, geologic, and ecologic zones and provinces in North America north of Mexico. We had been fortunate in having for study samples of licks used by all species of native North American ungulates save one—the musk-ox. Fortunately, Tener's report (1954) of a musk-ox lick on Ellesmere Island, Northwest Territories, filled this hiatus. It was evident from a review of the literature as well as from our own subsequent findings that only by comparing lick materials from a wide range of soil types and rock substrates would it be possible to arrive at a definitive understanding of the ecologic, physiological, and biochemical significance of salt licks in the lives of game animals. This chapter summarizes analyses of lick earths submitted by many contributors in Canada and the United States.

Lick localities

For the conterminous United States, the county in which a lick sample was taken is indicated in black in Figs. 6.1 and 6.2. The locations of licks in Canada and Alaska represented in the sample series are shown in Figs. 6.3 and 6.4, respectively. The specific location of each lick is noted in the site descriptions for all licks in App. 1. Reference to these maps confirms the broad geographic representation of the samples; represented are samples from the North Slope of the Brooks Range of Alaska southward to California and from Quebec southwestward to Georgia.

Geologic settings

The geologic setting of each lick was determined from literature sources and is described in App. 1. The age of the rocks giving rise to the earths eaten spans the Precambrian to the Holocene. These rocks are mostly sedimentary in origin. In the few cases where licks are situated in ingneous terrane, circumstances favor occurrence of calcium-bearing waters and precipitation of calcium salts. Sheep Creek (samples 244–49, 270) and Granite Creek (samples 276–79, 302–6) are examples where large ingneous rock bodies typify the mountainous range used by the Dall sheep, but their licks are situated in localized outcrops of calcareous materials, some of which may be metamorphosed sedimentary rocks. The lick (sample 398) along Lytle Creek, used by desert bighorn sheep in San Bernardino County, California, also occurs in a granitic terrain; but the lick earths they resort to are calcareous in nature. Evidently this lick is associated with a fault that has controlled groundwaters and from which calcium and magnesium carbonates have precipitated. The moose licks in the area of Lake Nipigon, Ontario (samples 384–93), and to the south at Nipigon Bay (samples 199–205) occur in a region dominated by outcrops of igneous and metamorphic rocks of the Canadian Shield. Here some rocks of the Sibley group are calcareous and it appears that moose have sought out these outcrops. In other cases lineaments in the intensely folded rocks have controlled groundwater, and licks occur in suitable places. To the southwest, on Isle Royale, rocks of age and lithology similar to the Nipigon area occur and the geologic relationships of the licks are similar to those at Nipigon Bay. The licks in Georgia and South Carolina are located in deeply weathered igneous and metamorphic rock areas. Certain licks (samples 119, 163, 164, 165, 168) in Colorado seem to be associated with sites of extinct hot-spring activity located in Tertiary volcanic rocks. Travertine deposits are common under these circumstances. Two licks (169, 173) from Clear Creek County, Colorado, occur in coarse-grained igneous rocks typical of the Front Range. A lick in Fauquier County, Virginia, occurs in an outcrop of Wissahickon schist. The water extract of this sample indicated that its sodium chloride content could be attributed to humans.

The sedimentary rock types found at licks are diverse. In the northeastern portion of the United States where Paleozoic rocks occur, there is no underlying similarity in type (i.e., shale, claystone, limestone, sandstone, or siltstone) nor is there preference in use of a particular formation or rock unit over great geographic distances. However, in several instances the same formation is sought out by deer; in Virginia, for example, licks (207, 208) in Martinsburg shale are about 51 km (32 mi) apart. The licks in the National Bison Range of Montana appear to be associated with rocks of the Belt series, which also are utilized by mountain goats about 117 km (73 mi) to the northeast. In the Brooks Range, Dall sheep are using rocks of the Lisburne group along Canning

River (280–301) and to the east in the upper reaches of Sheenjak River (274).

It is not surprising that licks occur at certain springs. In some cases the spring arises along a fault (e.g., 166, 372). Often when the geology of a lick is sought in the literature, the map suggests that the lick occurs at a contact between formations of differing permeabilities, a circumstance that sometimes leads to development of seeps and springs. Groundwater under these circumstances can carry salt, which probably is the origin of some of the licks in southeastern Ohio, eastern Kentucky, and adjacent West Virginia where the Pottsville formation is known to yield saline waters. More often these waters are of the calcium and magnesium bicarbonate type that precipitate calcium carbonate upon exposure at the ground surface, leaving the magnesium in solution from whence it can exchange onto clay minerals. In Chap. 8 we discuss the importance of magnesium and sodium in deer licks of northeastern United States. High levels of these elements tend to occur together in response to the solubility of their salts.

Texture of lick earths

The distribution of textures among the earth materials of the licks is shown diagrammatically in Fig. 7.1, using the textural classes of the USDA (Soil Survey Staff 1951:210–11). Although the silt-loam class is most abundant among the samples, sand-rich samples appear to be the most common. This character of the earth materials would seem to humans to lower the palatability of the material, not to mention injury from the very abrasive particles. However, Cooley and Burroughs (1962) found that the addition of 2% sand—the kind used in cleaning buildings—to high-concentrate beef cattle rations resulted in increases in gain and feed efficiency of about 5%. Smaller and larger amounts of sand gave lower gains. Their results (published only in abstract form) suggest that coarse material in the lick earths of the present study may not be detrimental but of benefit.

Estimates of material coarser than 2 mm were made. Sixty-three % of 268 lick-earth samples contained 5% or less of the coarse grades, and 78% contained 20% or less. Descriptions, as discernible megascopically, of material of this size are given in App. 1 with each site description. Many kinds of coarse materials are present, ranging from twigs and leaves to gastropod shells and a great variety of rocks. It is probable that animals select against some of these materials when eating the lick earths. Coarse materials that are friable—some shales, limestones, and marls—may be chewed and then swallowed. Bezoars or the so-called

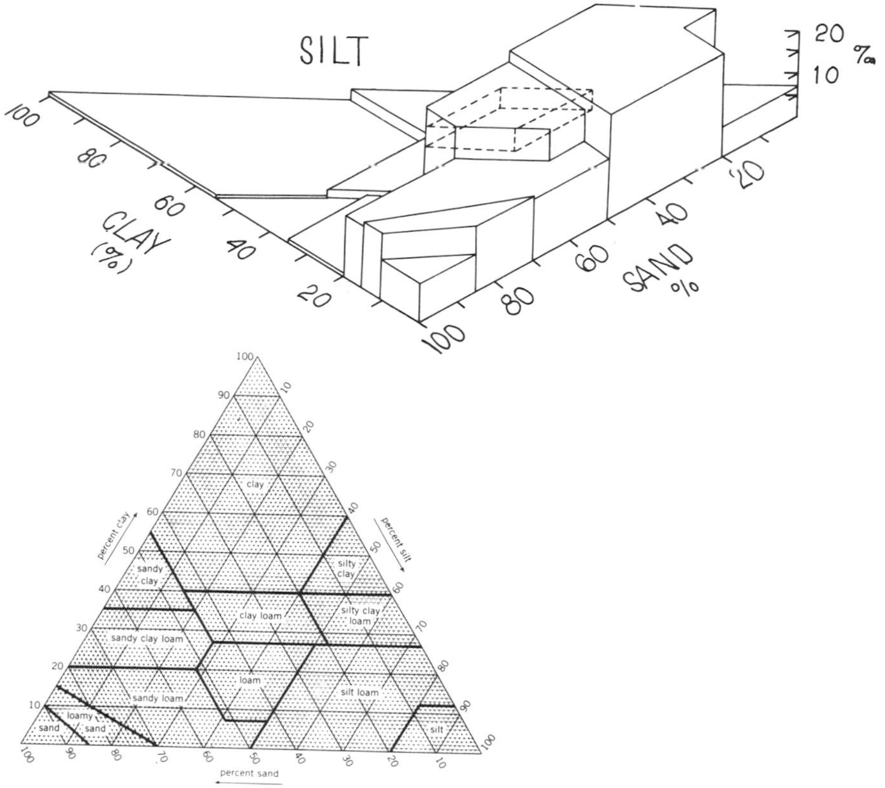

Fig. 7.1. Incidence of textural types among 268 lick earths (isometric drawing). Contents of sand, silt, and clay in each texture category are identified in the regular equilateral triangle.

madstones, found rarely in the rumens of ungulates, typically consist of a smoothed rock around which a mineral matrix from the rumen contents has precipitated. Perhaps these rocks were ingested during geophagy.

Miller et al. (1978) suggest that retention of silicious minerals in the gastrointestinal tracts of steers is related to texture of the earth material ingested, the coarser materials—evidently sand size—being retained (Table 5.1). These investigators found that most particulates were retained in the rumen, with lesser but significant amounts found in the omasum.

Clay mineralogy of lick earths

Samples were chosen to represent as wide a geographic representation as possible. The results of analyses of these samples for clay-size minerals (Table 7.1 and Fig. 7.2) indicate that lick earths tend to be rich in illite (a clay-size mica having more hydroxyl and less potassium in its molecule than the familiar micas seen in hand specimens) and kaolin. In samples occurring along the kaolin-illite line in Fig. 7.2, chlorite and/or vermiculite are common constituents (see Table 7.1). These latter minerals provide sources of magnesium and iron. The common occurrence of quartz suggests that some material slightly coarser than 2 μm was sampled, although quartz is not uncommon in the clay-size range, particularly in sizes near 2 μm; and its unequivocal assignment to the clay fraction would have involved different analyses.

The mineralogy of these samples is consistent with a hypothesis involving the importance of calcium and magnesium nutrition and geophagy. Kaolin displays a particular affinity for calcium (Marshall

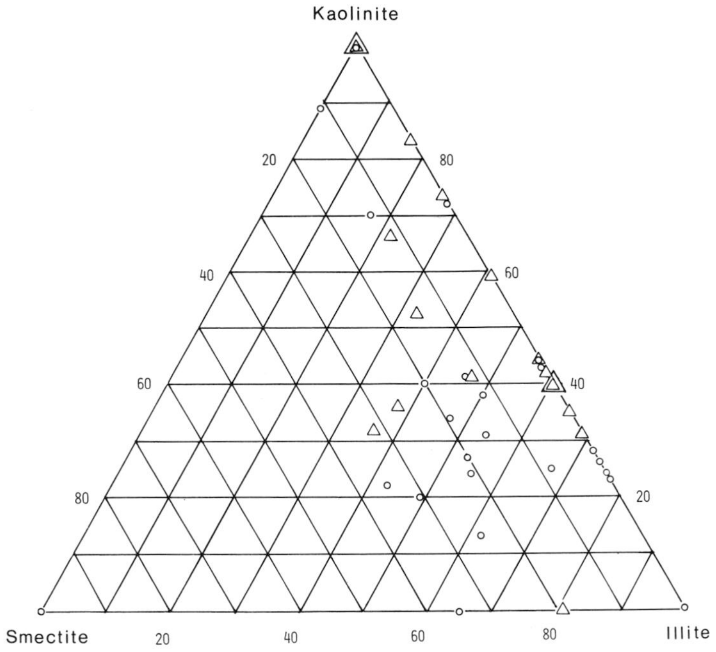

Fig. 7.2. Proportions of clay-mineral species in the less than 2 μm fraction of selected lick earths. Triangles indicate that chlorite and/or vermiculite also was a constituent. Superposed symbols indicate samples with same proportion.

Table 7.1. Mineralogy of the clay-size fraction from salt licks in North America.

State or province	Sample no.	Platelike silicates, %			Accessory minerals present			Remarks
		smectite	illite	kaolinite	calcite	dolomite	quartz	
Alabama	227	35		65				
Alaska	176	100						
Arkansas	132	15	54	31				
Alberta	238		100		✓	✓	✓	
	329		76	24				
	309		56	44			✓	
	413	32	36	32				swelling component is expanding chlorite and/or vermiculite
	218	8	67	25			✓	
	309		57	43			✓	
	312	11	tr	89	✓			
	237	12	50	38			✓	
	242	tr	72	28			✓	
British Columbia	234	12	47	41				also chlorite and/or vermiculite
	396		56	44	✓			also chlorite and/or vermiculite
	405		77	23	✓			
	400		58	42	✓	?		also trace of chlorite and/or vermiculite
Colorado	170	25	62	13	✓			
	165	21	55	24	✓			
Georgia	214		tr	100				also minor chlorite and/or vermiculite
Indiana	327	12	22	66				also chlorite and/or vermiculite
Kentucky	157		28	72			✓	
	147	26	38	36			✓	also chlorite and/or vermiculite
	153			100			✓	trace of chlorite and/or vermiculite
	152		63	37				also chlorite and/or vermiculite
Louisiana	197		27	73			✓	also minor chlorite and/or vermiculite
	194	13	17	70			✓	illite is poorly crystalline

Table 7.1 *(continued)*

State or province	Sample no.	Platelike silicates, %			Accessory minerals present			Remarks
		smectite	illite	kaolinite	calcite	dolomite	quartz	
Michigan	110	20	40	40				
Manitoba	148	31	49	20				
Minnesota	158	19		81			✓	also chlorite and/or vermiculite
Missouri	377	20	53	27			✓	
Mississippi	102	16	21	63			✓	also minor chlorite and/or vermiculite
	115	75		25				
Montana	141		60	40			✓	also vermiculite
	334		41	59	✓		✓	also vermiculite, expanding, in part
North Dakota	135	13	46	41				
Northwest Territories	100	tr	74	26	✓			expanding minerals interlayered
	112		60	40			✓	also vermiculite
Ontario	204	15	33	52			✓	minor chlorite or vermiculite, feldspars
Pennsylvania	185	tr	65	35				also chlorite or vermiculite
Tennessee	154			100			✓	
	233		17	83				also chlorite or vermiculite, illite weathered
Virginia	207		69	31				also chlorite or vermiculite
Wyoming	139	29	47	24	✓			weathered illite
	350	35	43	22			✓	

1964), holding it against leaching. Also, some of the illitic minerals and the vermiculite likely will reduce activity of potassium in the gastrointestinal tract, thereby affecting the interactions of potassium, calcium, and magnesium discussed later.

Evidence of a significant behavioral pattern in searching out a specific geologic formation is found in the clay-mineral analyses. Samples from western Kentucky (153, 157), western Tennessee (154, 195, 196), and Mississippi (102, 115) constitute a sequence of earths from the almost continuous outcrop of a band of Cretaceous rocks stretching from Kentucky southward to Mississippi, eastward through Georgia, and northward into east central North Carolina (see Fig. 2.5 for feature names including the word *lick* in this outcrop). In Mississippi, Alabama, and Georgia the rocks are sandy in character, comprising the Tuscaloosa and overlying Eutaw formations that have weathered to form the Fall Line Hills (Fenneman 1938:67). To the east and south of the Fall Line Hills in Mississippi and Alabama is the prairie known as Black Belt, which is formed on Selma chalk and perhaps represents the most continuous area of fertile upland soils in southeastern United States. No doubt deer have sought out these base-rich spots in a region of soils that is otherwise impoverished of bases.

Odor of lick earths

During the determination of physical characteristics of the samples the odor of certain samples, particularly when wet, was apparent. These odors are noted in the sample descriptions and are summarized into categories in Table 7.2. The earthy and argillaceous odors are explained by the presence of clay minerals that often possess this characteristic odor. Bear and Thomas (1964) identified a complex oil, steam-distilled from ground rock and clays, that bore the earthy odor; in fact, the authors reported a small perfume industry based on it in the state of Uttar Pradesh, India. Bear and Thomas speculated that the oil was formed by adsorption of unspecified organic compounds from the atmosphere and that subsequent changes in these compounds were brought about by the catalytic properties of the mineral surfaces. Sheep, particularly, use licks with earth-type odors.

The aroma of pineapple was associated with three licks spanning the distance from the Brooks Range in northern Alaska to San Bernardino County in southern California. In all likelihood, this odor is an excretion product of the sheep. An aroma described as sweet occurred in a Manitoba sample.

Sample 349 from Teton County, Wyoming, was in several respects the most interesting of all. Its bulk density is incredibly low, and

Table 7.2. Odors of earth materials from licks, animals using licks, and selected characteristics of the material. Full sample descriptions are in App. 1.

Odor	Sample no.	Species using lick	Remarks
Earthy			
earthy	169	elk, sheep	fine sand
earthy	275	sheep (?)	silt
earthy	286	sheep	clay loam, macerated rock
earthy	355	sheep	friable acicular laths
earthy	357	sheep	friable acicular laths
earthy	388	moose	silty clay loam, slippery
"newly plowed"	399	cattle	loam, hydrophobic
argillaceous	167	sheep, elk, deer	clay loam
argillaceous	176	sheep	clay (bentonite)
argillaceous	177	(sheep)	loam, a control sample for 176
argillaceous	178	sheep	loam
argillaceous	179	(sheep)	loam, a control sample for 178
argillaceous	304	sheep	sandy loam
argillaceous	400	sheep	silt
argillaceous	404	elk	fine sandy loam, below clay bank
argillaceous	406	goat	silt, weak odor
chalky	272	sheep	silt, weakly calcareous
musty	185	deer	loam, hydrophobic, weak odor
Fruity and sweet			
pineapple	281	sheep	sandy loam
pineapple	308	sheep	loam, hydrophobic
pineapple	398	sheep	clay loam
sweet	145	deer, elk	loam
molasses	349	elk, cooperator's horse	loam (?), strongly aggregated, hydrophobic
Algal	311	elk	peat, could be elk wallow
	378	deer	silt loam, buffalo wallow
	379	deer	silt loam
	382	deer	sandy loam
	383	deer	silt

Table 7.2 *(continued)*

Odor	Sample no.	Species using lick	Remarks
Rubbed tobacco	224	deer	abundant organic matter
	229	deer	abundant organic matter
	384	moose	abundant organic matter
	386	moose	silt, somewhat hydrophobic
	387	moose	silt loam
	389	(moose)	silt loam, faint odor, a control sample for 384
	390	(moose)	silty clay loam, slippery, weak odor, a control sample for 385
	391	(moose)	silt loam, somewhat hydrophobic, weak odor, a control sample for 386
	392	(moose)	loam, a control sample for 387
	393	(moose)	silt loam, a control sample for 388
Peatlike	235	elk	silt loam, weak odor
	412	moose, mule deer	silt, moose wallow below lick, weak odor
	413	moose, mule deer	silt loam, weak odor
Soybean oil	108	moose	loam, hydrophobic
Menthol	144	elk, deer	abundant organic matter
Dry dog food	150	moose, deer	peat
Sour	267	sheep	silty clay loam
Sour	279	sheep	sandy loam, hole in clay bank, faint odor
Carrots	356	sheep	loam, strongly calcareous
Amine	395	sheep, elk	loam, strongly calcareous, odor of decaying fish
Amine	396	sheep	silt loam, strongly calcareous, odor of decaying fish

combined with its hydrophobic character, the stable, angular aggregates float on water. Eventually the water is colored amber and later a dark brown. At this writing a sample of the aggregates remains unwetted and floating after 26 months in water. A distinct aroma of molasses rises from the sample. It is little wonder that when R. H. Johnson (who supplied the lick sample) visited the meadow, his horse ate this material.

We cannot help but reflect on the name of this locality—Upper Beauty Park—and speculate about the oasis it must represent to the ungulates that visit it and the enchantment that it and the surrounding Jackson Hole held for native Americans, early trappers, and explorers. The material must contain caramelized sugars. We have no idea how the material formed. Because of its unique character, the deposit and the material deserve study and perhaps protection from alteration and destruction.

The odor described as algal is the sharp odor associated with growths of freshwater algae. It is similar to the odor of fresh fish. Four of five samples displaying this odor were taken from north central Missouri; these moist sites may have had algal flora. The Manitoba site (sample 311) is possibly an elk wallow. Elk habitually roll on the spot on which they urinate. Perhaps the odor in sample 311 is an excretion product or a modification of one. W. R. Porath noted that one of the licks from Missouri (sample 378) was once a buffalo wallow.

The rubbed tobacco odors are undoubtedly related to vegetation at some stage of decomposition. These samples also are localized. One set comes from control samples and five associated moose licks at Lake Nipigon, Ontario. The other two are from adjacent counties in the Ridge and Valley Province of West Virginia.

Other odors include the oleaginous aromas of carrots, menthol, and soybean oil. We have no speculations about them. The sour and amine odors may be artifacts of use by the sheep, although sample 279 came from a hole in a clay bank that was accessible to only the animal's head. These sour-smelling samples are from Dry Creek Lick and Granite Creek Lick, which are about 100 km apart in the Alaska Range. The aroma of dry dog food may be derived from decomposition of the peat.

The part that odor plays in initiating a lick is unknown. Note the geophagous activity of deer (R. Kea, pers. commun.) in Kalkaska County, Michigan, where a burial pit for animals and materials contaminated with polybrominated biphenyls was carefully filled and then sealed with a layer of bentonite—perhaps from the Wyoming locality; then topsoil was spread over the clay and grass was planted on the site. Recently deer have been digging holes and eating the topsoil and exposed clay cap. How did the deer find this clay? Were they searching out the argillaceous odor? Soils in this area are coarse textured and deer would not normally come into contact with clay, which implies an instinctive behavior toward finding and eating the bentonite.

Several samples that contained fecal material were received from cooperators. Certainly urine, feces, and scent left at a lick would serve as odor attractants to other animals in the vicinity, thereby increasing lick use.

Chemical composition of lick earths

OVERVIEW OF AMMONIUM ACETATE EXTRACTS. We are not aware of any compilations of large data sets with which to compare the means for calcium, magnesium, sodium, and potassium extracted. However, in Chap. 8 we present the calculated mean contents of extractable bases reported for soils from a number of states (Tables 8.7, 8.10, 8.12, 8.18). The means for calcium, for magnesium, and for potassium in licks compare favorably with those from soils derived from calcareous rocks in West Virginia, the soils on the late Pleistocene lake plain of northwestern Ohio, the soils of northeastern Illinois (strongly influenced by medium- and fine-textured Wisconsinan tills), the soils of northwestern Illinois (developed in late-Wisconsinan loess), and the soils of the Dakotas (developed under moisture conditions favoring accumulation of salts in or near the surface). All these areas produce large deer (see Chap. 8). In northwestern Ohio, northern Illinois, and the Dakotas, deer resort to few licks; evidently the high soil fertility of these areas obviates their need for seeking out highly mineralized sites.

OVERVIEW OF WATER EXTRACTS. Conductivity of water extracts provides a ready evaluation of the easily soluble minerals present. The geometric mean conductivity amounted to 1.82 mmho/cm and a median of 1.56 mmho/cm (Fig. 7.3). For comparison, data for Iowa soils were summarized; they represented levels expected in a humid temperature area of high soil productivity. The median concentration for the Iowa samples of A, B, and C horizons was about 0.4 mmho/cm. In contrast, surface soils of the Black Glaciated Plains of the Dakotas (Austin 1965), an area of Chernozem soils (more recently classified as Borolls and less extensive Udolls), have a median conductivity of almost 0.9 mmho/cm.

The plots of these sample distributions on probability paper suggest that the Iowa sample is nearly normal, whereas the Plains sample and particularly the lick earths represent more complicated assemblages. The lick earths are also more saline than soils from these two regions.

Sodium is the most abundant of metals analyzed, but even this element was present in the lick samples in low amounts, the geometric

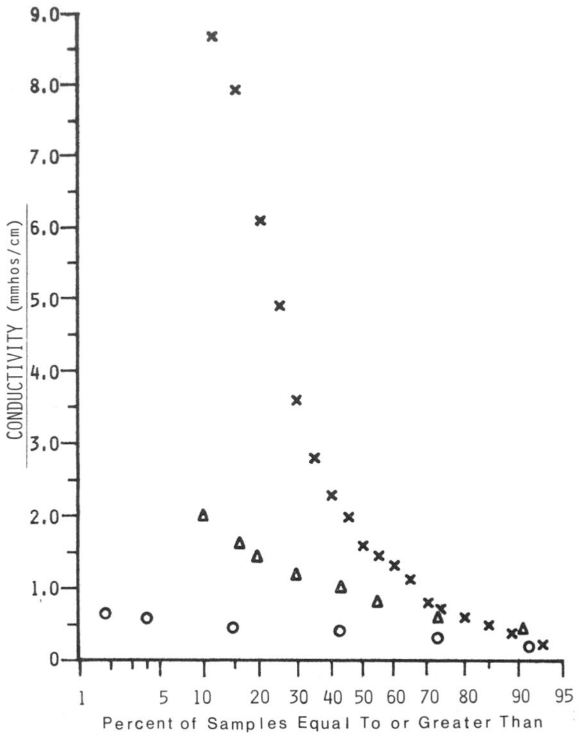

Fig. 7.3. Distribution, plotted on a probability scale, of conductivities of saturation paste extracts of 136 lick earths (crosses), 31 surface soils in the Black Glaciated Plains of the Dakotas (triangles), and samples of 19A, 19B, and 18C horizons of Iowa soils (circles).

mean equaling 0.58 me/100 g soil or 13.3 mg/100 g soil. Sulfate appears to be about three times more common than chloride, but all of the analytes varied substantially. Water in contact with the earths tended to be nearly neutral.

ORGANIC CARBON CONTENT. To compare lick materials with natural darkened surface horizons, determinations of organic carbon were made of selected samples. Dark samples (exclusive of the peat samples) with values of 3 and 4 and chromas of 2 and 3 in the Munsell system were chosen at random from a list of licks. Among these selected samples, organic matter contents range from 0.40 to 31.9% (Table 7.3). This range might be expected in such a diverse group of samples from widely different soil-development environments. These results also confirm that in many instances animals exploit the organic-enriched surface horizon of soils. Ruminants apparently can ingest large quantities of soil organic materials—humic and fulvic substances and other complex polymers of natural origin—without apparent physiologic stress or subsequent pathology.

Table 7.3. Organic matter contents of dark-colored lick earths. Samples are divided into mineral and organic earths based on a content of 20% organic matter. Deer means white-tailed unless otherwise noted.

Sample type and state or province	Sample no.	Munsell color in moist condition	Texture	Organic matter	Species using lick
Mineral earths					
Pennsylvania	187	dark grayish brown	silt loam	5.80	deer
Michigan	104	brown to dark brown	sand	1.17	moose
	110	very dark grayish brown	loamy sand	2.25	moose
Ontario	203	brown to dark brown	loam	4.21	moose
Tennessee	196	dark brown	silty clay	0.40	deer
Missouri	377	very dark grayish brown	silt loam	3.41	deer, elk[a], bison[a]
	380	very dark grayish brown	silt loam	2.44	deer
	382	very dark grayish brown	sandy loam	1.08	deer
Mississippi	115	brown to dark brown	loam	1.44	deer
Manitoba	149	very dark grayish brown	silt	10.78	elk, moose
Colorado	116	dark brown	loamy sand	1.17	deer, elk, sheep
Organic earths					
West Virginia	230	very dark grayish brown	mucky loam	26.0	deer
Minnesota	415	black	peat	77.5	moose
Ontario	201	grayish black	peat	49.7	moose
North Dakota	134	very dark grayish brown	mucky sandy loam	31.9	deer[b]
Manitoba	150	very dark brown	peat	58.8	moose, elk
Wyoming	349	brown to dark brown	mucky loam	25.0	elk
Alberta	310	very dark grayish brown	peat	84.4	moose, deer
	311	very dark grayish brown	mucky silt loam	20.0	elk
	315	very dark grayish brown	peat	62.9	moose, deer[c], horse
	335	dark reddish brown[d]	peat	75.4	moose, deer[c], horse
	322	dark reddish brown	peat	78.4	moose, elk, deer[b]
Northwest Territories	339	dark reddish brown	peat	86.5	moose
	340	dark brown	peat	81.3	moose
Alaska	175	very dark grayish brown	peat	75.7	moose

[a]Used by elk and bison in early historic times.
[b]Mule deer.
[c]Mule and white-tailed deer.
[d]Color in dry condition.

Ten of the lick samples, classified as peats, were ashed at 550°C to determine organic matter content. This procedure is not specific because various oxidation, dehydration, and dehydroxylation reactions of inorganic substances obscure the true gravimetric loss of organic matter through oxidation. Also, some of the organic matter may not be pyrolyzed at 550°C. In spite of these shortcomings, pyrolysis does give an idea of the weight proportion of organic material (Table 7.3). The average organic matter content for these 10 peat samples was 73.1% (coef. var., 16.4%) and range was 49.7–86.5%. All the materials were used by moose except for the lowest sample, which was from an elk lick in Alberta.

SULFUR AND CALCIUM CARBONATE CONTENTS. The distribution of total sulfur content of 262 lick samples is depicted on the probability scale of Fig. 7.4. At least two populations are indicated. There are 85 samples greater than 1000 ppm sulfur, 36 of which are from Alaska—most are from the lick complex at Dry Creek in the Alaska Range and along Canning River in the Philip Smith Mountains. The geometric mean sulfur content (i.e., total sulfur) for the 262 samples is 630 ppm, which is about equivalent to the 700 ppm given by Bowen (1966:200) for soils. Because the mean concentration is near the few average concentrations available for comparison and because of the large number of lick earths that are low in sulfur, we now reject the hypothesis

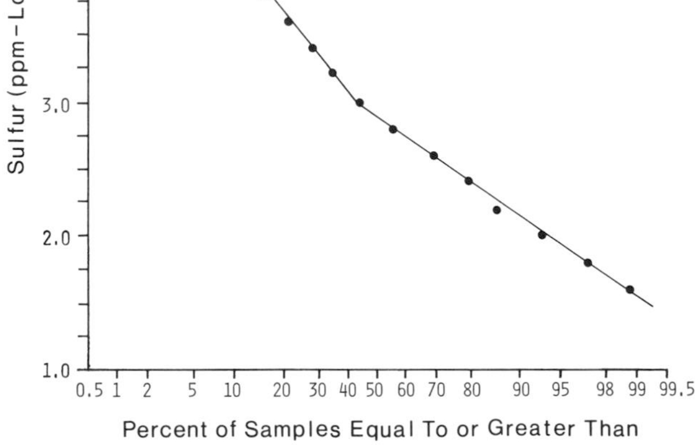

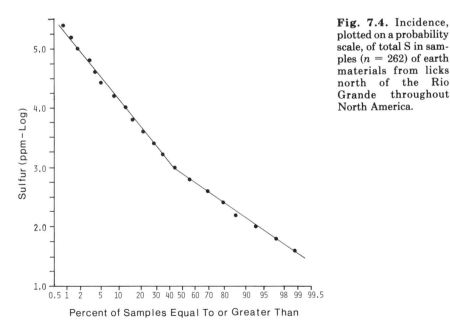

Fig. 7.4. Incidence, plotted on a probability scale, of total S in samples (n = 262) of earth materials from licks north of the Rio Grande throughout North America.

that sulfur is a major attraction in most licks. However, sulfur salts are an important constituent in some licks, notably in the bighorn ranges of Alberta and Alaska; here sulfur salts must contribute importantly to the sulfur nutrition of these animals.

Of the 265 lick samples processed for calcium carbonate equivalent, 113 contained 0.1% or more, the minimum detectable limit in this study. This characterization of lick samples is important in the explanation of lick use. We did not distinguish calcite and dolomite in the samples. This proved to be unfortunate, since we hypothesize that magnesium is one of the key elements explaining geophagous behavior. In the first approximation, the carbonates can be considered as either primary or secondary constituents of the lick earths. Primary occurrences are those in which the parent material or unaltered rock at the site is eaten. Secondary occurrences are those arising from precipitation from waters issuing from springs and seeps or evaporation of carbonate-bearing surface waters. Both occurrences are important sources and are discussed below and noted in the geologic interpretation for each lick in App. 1.

LICKS HIGH IN EXTRACTABLE SODIUM. Arbitrarily placed in this category are those samples with sodium contents of $\geq$ 5 me/100 g earth material as determined in the ammonium acetate extract. Thirty-six samples meet this criterion (Table 7.4) and their occurrence is literally continentwide. The accumulation of sodium in soil under temperate conditions and a freely draining topography is unusual. Generally sodium is lost readily by leaching, and sodium-bearing silicate minerals are among the least stable silicate minerals in an acidic weathering environment. In the eastern portion of the continent, sites in Mississippi (sample 115), Michigan (sample 104), and West Virginia (sample 215) high in extractable sodium were associated with springs. The licks in Missouri (samples 378, 383) are in low positions on the landscape and perhaps are under persistent groundwater effects. The two licks from Georgia (samples 209, 213) are unusual in that the sodium content is appreciable, but textures of silt loam and loam and the yellow and red chromas in sample 209 suggest a free-draining, oxidized environment. The Michigan sample (104) is from an area influenced by water from an oil well, and judging from the near equivalence of sodium and chloride ions, the material bears salt. The two Louisiana samples (189, 194) also appear to be salt bearing. A sample from Lake Nipigon, Ontario (388), contains substantially more sodium than the other four licks in the area. Both chloride and sulfate are substantially less than the 2.2 me sodium/100 g found in the saturation extract. The lick evidently occurs at or very near the shore. Origin of the sodium in this sample is not clear. Sample 372 from Illinois comes from near a spring that once was used in

Table 7.4. Selected chemical characteristics of earth materials (me/100 g) from licks with ≥ 5 me Na extractable with NH_4OAc. Deer means white-tailed deer.

State or province	Sample no.	Ammonium acetate extract			Water extract			$CaCO_3$	Species using lick
		Ca + Mg	Na	(Ca + Mg)/Na	Na	Cl	SO_4		
Eastern area									
Mississippi	115	17.6	5.37	3.3	2.72	0.21	3.77	4	deer
Georgia	209	13.5	7.78	1.7	0.48	0.26	0.03	0	deer
	213	12.8	8.97	1.4	0.27	0.18	0.02	0	deer
Louisiana	189	19.6	6.36	3.1	1.46	1.23	0.18	2	deer
	194	2.79	14.6	0.2	11.2	12.7	2.45	0	deer
West Virginia	215	34.3	6.52	5.3	5.07	5.46	0.82	30	deer
Illinois	125	4.33	6.04	0.7	2.21	0.31	1.30	4	deer
	372	67.7	28.7	2.4	19.3	16.5	3.56	126	deer
Missouri	378	23.7	12.1	2.1	6.84	11.2	0.85	6	deer, elk[a], bison[a]
	383	15.5	25.9	0.6	22.1	33.2	2.75	0	deer
Michigan	104	22.3	5.03	4.6	4.22	4.69	0.49	120	deer
Ontario	388	42.8	5.22	8.2	2.22	0.63	0.09	346	moose
Western area									
Colorado	117	5.05	5.03	1.0	0.58	0.26	0.24	0	deer[b], elk, sheep
	118	20.1	17.9	1.1	2.97	0.13	0.16	0	deer[b], elk, sheep
North Dakota	133	30.1	46.4	0.6	27.7	1.00	36.9	0	mule deer, sheep
	134	2.1	25.6	2.1	13.8	0.31	19.3	0	mule deer, sheep
	135	40.3	9.09	4.4	2.98	0.07	5.84	160	mule deer, sheep
Wyoming	346	5.34	10.5	0.5	2.05	0.05	1.38	16	antelope
Montana	141	4.21	21.2	0.2	2.72	1.38	1.06	0	bison
	142	50.9	21.5	2.6	2.18	0.72	3.81	72	bison
	333	18.8	5.84	3.2	0.37	0.001[c]	0.001[c]	0	elk
California	155	17.2	27.3	0.6	n.d.	n.d.	n.d.	0	deer[d], raccoon, birds
Oregon	222	80.6	14.4	5.6	5.33	10.3	1.66	48	mule deer
Manitoba	127	22.9	14.2	1.6	8.87	11.2	1.14	0	deer, moose
	143	118.0	6.39	18.6	3.65	0.57	8.39	374	deer, elk
	150	84.9	45.6	1.9	22.0	9.73	12.9	645	elk, caribou

Table 7.4 *(continued)*

State or province	Sample no.	Ammonium acetate extract			Water extract			$CaCO_3$	Species using lick
		Ca + Mg	Na	(Ca + Mg)/Na	Na	Cl	SO_4		
Alberta	238	82.4	6.94	11.9	3.98	0.15	19.9	659	sheep
	310	107.0	21.2	5.0	14.5	1.09	14.3	0	mule deer, elk, moose
	315	106.0	10.5	10.0	2.71	0.41	0.33	0	mule deer, elk, moose, horse
	414	18.3	5.09	3.6	2.17	0.04	3.54	30	mule deer
British Columbia	396	88.5	5.48	16.2	4.01	0.23	43.1	324	moose, Stone sheep
Alaska	175	87.5	7.14	12.3	3.89	3.86	1.22	0	moose
	176	38.4	24.4	1.7	7.07	0.001[c]	0.62	16	sheep
Northwest Territories	100	54.0	15.3	3.5	11.0	10.6	7.05	753	caribou, moose
	339	86.6	11.5	7.5	n.d.	n.d.	n.d.	44	moose
	340	99.4	33.7	3.0	n.d.	n.d.	n.d.	16	moose

[a]Formerly used by these animals.
[b]Probably mule deer.
[c]Less than 0.01 me Na/100 g.
[d]Species uncertain.

salt manufacture. The other sample from Illinois (125) appears to have been influenced by sodium sulfate waters.

Samples from drier regions of the continent would be expected to contain sodium salts. Inspection of the data in Table 7.4 indicates that the levels of sodium in both saturation-paste extract and ammonium-acetate extracts are of the order of magnitude of those found in the east. Three samples (133, 134, 135) from North Dakota probably arise from the same bentonite-bearing formation—the Sentinel Butte member of the Fort Union formation. Two samples (339, 340) from the Northwest Territories were taken at a hot spring. Samples from Alberta (414), Manitoba (127, 143), California (155), and Oregon (222) also come from the vicinity of springs. The single lick used by antelope, 346 from Wyoming, was taken at an artesian well.

Sodium does not represent the only cation of physiological importance that is readily available in these samples. We have included in Table 7.4 the ratios of extractable calcium plus magnesium to sodium. In six instances, the sum of calcium and magnesium is less than sodium. Gypsum, calcite, dolomite, and other salts of calcium and magnesium are soluble to some extent in ammonium acetate; and these compounds, particularly the carbonates, would be expected to yield calcium and magnesium assimilable by ungulates. Among these 33 high sodium samples are 18 samples with $> 0.2\%$ calcium carbonate (4 me $CaCO_3$/100 g), the level representing the lower detection limit in the presence of organic matter. If the calcium carbonate levels are considered with respect to those of sodium, some of the samples contain levels of carbonate that must be of nutritional significance and physiological importance in geophagy.

Lick composition by species use

Numerous data for lick composition are scattered through a variety of publications and agency reports, some of which are not widely circulated. The interpretation of these data is often difficult because the methods of analyses are either not described or not noted. Much of this information is summarized in the following discussions. Results of analyses of water extracts of earths or waters from licks are presented in Table 7.5; and data for ammonium acetate, acid, and total analyses of lick earths are given in Table 7.6. Most data for soil and earth analyses here are reported on the equivalence basis because of the importance placed on extractable elements later in the discussion.

The central tendency and variation of analytes in lick earths analyzed are given in Table 7.7. The variation found is indicative of the great diversity of substrates exploited by ungulates in seeking to balance

Table 7.5. Contents (mean and standard error) of Ca, Mg, Na, and K in waters or water extracts of soils from licks in United States and Canada. Livingstone's estimates for contents of these bases in river water are given for comparison.

State, province, or territory	Species using licks	n	Extractable cations (me/100 g) Ca	Mg	Na	K	Type of analysis	Source
Quebec	moose	19	9.09 ± 0.01	0.42 ± 0.06	0.92 ± 0.23[a]	0.004 ± 0.001	water	Bouchard 1970
Ontario	moose	7	1.8 ± 0.5	0.31 ± 0.07	1.9 ± 0.7	n.d.	water	Peterson 1953
Ontario	moose	4	5.6 ± 2.6	2.9 ± 1.7	7.1 ± 3.6	0.23 ± 0.04	water	Fraser and Reardon 1980
Ontario	moose and white-tailed deer	13	13.0 ± 10	0.8 ± 0.3	4.0 ± 1.7	0.16 ± 0.03	water	Chamberlin et al. 1977
Alberta and British Columbia	various	11	12.0 ± 3.0	n.d.	n.d.	n.d.	soil	Cowan and Brink 1949
Yukon Territory	barren-ground caribou	4[b]	27.0 ± 4.0	4.1 ± 1.2	9.0 ± 2.0	0.32 ± 0.08	soil	Calef and Lortie 1975
Northwest Territory[c]	musk-oxen	1	5.5	9.1	235.0	tr.	soil	Tener 1954
Wyoming	mountain sheep	1	3.2	0.13	25.9	n.d.	soil	Beath 1942
Idaho	moose, elk, mule and white-tailed deer	4	0.34 ± 0.13	0.05 ± 0.04	1.7 ± 0.2	0.29 ± 0.06	water	Mayland and Jimeno 1979[d]
Indiana	white-tailed deer	12[e]	n.d.	n.d.	3.1	.12	water	Weeks and Kirkpatrick 1976
Average river water			0.75	0.34	0.27	0.06	water	Livingstone 1963

[a]For 16 samples the statistics are 0.92 ± 0.23
[b]Samples from one site.
[c]Analyses are assumed to be of a water extract.
[d]Pers. commun. H. F. Mayland and G. B. Jimeno 1979
[e]Weighted average for April, May, and June.

Table 7.6. Contents (mean and standard error) of Ca, Mg, Na, and K either extractable from earth materials of licks or determined as total analyte.

State, province, or territory	Species using licks	*n*	Extractable cations (me/100 g)				Type of analysis	Source
			Ca	Mg	Na	K		
Quebec	moose	15	12 ± 3	2.6 ± 0.7	0.58 ± 0.15	0.25 ± .06	acetate[a]	Bouchard 1970
Ontario	moose, white-tailed deer	5	285 ± 76	202 ± 44	64 ± 5	38 ± 4	total	Chamberlin et al. 1977
Indiana	white-tailed deer	13	6.2 ± 1.1	2.1 ± 0.3	1.7 ± 0.2	0.17 ± .02	acetate	Weeks 1978
Montana	various	17	19 ± 3	8.3 ± 1.3	3.5 ± 1.3	0.86 ± .24	acetate	Stockstad et al 1953
Montana	elk	4	294 ± 421	100 ± 26	32 ± 11	n.d.	total	Knight and Mudge 1967
Montana	mountain goat	10	80 ± 43	12 ± 6	0.90 ± 0.28	2.1 ± .9	Acid[b]	Singer 1975
Alberta	mountain goat	1	821	194	...	2.7	total	McCrory 1967
British Columbia	mountain goat	11	62 ± 59	n.d.	2.9 ± 7.0	n.d.	acetate	Herbert and Cowan 1971
Alberta and British Columbia	various	11	201 ± 28	n.d.	n.d.	n.d.	acid[c]	Cowan and Brink 1949

[a] NH_4OAc.
[b] 0.1 N HCl.
[c] 1.0 N HCl.

Table 7.7. Values of GM and GD for selected elements, pH, and conductivity in NH_4OAc and water extracts of earth materials from licks, and total S and $CaCO_3$ equivalents. Licks are from throughout North America north of the Rio Grande.

Analyte and unit and measure	*n*	GM	GD
Acetate extract			
Ca (me/100 g)	137	13.3	3.86
Mg (me/100 g)	144	3.6	6.40
Na (me/100 g)	144	2.0	3.94
K (me/100 g)	144	0.41	2.40
Water extract			
Ca (me/100 g)	137	0.21	7.62
Mg (me/100 g)	137	0.13	9.75
Na (me/100 g)	137	0.58	5.64
K (me/100 g)	137	0.02	3.86
SO_4 (me/100 g)	137	0.35	10.4
Cl	137	0.12	10.6
Cond. (mmho/cm)	136	1.82	4.10
pH	137	6.9 [a]	1.5 [b]
Total analysis			
S (%)	262	0.063	5.79
$CaCO_3$ (%)	113 [c]	5.1	5.11

[a] Arithmetic mean.
[b] Standard deviation.
[c] Samples having 0.1% or more $CaCO_3$.

their mineral nutrition. This variation, in turn, encourages scrutiny of the sample data for subsets. In later discussions the relation of chemical composition of earth licks to the species or species assemblage of ungulates using the licks is evaluated. The following discussions are confined to treatment of the overall averages. Data for each lick are presented in App. 2.

MOOSE LICKS

Previous studies. Bouchard (1970) analyzed the waters and soils of moose licks in the Matane Reserve, southeastern Quebec. The waters (Table 7.5) and ammonium acetate extracts (Table 7.4) of the earth materials were analyzed for calcium, magnesium, potassium, sodium, iron, and manganese and determinations for pH, organic carbon, and nitrogen were made on the soils. The waters, which were uniformly alkaline, were dominated by sodium and magnesium, and calcium and potassium were conspicuous for their low levels. Ammonium acetate extracts of the soils removed substantial amounts of calcium. Samples 190, 191, 192, and 193 (App. 1) came from the same areas in the Matane.

Peterson (1953) reported that moose licks are rare in Ontario. He gave data (Table 7.5) for the supernatant water of seven licks on St. Ignace Island in Lake Superior. These results do not show any consistent pattern, as Peterson concluded (p. 21). Sodium concentrations were reported for only three of the licks and their absence in the others was not explained. Sample 204 (App. 1) came from St. Ignace Island.

Chamberlin et al. (1977) reported the results of their sampling of 13 moose and white-tailed deer licks in the Nipigon lake and bay areas and the notorious lick known as the "moose graveyard" near Kirtland Lake, Ontario. They sampled lick waters and waters from adjacent areas to represent comparative materials, and they took soil samples from the licks and control samples from nearby areas. Sodium and calcium were 47 and 20 times, respectively, more concentrated in the lick waters and nitrogen and phosphorus were about 20 times higher. The latter elements no doubt reflected excreta in the waters, and therefore some amount of sodium and calcium should also be attributed to this source. In their summary, the authors did not use data for three licks because they felt that waters used by animals were diluted by streams traversing the lick. If the McIntyre Bay Lick, which contains the highest sodium and calcium levels of all licks, is also removed from the calculation, sodium and calcium in the remaining lick waters are, on the average, 28 and 3.9 times, respectively, over that in control waters. Analytical data for acid digestates of soils from licks and their respective control sites do not reveal a pattern of preferential concentration of any element in the earth material of the lick. The authors concluded that "this study does not give conclusive evidence of a single factor which attracts moose and deer" (p. 213). They do suggest that perhaps the clay soil common to almost all the sites may be sought. The glacial sediments of the region tend to be rich in sand and the clay-rich nature of the licks is a strong textural contrast. These fine textured materials also tend to concentrate cations, thus contributing to the higher salt (in the strictest sense) concentrations in the waters of the lick sites. Indeed, ingestion of the clayey sediments will carry a load of calcium, magnesium, potassium, and sodium not available when the sandy materials are eaten.

Fraser and Reardon (1980) have reported data (Table 7.5) for the composition of water in four licks on the Sibley Peninsula, southwestern Ontario. These licks are in settings similar to and not far west of several licks reported by Chamberlin et al. (1977) for the Nipigon Bay area. Like those licks, sodium is more concentrated (about 50 times) in lick waters than in nearby streams. Magnesium, calcium, and potassium are about 5, 6, and 13 times, respectively, more concentrated in the lick water. As described later, cafeteria trials in the vicinity of one of the licks resulted in a clear choice of sodium ion, which (in view of the concentration factor of sodium in lick water relative to local stream water) led the authors to conclude that sodium was sought.

Table 7.8. Composition of NH_4OAc and saturation paste extracts and total S and $CaCO_3$ equivalent of earth materials from licks used exclusively by moose. Concentrations are expressed as GM and GD on a dry-earth basis.

Analyte and unit of measure	*n*	GM	GD
Acetate extract			
Ca (me/100 g)	19	22.67	2.76
Mg (me/100 g)	19	3.19	2.77
Na (me/100 g)	19	1.68	3.65
K (me/100 g)	19	0.42	2.65
Water extract			
Ca (me/100 g)	16	0.42	4.09
Mg (me/100 g)	16	0.16	3.97
Na (me/100 g)	16	0.45	3.09
K (me/100 g)	16	0.03	2.94
SO_4 (me/100 g)	16	0.35	4.58
Cl (me/100 g)	16	0.06	12.48
Cond. (mmho/cm)	16	1.26	2.85
pH	16	6.88[a]	1.18[b]
Total analysis			
S (%)	24	0.058	4.17
$CaCO_3$ (%)	11	1.81	7.50

[a]Arithmetic mean.
[b]Standard deviation.

Present findings. Sixteen licks located in the Matane, Quebec, area; in the Nipigon lake and bay areas, Ontario; on Isle Royale, Michigan; in northeastern Minnesota; along the South Nahanni River, Northwest Territories; and on the Kenai Peninsula are included in this evaluation. Both water and acetate extracts and total sulfur and calcium carbonate equivalents were determined for this sample set. All these licks were reported by cooperators to be used exclusively by moose. The geometric means and deviations for the various analytes except the water-soluble cations are presented in Table 7.8. Compared with the mean (Table 7.7) of all licks, moose licks tend to have more extractable calcium and less magnesium and sodium. Because the soils of bog and fen habitats sought by moose tend to be neutral or acid, the low carbonate levels of these lick samples are not surprising; indeed, the presence of carbonates in 11 of the 19 samples may primarily explain their use. Sulfate, pH, and total sulfur are close to the means for all licks. An idea of the pH and extractable calcium and magnesium in peat can be gained from the averages for the limited samples from northern Wisconsin in Table 8.17. Given the differences between wood and moss peats, it cannot be stated that the lick earths differ in calcium and magnesium contents from these surface horizons; however, pH of the licks may be higher.

WHITE-TAILED DEER LICKS

Previous studies. Weeks and Kirkpatrick (1976) provided analytical data for water in wet licks (Table 7.5) and Weeks (1978) provided data for soil from earth licks (Table 7.6) used by white-tailed deer in Martin County, Indiana. Also, Weeks and Kirkpatrick (1978) sampled soils of roadside, old-field, and woodland habitats and compared these situations with the licks. Sodium and magnesium were significantly greater ($P <$ .001 and $<$ .01, respectively) in the lick earths than in the earths from nonlick habitats. The authors attributed geophagy by deer in Martin County to the need for sodium, exacerbated by potassium-rich forage in the spring.

Weeks generously provided the data for extractable calcium, magnesium, sodium, potassium, water soluble sulfate, chloride, salts and pH, and available phosphorus for each site. In analyzing these data, we combined the 20 nonlick woodland and old-field samples and compared them with the 15 lick samples by discriminant analysis. Log transformation of these data except for pH resulted in significant univariant F-ratios for all variables at the 5% level. Using Rao's V-statistic for choice of the seriatim arrangement of variables, all the variables made significant contributions to the discriminant function, entering in order: sodium, calcium, pH, potassium, salts, magnesium, sulfate, and available phosphorus. The function explained, as measured by the $\hat{\omega}^2$-statistic (Tatsuoka 1970), over 96% of the separation of the groups, and consequently, a posteriori classification resulted in 100% of the sites being correctly assigned. From this evaluation, the licks clearly represent unique positions in the Indiana landscape, being sodium rich and alkaline. Weeks provided us with samples of several of the licks (see App. 1, samples 324–28), two of which were found to contain a trace of carbonates.

Present findings. Seventy-five analyses of water extracts and 76 ammonium acetate extracts of lick earths were obtained for licks used either exclusively by deer or frequented also by other ruminants, usually elk. Both mule and white-tailed deer are included in this characterization. The geometric means and deviations for these determinations, together with data for total sulfur and calcium carbonate equivalent for 20 additional licks, are presented in Table 7.9. Graphs for the probability distribution of calcium, magnesium, sodium, and total sulfur are presented in Figs. 7.5, 7.6, 7.7, and 7.8, respectively. Only the water extracts for potassium, magnesium, and chloride do not fit a log-normal distribution. The data possess substantial variability and exhibit a positively skewed distribution except for acetate-extractable potassium. Reference to the data for water extracts indicates a low average level of soluble salts. Ammonium acetate extracts substantially more bases than water. The mean contents of the four bases determined by ammonium

Table 7.9. Values of GM and GD for selected elements in NH_4OAc and water extracts; conductivity, pH, and total S and $CaCO_3$ equivalent of earth materials from deer licks.

Analyte and unit of measure	*n*	GM	GD
Acetate extract			
Ca (me/100 g)	76[a]	6.44	4.04
Mg (me/100 g)	76	2.27	3.40
Na (me/100 g)	76	2.15	3.88
K (me/100 g)	76	0.31	1.47
Water extract			
Ca (me/100 g)	75[a]	0.12	8.45
Mg (me/100 g)	75	0.06	9.31
Na (me/100 g)	75	0.62	4.95
K (me/100 g)	75	0.02	3.78
SO_4 (me/100 g)	75	0.22	9.59
Cl (me/100 g)	75	0.19	8.45
Cond. (mmho/cm)	75	1.79	3.88
pH	75	6.36[b]	1.56[c]
Total analysis			
S (%)	96[d]	0.034	4.11
$CaCO_3$	28[e]	1.78	5.63

Note: Extract data, except for conductivity and pH, were calculated to soil-weight basis. Licks were used by white-tailed and/or mule deer or simply deer; 8 licks in western North America were also used by elk or moose and 3 by bighorn sheep.

[a]64 white-tailed deer licks.
[b]Arithmetic mean.
[c]Standard deviation.
[d]79 white-tailed deer licks.
[e]20 white-tailed deer licks.

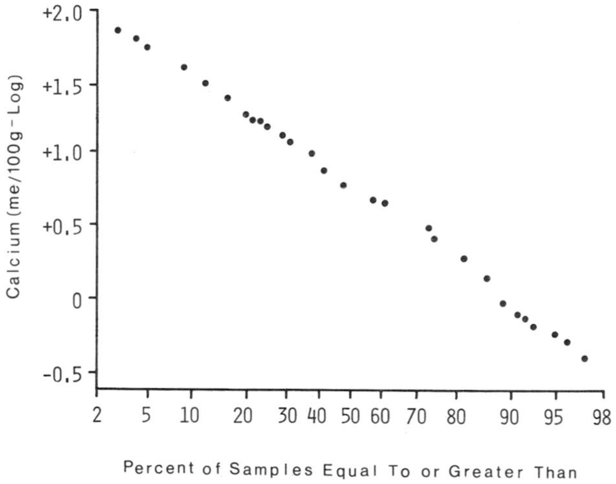

Fig. 7.5. Incidence of calcium extractable with ammonium acetate in earth materials, plotted on a probability scale, for 76 deer licks.

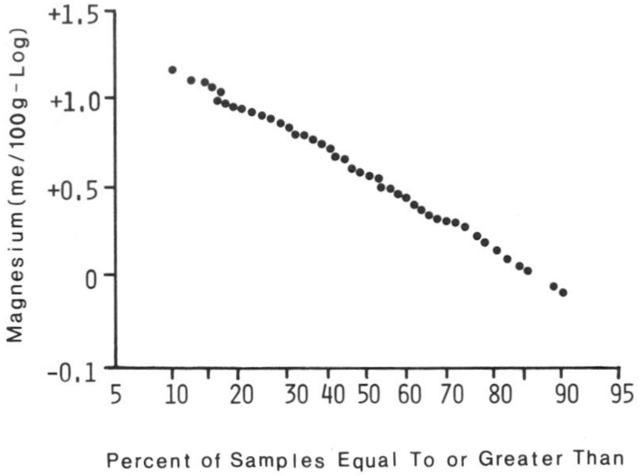

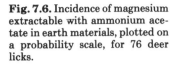

Fig. 7.6. Incidence of magnesium extractable with ammonium acetate in earth materials, plotted on a probability scale, for 76 deer licks.

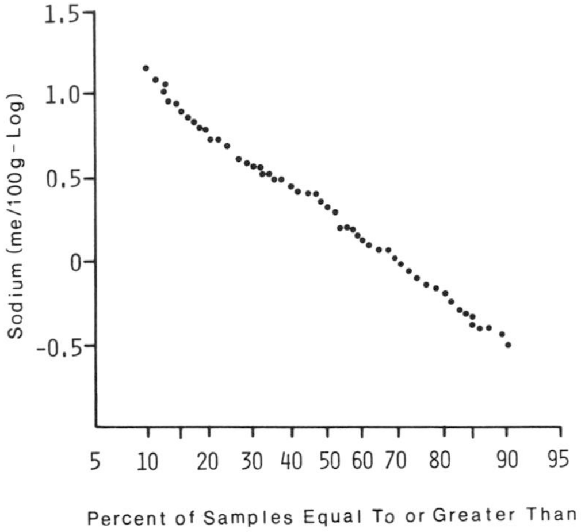

Fig. 7.7. Incidence of sodium extractable with ammonium acetate in earth materials, plotted on a probability scale, for 76 deer licks.

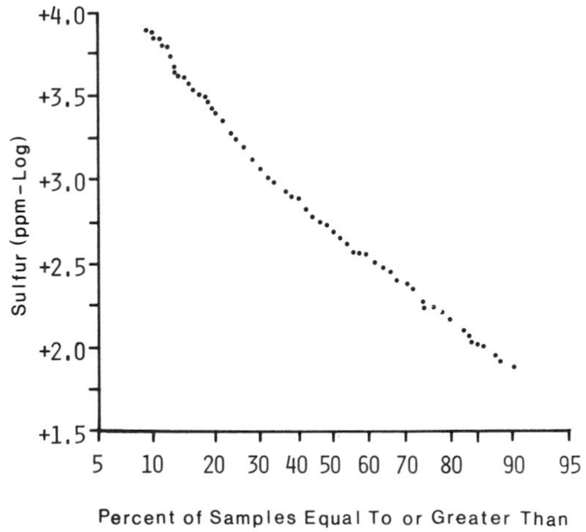

Fig. 7.8. Incidence of total sulfur in earth materials, plotted on a probability scale, for 96 deer licks.

acetate extract compare closely with levels found by Weeks (1978) in Table 7.6 for licks used by white-tailed deer in southern Indiana. The mean sulfur content of 340 ppm is probably lower than the mean abundance of sulfur for soils in North America; therefore we must conclude that deer using licks are seeking primarily constituent(s) other than sulfur. Of the 96 licks available for study, 28 contained carbonates with a mean of 1.78%, equivalent to 0.71 g Ca/100 g of earth if the carbonate is calcite or perhaps aragonite.

DEER LICKS IN LOW-CALCIUM AREAS. Analyses were performed on earths from 55 licks of the samples in our collection. Licks used by white-tailed deer in the area east of the 100th meridian were chosen for comparison of chemical characteristics with surrounding soils. Data for calcium, magnesium, sodium, and potassium extracted with ammonium acetate were available in compilations of analytical data contained in Soil Survey Investigations Reports published by the *USDA Soil Conservation Service* for states listed in Apps. 6 and 7. Data for counties where the licks occurred were so rare that data for sites in adjacent counties often had to be accepted. Using the exchangeable calcium map (Lane and Sartor 1966) for United States surface soils as a guide, we used only lick samples from counties having predominantly < 10 me Ca/100 g of soil in this evaluation. Thus lick earths from potentially base-rich localities were excluded. Physiographic, geologic, and soil maps were consulted in an attempt to confine the comparative data to soils that developed from similar parent material. Data for surface or A horizons; clay-enriched or B horizons (usually B22); and, in a few cases, the C horizon or parent material were extracted from the reports. These horizons were chosen because licks in this region most often are described as occurring in moist surface soil or in subsoil exposed through erosion, cultural practices such as road construction, or digging by the deer. Sample statistics for licks, topsoil, and subsoil groups are presented in Table 7.10.

Discriminant analysis was used to determine the variates that best

Table 7.10. Values of GM and GD (me/100 g) for extractable Ca, Mg, Na, and K in white-tailed deer licks and associated soils in eastern United States in areas of $<$ 10 me exchangeable Ca/100 grams topsoil (map of Lane and Sartor 1966). Lick samples used are listed in App. 6.

Sample		Ca		Mg[a]		Na[a]		K	
type	*n*	GM	GD	GM	GD	GM	GD	GM	GD
Lick	48	4.05	3.50	2.03a	2.98	1.49a	3.50	.25	2.02
Topsoil	82	2.54	4.08	0.72b	3.60	0.97b	1.57	.21	1.96
Subsoil	78	2.13	4.72	1.09c	3.86	0.13b	3.04	.19	1.90

[a]Means not followed by the same letter are significantly different ($P \leq .05$).

distinguish lick earths from surface and subsoil horizons of the area. The null hypothesis states that no difference exists among the group centroids in discriminant space. Because the data for each element in the lick samples were found to be log normally distributed by the Kolmogorov-Smirnov statistic, the discriminant analysis was carried out on log-transformed data. At the outset, univariate F-ratio tests indicated that significant differences occurred for extractable magnesium and sodium. The within-groups correlation matrix yielded r values of .57 or less, and only magnesium displayed any covariation—weak as it was—with the other three variates. The stepwise discriminant procedure that maximizes Rao's V-statistic was used. Results of the analysis are given in Table 7.11. Sodium and magnesium represented the greatest discriminating power, which is not surprising because the variance test had indicated that differences for these elements existed among the groups. Potassium and calcium, in that order, also contributed significantly to discrimination of the lick and soil samples. Comparison of the standardized discriminant function coefficients indicates sodium was about six times as important as magnesium in contributing to group separation for the first function. Calcium and potassium were about one-third and one-fourth as effective, respectively, as magnesium. Magnesium was most important in the second function followed by calcium and potassium, which were about equally important; sodium was least important. A posteriori classification of the samples provides insight into the kinship of the lick samples with surface soils and the subsoils. Results of this reclassification are given in Table 7.12. These

Table 7.11. Normalized and standardized coefficients for variables in discriminant functions of deer licks and soils with extractable Ca < 10 me/100 g of surface soil. Sample statistics for the materials are presented in Table 7.10.

Variable[a]	Normalized vector		Standardized vector	
	Y1	Y2	Y1	Y2
Ca	0.098	1.131	0.061	0.709
Mg	0.303	−2.262	0.174	−1.299
Na	−1.667	0.334	−1.104	0.221
K	0.140	2.850	0.041	0.834
Constant	−1.204	1.753	...	...

Function	Y1	Y2
Eigenvalue	1.789	0.066
Contribution (%)	96.45	3.55
χ^2	221	13
df	8	3
P	< .000	.005

[a]Taken as log transformations.

Table 7.12. Prediction of group membership in a posterior classification of white-tailed deer licks and soils in eastern United States.

		Predicted group (%)		
Actual group	n	lick	surface soil	subsoil
Lick	48	81.2	4.2	14.6
Surface soil	82	0.0	75.6	24.4
Subsoil	78	11.5	34.6	53.8

data seem to indicate that deer are seeking specific locations on the landscape that are likely to be more similar to subsoils than to surface soils. Considering the leached nature of soils of this eastern humid temperate region, it is expected that none of the surface soils were misclassified into the lick group, which is comparatively base rich. The substantial misclassifications between surface soils and subsoils reflect the great diversity in soil development and parent material in the samples that represent half a continent.

The results of this comparison support the conclusions of Weeks and Kirkpatrick (1976) that during spring, white-tailed deer are under stress to supplement a sodium-deficient state that we believe to be transient. The importance of magnesium, subrogated to sodium to some degree, supports the hypomagnesemia hypothesis put forth in this book. But either sodium or magnesium supplemented from licks alleviates the same physiological stress—high potassium and low alkaline earth intakes associated with the spring flush of grasses (see Chap. 9). The discriminant analysis emphasizes the unique ability of ungulates to seek areas that do not have close affinities with surface soils and subsoils.

DEER LICKS IN HIGH-CALCIUM AREAS. White-tailed deer licks in Missouri, Arkansas, and Mississippi were compared with surface soils and subsoils in the same manner as lick samples from low calcium areas. The concentration base-line was again 10 me Ca/100 g soil (Lane and Sartor 1966). The sample was small, involving three licks from north central Missouri, two licks from the Mississippi flood plain in Arkansas, and two licks from the Selma chalk belt in extreme east central Mississippi (Table 7.13). Data for 9 surface soils and 18 subsoils and parent materials found in Soil Survey Investigations Reports for Missouri, Arkansas, and Mississippi were used. Log transformations of the data were made for computation. Analysis of variance of the individual variates indicated that at least two sodium populations existed. The within-group correlation matrix indicated that, like the low calcium sample, little correlation existed among elements. In the discriminant analysis, sodium, magnesium, and potassium contributed significantly to the

Table 7.13. Values of GM and GD (me/100 g) for extractable Ca, Mg, Na, and K in white-tailed deer licks and associated soils in eastern United States in areas of > 10 me exchangeable Ca/100 g topsoil (map of Lane and Sartor 1966). Lick samples used are listed in App. 7.

Sample type	n	Ca GM	Ca GD	Mg GM	Mg GD	Na[a] GM	Na[a] GD	K GM	K GD
Lick	7	8.99	2.76	3.12	2.32	3.92a	4.01	0.37	2.56
Topsoil	9	7.41	2.03	3.67	1.33	0.11b	1.52	0.36	1.76
Subsoil	18	5.49	3.42	5.06	1.50	0.32b	2.75	0.35	1.67

[a]Means not followed by the same letter are significantly different ($P \leq .01$).

Table 7.14. Normalized and standardized coefficients for variables in discriminant functions of deer licks and soils with extractable Ca > 10 me/100 g surface soil. Sample statistics for the materials are presented in Table 7.13.

Variable[a]	Normalized vector Y1	Normalized vector Y2	Standardized vector Y1	Standardized vector Y2
Ca	−0.003	1.397	−0.001	0.645
Mg	2.210	−4.404	0.504	−1.005
Na	−1.361	−0.634	−0.938	−0.437
K	−0.677	1.033	−0.180	0.274
Constant	−2.215	1.814	...	...

Function	Y1	Y2
Eigenvalue	2.852	0.298
Contribution (%)	90.55	9.45
χ^2	47	7.7
df	8	3
P	< .000	.053

[a]Taken as log transformations.

discriminating power of the function (Table 7.14). The first function was highly significant, contributing 90% of the differentiation. The second function barely missed significance at 5%. The overall discriminatory power, as measured by the $\hat{\omega}^2$-statistic (Tatsuoka 1970:48), indicated that 78.2% of the variability in the function was due to group differences. The importance of sodium and magnesium in this small sample of licks from areas that are relatively base rich suggests that deer in these states were seeking out licks for the same reasons as deer in areas of eastern states with soils that have, on the average, exchangeable levels of < 10 me Ca/100 g. The posterior classification of these licks and soils (Table 7.15) followed a qualitative and quantitative pattern similar to that found in the base-poor sample (compare Tables 7.12, 7.15).

Table 7.15. Prediction of group membership in a posterior classification of white-tailed deer licks and soils in eastern United States where surface soils average > 10 me/Ca/100 g surface soil.

		Predicted group (%)		
Actual group	*n*	lick	surface soil	subsoil
Lick	7	71.4	0.0	28.6
Surface soil	9	0.0	88.9	11.1
Subsoil	18	0.0	27.8	72.2

ELK LICKS

Previous studies. Knight and Mudge (1967) analyzed four licks used primarily by elk in the Sun River basin of west central Montana. Their analyses were for total elemental contents, an approach commonly used in geological investigations. They also sampled the same lick-bearing geologic formations adjacent to the licks and rocks of similar character at some unspecified distance from the Sun River area. The authors concluded (p. 295): "Sodium carbonate is the only compound that occurs in larger proportions in the units containing licks than in units where there are no licks." Sodium carbonate is important in the African licks described by French (1945). We selected calcium, magnesium, and sodium from the analytes they reported and subjected the data to stepwise discriminant analysis. In spite of the very small sample size, discrimination between lick and nonlick materials was significant; sodium contributed most to the discrimination and was followed by magnesium and calcium, neither of which added substantially to separation (as measured by Rao's V) of the groups. When sodium, calcium, and carbon dioxide (a measure of carbonate) were tested and three groups that were equivalent to licks, associated rock units, and distant rock units were tested, the resulting two discriminant functions had less discriminating power. The first function accounted for almost 92% of the combined discrimination and, as measured by Wilks's λ, was significant at the .10 level. A posteriori classification indicated that one lick was misclassified, being placed in the distant rock unit category ($P = .59$); and substantial misclassification (70%) was found between the adjacent and distant rock units. This poor separation is to be expected in the effort to sample rocks of similar lithologic character at some distance to the licks. This analysis of the Sun River licks agrees with Knight and Mudge's conclusion that sodium was the element sought.

Present findings. Ten licks of those for which we have analytical data were used exclusively by elk. Six of these licks were located in Wyoming,

Table 7.16. Composition of NH_4OAc and saturation-paste extracts of earth materials from licks used exclusively by elk. Concentrations are expressed as values of GM and GD on a dry-earth basis.

Analyte and unit of measure	*n*	GM	GD
Acetate extract			
Ca (me/100 g)	9	24.2	2.55
Mg (me/100 g)	9	3.40	2.32
Na (me/100 g)	9	0.86	3.54
K (me/100 g)	9	0.66	2.92
Water extract			
Ca (me/100 g)	9	0.13	6.97
Mg (me/100 g)	9	0.068	4.78
Na (me/100 g)	9	0.18	2.60
K (me/100 g)	9	0.018	3.39
SO_4 (me/100 g)	9	0.056	11.2
Cl (me/100 g)	9	0.017	9.12
Cond. (mmho/cm)	7[a]	1.65	4.69
pH	9	7.32[b]	1.65[c]
Total analysis			
S (%)	10	0.027	3.33
$CaCO_3$ (%)	8	8.63	2.50

[a]Two samples were 1 to 5 water extracts.
[b]Arithmetic mean.
[c]Standard deviation.

two in Alberta, and one each in Manitoba and British Columbia. The geometric means and deviations for analytes in materials from these licks are presented in Table 7.16. Water extracts tend to be low in salts, and of those present, sodium and calcium predominate. These elk licks tend to be calcareous, 8 of the 10 having a mean content of 8.6% calcium carbonate equivalent. This characteristic is reflected in the contents of ammonium acetate extracts, which contained a mean of 24.2 me Ca/100 g. Magnesium was less abundant at 3.40 me/100 g, and sodium and potassium were considered to be similar at 0.86 and 0.66 me/100 g, respectively. Compared with the grand means for all licks (Table 7.7), the materials used by elk tend to be higher in calcium and lower in water extractable sulfate and chloride and in total sulfur.

CARIBOU LICKS

Previous studies. Calef and Lortie (1975) studied a lick used by barren-ground caribou located about 21 km (13 mi) up the Firth River, Yukon Territory, from where it enters the sea. Six samples were taken, five of which were from the lick. Water extracts were made and analyzed for the common cations and anions. All the samples were calcareous—the calcium carbonate equivalent ranging from 8.7 to 17.4% (172 to 345 me/100 g). Soluble salts increased with the estimated use of the portion of the lick sampled and substantial accumulation of salt occurred at the surface. A crust gave 101 me Na/100 g, 96 me Cl/100 g and a conductivity of 94 mmho/cm. Below the crust the earthy material ranged from 22.8 to 39.3 me Ca/100 g and 0.18 to 26.4 me Na/100 g, with calcium always being more abundant. Sulfates ranged from 0.03 me/100 g in the little used portion to 1.83 me/100 g in the uppermost 1.3 cm of the heavily used portion of the lick; however, in 15-cm-deep samples, sulfate amounted to only 0.05 and 0.06 me/100 g in moderate- and heavy-use areas, respectively. Unlike chloride and the metals, sulfate is reported in parts per million and it is assumed that the data are on a weight per soil basis. Unusual among analyses of licks is the report of nitrate, which amounts to 3.6 me/100 g (2230 ppm) in the 1.3 cm layer referred to above. In the low- and moderate-use areas, nitrate is about 0.01 me/100 g. No doubt its concentration indicates contamination with excreta, but oddly, phosphorus contents (which are $\leq$ 1 ppm) do not follow a similar trend. Calef and Lortie (p. 241) concluded that sodium, chloride, and sulfate were "logical candidates" for the sought-after nutrient or nutrients at this lick. They cite the work of Skoog (1968), who studied the Forty-Mile, Nelchina, and McKinley caribou herds and noticed licks. He speculated that calcium and phosphorus were likely to be needed because of the low levels of these nutrients in lichens, the principal winter food. Calef and Lortie did not attach importance to the calcium levels in the Firth River lick.

Present findings. Five licks used by woodland caribou were sampled (Table 7.17). In each case the lick was also used by moose; at two licks in the Rocky Mountains of northwestern Alberta mountain sheep and goats also ate the earths. Except for a large amount of sodium in a lick on Caribou Flats along the Keele River, Northwest Territories, calcium and magnesium occur in large amounts in these five lick samples. Magnesium sulfate is an important constituent in the two licks from the Northwest Territories; in both cases, rocks of the Sekwi formation appear to be utilized. Sodium chloride comprises about 0.5% of the Caribou Flats sample. The analyses of these five samples indicate that woodland caribou seek out alkaline materials rich in magnesium and calcium.

MUSK-OX LICKS

Previous studies. Tener (1954) described a musk-ox lick in white soil exposed on both banks of a stream flowing between Romulus Lake and Slidre Fiord on Ellesmere Island. He recorded the results (Table 7.5) of what appear to be a water extract on the earth; sodium salts are common

Table 7.17. Composition of NH_4OAc and saturation-paste extracts and total S and $CaCO_3$ equivalent of caribou licks; values of GM and GD for these also are provided.

Province or territory	*n*	Acetate extract Ca	Mg	Na	K	Water extract Ca	Mg	Na	K	SO_4	Cl	Cond.	pH	Total S	$CaCO_3$
		me/100 g				me/100 g						mmho/cm		%	
Northwest Territories	100	47.16	6.79	15.33	0.52	3.40	2.80	11.04	0.22	7.05	10.61	17.15	7.6	0.76	37.7
Northwest Territories	112	3.09	15.67	2.59	0.39	1.14	8.96	2.10	0.07	12.63	0.09	9.21	4.5	0.32	0.0
British Columbia	234	25.20	2.49	0.39	0.24	0.28	0.16	0.12	0.01	0.26	0.03	0.96	7.8	0.090	4.3
Alberta	309	13.35	5.80	0.28	0.59	0.09	0.07	0.03	0.008	0.14	0.01	0.61	7.1	0.097	1.5
Alberta	329	71.36	12.22	0.76	0.26	1.00	4.70	0.15	0.02	6.10	0.15	7.96	7.4	5.10	21.4
GM		20.4	7.15	1.27	0.38	0.63	1.06	0.42	0.03	1.82	0.13	3.74	6.88[a]	0.40	8.49
GD		3.42	2.05	5.11	1.50	4.06	8.68	10.93	4.04	8.06	14.27	4.42	1.36[b]	5.32	4.38

[a]Arithmetic mean.
[b]Standard deviation.

Table 7.18. Composition of NH_4OAc and saturation-paste extracts and total S and $CaCO_3$ equivalent from bison licks on the National Bison Range, Montana.

Sample no.	Acetate extract Ca	Mg	Na	K	Water extract Ca	Mg	Na	K	SO_4	Cl	Cond.	pH	Total S	$CaCO_3$
	me/100 g				me/100 g						mmho/cm		%	
140	2.79	4.91	3.45	.40	.005	0.005	0.52	.005	0.25	0.50	1.42	7.5	.017	2.2
141	2.17	2.04	21.20	.80	.03	0.06	2.72	.04	1.06	1.38	0.65	8.3	.044	0.0
142	44.91	5.96	21.53	.50	.89	1.16	2.18	.02	3.81	0.72	5.70	7.8	.394	3.6

in this material, there being 9.88% (166 me/100 g) sodium sulfate and 5.59% (103 me/100 g) sodium chloride. Occurrences of these salts, particularly sodium sulfate as the mineral thenardite, are common in the Arctic (Tedrow 1977). We do not know if these salts gave the soil a white color or if calcium carbonate also was present. From the data it appears that the soil was eaten for its sodium content and perhaps common salt acted as the main attractant. Tener (1965) found soils in the Thelon Game Sanctuary and Lake Hazen, Northwest Territories, musk-ox summer ranges to be neutral or mildly alkaline, with especially large amounts of alkaline earths in the Lake Hazen area where soils may be calcareous.

BISON LICKS

Previous studies. We have discussed the literature of bison licks (Chap. 2). Earth at Big Bone Lick in Kentucky evidently is rich in calcium sulfate. We are not aware of comprehensive analytical data for any of the eastern licks used by bison.

Present findings. Three samples (140, 141, 142) of earths eaten by bison on the National Bison Range, Montana, were analyzed (Table 7.18). They are composed of mildly to moderately alkaline materials, but the two samples taken along Mission Creek are notably rich in sodium with 21 me/100 g. The other sample, from about 3 km to the southwest, contains about 3.5 me Na/100 g. Carbonates occur in two samples. Calcium and magnesium are quite variable, with calcium reaching almost 45 me/100 g in sample 142 and magnesium almost 5 me/100 g in sample 140. The water extracts of these samples reflect a great range of salt content with conductivity varying between 0.6 and 5.7 mmho/cm. Both sulfate and chloride salts are important, although total sulfur content averages 1500 ppm (skewed by 3940 ppm in sample 142).

These results are consistent with an interpretation of outcrop of Belt Series rocks that consist, in part, of magnesium-rich carbonates. The origin(s) of the substantial content of sodium in these samples is not clear. Its abundance suggests that these earths are sought for their sodium contents. However, this evaluation may be simplistic, since the collector of sample 142 (see App. 1), Babe May, noted that bison eat the earths "in spite of having mineral blocks available to them year around."

PRONGHORN LICKS

Previous studies. We have discussed the contradictory literature regarding geophagy by the pronghorn. We know of no analytical data.

Present findings. The single sample of a medium sand (346) provided by Dave Lockman from a site adjacent to an artesian well at Clay Butte in Sweetwater County, southwestern Wyoming, is strongly alkaline and rich in sodium (App. 1). Sulfate and probably bicarbonate are the important anions; chloride is insignificant. It appears that this earth material is used for its sodium content and alkalinity.

MOUNTAIN GOAT LICKS

Previous studies. Data for licks used solely by goats have been reported by McCrory (1967) for the Athabasca Falls lick in Jasper National Park, Alberta (Table 7.6); by Vaughan (1975) for licks in the Wallowa Mountains, Oregon (Table 7.19); and by Singer (1975) for licks in Glacier National Park, Montana (Table 7.6) These licks are especially rich in calcium and the Alberta lick is also high in magnesium. McCrory (pp. 10–11) concluded that goats were eating the earth at Athabasca Falls for its calcium content. Vaughan noted goats licking a rock that appears to us to contain calcium carbonate (see Table 7.19).

Present findings. Data for the six licks sampled for this study and for three licks in northeastern Oregon that were investigated by Vaughan (1975) are presented in Table 7.19. Except for samples 329 and 405, which appear to contain epsomite, the water extracts were especially low in salts, particularly in sodium forms. However, ammonium acetate extracted substantial amounts of calcium from all the samples except 163; the calcium content was greater in seven of the nine goat lick samples studied than in the mean calcium for all licks (Table 7.7). In six samples, magnesium was greater than the mean for all licks, but all the sodium levels were less than the general mean. Potassium levels exceeded the general mean in only two samples.

Total sulfur content of these licks was quite variable, but notably high—5.1%—in sample 329 from Muskwa Range in northern Alberta, where goats (with moose, elk, caribou, and sheep) use a spring bearing sulfur- and calcium-rich water. When carbonate data were available, the rather high levels found suggest that this material is sought. It appears that calcium and perhaps magnesium are the major mineral elements sought by mountain goats.

MOUNTAIN SHEEP LICKS

Previous studies. Both Allred and Beath (in Honess and Frost 1942: 86–87) present data for bighorn sheep licks in the Gros Ventre Mountains of western Wyoming. The data are incomplete and, in part, semiquanti-

Table 7.19. Composition and characteristics of earth materials eaten by mountain goats. Data for Oregon licks are from Vaughan (1975).

State or province	Sample no.	Acetate extract				Water extract								Total	
		Ca	Mg	Na	K	Ca	Mg	Na	K	SO_4	Cl	Cond.	pH	S	$CaCO_3$
		me/100 g				me/100 g						mmho/cm		ppm	%
Colorado	163	7.86	3.66	1.59	.22	0.08	0.06	.42	.01	0.15	.09	1.56	7.9	110	0.1
Montana	334	30.69	2.06	0.22	.19	0.03	0.02	.01	.004	0.01	.005	0.34	8.2	516	26.8
Alberta	309	13.35	5.80	0.28	.59	0.09	0.07	.03	.008	0.14	.01	0.61	7.1	970	1.5
Alberta	329	71.36	12.22	0.76	.26	1.00	4.70	.29	.02	6.10	.15	7.96	7.4	51,000	21.4
British Columbia	405	39.55	8.60	1.39	.31	0.40	2.19	.77	.08	2.78	.37	4.82	8.6	676	33.6
British Columbia	406	26.32	3.17	0.42	.16	0.02	.09	.12	.006	0.08	.001	0.81	8.4	49	41.3
Oregon[a]		27.8	7.25	0.27	.49							0.45			
Oregon[b]		35.3	5.6	0.10	.25							0.15			
Oregon[c]		30.6	0.43	0.04	.04							n.d.			

[a]Chief Joseph Mountain.
[b]Hurricane Divide.
[c]Sacajawea Peak (rock that is licked).

tative. Allred gives analyses for 5 licks: in these, sodium ≤ 0.4% (17.3 me/100 g); sulfate is 0.64% (13.3 me/100 g), 3.98% (82.8 me/100 g), and trace (slight and large amounts); calcium oxide is given as 1.0% (35.7 me/100 g), 3.1% (111 me/100 g), and strong in 3 licks and calcium carbonate is given as slight and 48.3% (966 me/100 g) in the other 2; and magnesium oxide is given as trace, small amounts, and 0.8% (66 me/100 g). Much importance was attached to phosphorus (as P_2O_5) in the licks; the data are complete and are given as 0.2, 0.1, 0.209, 0.134, and 0.207% (14.1, 7.0, 14.7, 9.4, and 14.6 me/100 g, respectively). It is not clear how the analyses were performed, but it is likely that a water extract was used. Acid extracts of 3 of the licks yielded 5.56, 7.32, and 60.65% solubles (the latter from the lick with 48.3% calcium carbonate).

Beath (1942) alludes to analyses of 12 licks used by sheep that evidently were in the Gros Ventre area, certainly in Wyoming. From these data he selected one lick to represent the typical salts present (see Table 7.5.) He also describes three additional licks pictured in his report. One of these was used for its phosphorus content; in this lick 16% of the water-soluble salts are phosphorus species. Some calcium carbonate (reported as lime) but no chloride was present. In another lick he says that "40% of the salts given up by treating this earth lick with water is lime carbonate." Sodium chloride was present in the 1% (17.1 me/100 g) to 2% (34.2 me/100 g) range. Calcium carbonate is relatively insoluble in most waters and will bring a weakly acidic water to about pH 8.1 rapidly; therefore Beath's description of this sample is not clear. Finally, he presented a picture (Honess and Frost 1942: Fig. 30) of a nose of a hill in the Gros Ventre Mountains that has been extensively eaten into by elk. For this lick he reported 12% (240 me/100 g) calcium carbonate (again given as lime), and he was specific in citing that no other minerals were found. Beath did not come to any conclusions regarding a commonality of composition in these licks.

Present findings. Levels of bases derived by ammonium acetate extract from 18 licks used exclusively by sheep are reviewed, and of these licks, data for 14 saturation paste extracts were obtained (Table 7.20). These samples, well distributed over the geographic range of the mountain sheep, include sites from Alaska to California and east to Colorado. The Dall, Stone, Rocky Mountain, and desert subspecies are represented in what might, in big-game hunter jargon, be called the "grand slam" of the lick samples. Compared with the means for acetate extracts of the combined lick sample (Table 7.7), sheep licks are enriched in calcium and magnesium and results for the water extract suggest even higher soluble salt levels; however, substantial variability occurs. Sodium is near the grand means in both water and acetate extracts and sulfate is

Table 7.20. Composition of NH_4OAc and saturation-paste extracts of earth materials from licks used exclusively by mountain sheep. All concentrations are expressed as values of GM and GD on a dry-earth basis.

Analyte and unit of measure	*n*	GM	GD
Acetate extract			
Ca (me/100 g)	18	31.6	1.82
Mg (me/100 g)	18	6.97	2.87
Na (me/100 g)	18	2.03	3.03
K (me/100 g)	18	0.54	2.72
Water extract			
Ca (me/100 g)	14	0.61	4.21
Mg (me/100 g)	14	0.66	11.2
Na (me/100 g)	14	0.81	9.74
K (me/100 g)	14	0.066	3.23
SO_4 (me/100 g)	14	1.48	10.4
Cl (me/100 g)	14	0.16	7.30
Cond. (mmho/cm)	9[a]	5.66	4.56
pH	14	7.66[b]	1.29[c]
Total analysis			
S (%)	16	0.089	5.44
$CaCO_3$ (%)	15	10.8	4.97

[a]Five samples were 1 to 5 water extracts.
[b]Arithmetic mean.
[c]Standard deviation.

evidently more common than chloride. Total sulfur is near the grand mean, but calcium carbonate equivalent is about two times higher in the sheep licks.

The sample (123) of washings from concrete trucks in Colorado supplied by R. E. Keiss contains 15.6% calcium carbonate equivalent and the total sulfur content is 401 ppm. Although these are the only analyses that we performed, it appears that the washings are being eaten for their calcium content and perhaps for the buffering effect of the carbonate and alkalinity in the concrete.

Russo (1956), in his study of desert bighorns in Arizona, never found evidence of geophagy; but on one occasion he observed a ram strike a rock and knock off a fragment, which he chewed (Russo could hear the grinding at 15 m distance). Quoting Russo (p. 51): "The rock was later inspected and no other sign of use was found. A thin strata of lime ran through the formation, and a portion of this had been chipped off by the ram."

Sulfate salts of calcium and magnesium occur in a number of the sheep licks, the Stone sheep licks (396, 397) in northern British Columbia being particularly noteworthy. In our study we encountered a geologic report by Eckhart (1951) regarding gypsiferous deposits on Sheep

Mountain, which is about 145 km northeast of Anchorage, Alaska. Here pods of gypsum occur in altered Jurassic lavas (Talkeetna formation). A water course in the area is named Caribou Creek.

LICKS USED BY SEVERAL SPECIES

Previous studies. Cowan and Brink (1949) studied licks used by bighorn sheep, mountain goat, moose, elk, caribou, mule deer, and white-tailed deer in the Canadian Rocky Mountain parks of Alberta and British Columbia. Water and 1.0 N HCl extracts were made of the samples (Tables 7.5, 7.6). The average water-soluble substance was 0.53% ($n = 11$) with a range of 0.16 to 1.19%, and the average acid-soluble material was 22.11% ($n = 10$) with a range of 3.65 to 45.32%. These extracts were subjected to spectrographic analysis, and contents of 18 metals were estimated semiquantitatively. The results did not indicate that any one of the constituents was sought. Phosphorus, calcium, and chloride were determined quantitatively. The average phosphorus content was 0.016% in the acid extract, and calcium content of the water and acid extracts amounted to 0.25% and 4.02%, respectively. Chloride in the water extract amounted to 0.041%. At two sites, the authors failed to find substantial differences in phosphorus and chloride contents between lick and nearby nonlick samples. Cowan and Brink concluded that neither salt nor phosphorus was sought but trace elements might be sought. They suggested that preference tests might be more productive than analyses of the licks.

Stockstad et al. (1953) studied 17 earth licks and 1 aqueous lick in western Montana that were used by mountain goat, elk, and mule and white-tailed deer; however, the authors did not identify the animals using specific licks in their tabulation. Their study also included control samples taken near the licks. The averages and standard errors of means for the data in their Table 3 are given in Table 7.6. Stockstad et al. inferred from their data that only magnesium fulfilled the requirement of being more concentrated in lick materials than in control samples and therefore represented the sought-after nutrient. However, cafeteria trials (discussed later in this chapter) did not reflect a selection for magnesium, so the authors discounted the importance of this element. They concluded that sodium was the needed element, evidently because of the clear preference for sodium in the cafeteria trials, although they pointed out that 2 of the licks contained lower sodium than their associated control samples. When these preference data are subjected to variance analysis, pH and sodium are significantly different and magnesium just misses significance at the 5% level. When the data for pH; sulfate; and chloride and exchangeable sodium, potassium, calcium, and magnesium are treated by stepwise discriminant analysis, only pH and sulfate—in that order—serve to distinguish lick from control soils. Subsequent classification of licks is perfect and only 2 of the 17 controls were misclassified.

Present findings. The relation of species group to lick composition was studied to determine if there was a species affiliation with a particular elemental combination. Licks of the following species were compared by discriminant analysis: caribou, deer, and elk licks combined versus moose licks versus goat and sheep licks combined. The analysis was made using total sulfur, sulfate, chloride, exchange calcium, magnesium, sodium, potassium, calcium carbonate equivalent, and pH. All data except pH were transformed to their logarithmic equivalents. Because calcium carbonate did not occur in some of the samples (which therefore would not permit transformation), these samples were assumed to contain 0.001% calcium carbonate to allow the logarithmic transformation for discriminant comparison. The groupings were made to compare the cervids (deer, elk, and caribou), which for the purpose in mind might be assumed to have roughly similar physiological characteristics, with (1) moose, which occupy a fairly specific niche associated with organic soils, the willow ecotone, and aquatic plants; and (2) mountain sheep and Rocky Mountain goat (recognizing that the latter is an antelope). The handling and preparation of the lick samples had suggested that the gross physical characteristics of the samples had affinities with this three-way classification.

The geometric means and deviations of the three groups are presented in Table 7.21. Analyses of the variances of these data indicate

Table 7.21. Composition of lick materials in groupings used in a discriminant comparison of white-tailed deer, elk, and caribou ($n = 79$) with moose ($n = 16$) and mountain sheep and Rocky Mountain goat ($n = 18$). Except for pH, sample statistics are presented as values of GM and GD.

Analyte and unit of measure	Deer, elk, and caribou		Moose		Sheep and goat	
	GM[a]	GD	GM[a]	GD	GM[a]	GD
$CaCO_3$ (%)**	0.03a	52.9	0.02a	70.0	3.36b	31.3
Ex. Ca (me/100 g)**	6.87a	4.12	19.94b	2.70	27.85b	1.88
Ex. Mg (me/100 g)**	2.32a	3.12	2.67a	2.72	5.69b	2.72
Ex. Na (me/100 g)	1.65	3.52	1.14	2.56	1.67	3.56
pH**	6.44a	1.58	6.88b	1.18	7.80c	1.17
SO_4 (me/100 g)*	0.16a	9.04	0.35b	4.58	0.87a	12.10
Cl (me/100 g)	0.14	10.10	0.06	12.48	0.10	9.51
Ex. K (me/100 g)	0.32	2.51	0.33	2.19	0.42	2.70
S (ppm)*	296a	3.74	488b	3.66	770c	5.49

*$P \leq .05$.
**$P \leq .01$.
[a]Means followed by the same letters are not significantly different by univariate F-test and Duncan's multiple-range test.

Table 7.22. Normalized and standardized coefficients for variables in discriminant functions that distinguish among combined (1) deer, elk, and caribou; (2) moose; and (3) sheep and goat licks. Sample statistics for the groupings are presented in Table 7.21.

Variable	Normalized vector Y1	Normalized vector Y2	Standardized vector Y1	Standardized vector Y2
S (total)	0.237	0.179	0.146	0.110
SO_4	0.286	−0.054	0.279	−0.053
Cl	−0.190	0.056	−0.193	0.057
Ca (ex.)	0.930	−2.340	0.560	−1.410
Mg (ex.)	0.125	1.662	0.062	0.821
Na (ex.)	−0.639	0.118	−0.338	0.062
K (ex.)	−0.359	0.236	−0.142	0.094
$CaCO_3$	−0.015	0.704	−0.028	1.310
pH	0.262	−0.111	0.405	−0.171
Constant	−3.417	2.878	...	...

Function	Y1	Y2
Eigenvalue	0.332	0.296
Contribution (%)	52.83	47.17
χ^2	58	27
df	18	8
P	$<$ .000	.001

[a]Except for pH, variables are log transformations.

highly significant differences for extractable calcium and magnesium, calcium carbonate equivalent, and pH. Sulfate extracted in the saturation paste extract and total sulfur were significantly different at the 5% level.

Results of the discriminant analysis are presented in Table 7.22. Calcium carbonate equivalent, exchangeable calcium, and magnesium entered, in that order, into the function in a significant fashion. The other variates did not contribute to significant separation of the group centroids. The two functions contributed nearly equally to discrimination, and the overall contribution to discrimination as measured by the $\hat{\omega}^2$ statistic was 40.8%. The higher exchangeable calcium contents of the moose and sheep plus goat groups contributed to a high positive score on the first function and thus was important in distinguishing these licks from the deer licks. Similarly, calcium was important in separation of moose licks on the second function.

The posterior classification (Table 7.23) reflects the substantial variance for deer licks and demonstrates how deer might comfortably use these other licks as they do. The moose and sheep plus goat groups contain less variance and reclassify much better than do the other two groups.

Among the samples received, 27 represented licks used by from two to five species, the groupings of which did not correspond to the

Table 7.23. Prediction of group membership in a posterior classification of combined deer, elk, and caribou licks (group 1); moose (group 2); and combined goat and sheep (group 3) licks.

Actual group	n	Predicted group (%) 1	2	3
1	79	58.2	19.0	22.8
2	16	6.3	75.0	18.8
3	18	11.1	11.1	77.8

three-way classification described above. These samples were classified using the same discriminant functions used for the three-way grouping in Table 7.22. The positions of these licks in discriminant space are shown in Fig. 7.9. Eighteen (2/3 of the 27 licks) were used by two species, for example, deer and moose. Of the 9 licks that were misclassified, 7 were used by two species—close to the 6 that would have been expected purely on the basis of statistical chance. This classification procedure provides further evidence as to the nonspecific relationship of the earth materials in licks to a variety of ungulate species.

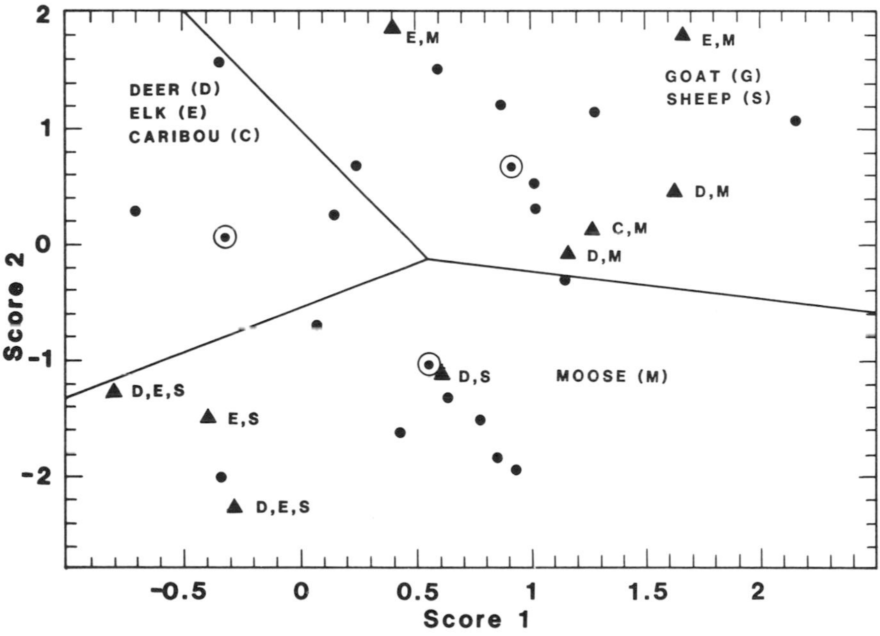

Fig. 7.9. Classification of licks used by several species. Circles indicate that at least one species in the classification group occurred in the field for that species; triangles indicate that none of the species in the test classification occurred in the field to which it was assigned by the function.

The use of discriminant analysis to differentiate these arbitrary groupings has emphasized the importance of calcium in earth materials of licks, either in the extractable form or as a calcium carbonate. A priori, this finding might be anticipated, given the habitats of the species groupings. Thus, the deer lick sample comes primarily from eastern United States where acidic soils are common; moose occupy habitats where organic soils tend to be acid; and sheep and goats inhabit areas where soils are often thin or unweathered rock exposures are common. And, as might also be anticipated, the average level of each variable (Table 7.21) reflects the more general or regional geochemical character rather than representing some mineral artifact of the habitats of the species groupings. We cannot say, for example, that mountain sheep and goats, because of some inherent specialized physiological needs, are using lick-earth materials richer in alkaline earths than those used by cervids. Rather, the deer in eastern United States likely would prefer calcium to magnesium carbonate–rich licks if they were available as, indeed, they do where travertine or tufa occur. However, as pointed out in Chap. 9, deer alleviate the same physiological stress induced by the high potassium and low alkaline earth levels in the spring flush of vegetation by their use of sodium- and magnesium-rich licks.

Artificial licks

There is no doubt that wild ruminants can be attracted to sodium chloride. Common salt has evidently been used for many years in attracting and baiting animals for hunting purposes. The ease of attraction and fidelity to these sites by wild ungulates has resulted in the universal outlawing of the practice. The data (Table 7.24) for two sites in Pennsylvania and one in Manitoba together with the data (provided by H. P. Weeks) for four sites in Indiana are of special interest because of levels of salt occurring in the soils. No records indicate when the sites were salted except for the Manitoba lick, so the extent of leaching by rain and snow melt is unknown. D. Bigelow, F. Anderson, and L. Dubray, who sampled the Manitoba site, reported that as much as 22.7 kg (50 lb) of common salt was spread on 9.3 m^2 (100 ft^2) annually from 1953 to 1976, the year of sampling. This amount of salt uniformly distributed through soil to a depth of 5 cm (assuming a soil density of 1.3 g/cm^3) creates a concentration of 38,000 ppm for a single addition. The level of 390 ppm found suggests clearly that much of the applied salt had been leached, consumed, or both.

The level of 0.5 me NaCl/100 g of the Manitoba lick sample, which is similar to levels of the Pennsylvania samples, is equivalent to 290 mg NaCl/1000 g of soil, or 290 ppm. The Manitoba sample evidently is

Table 7.24. Contents of selected elements (me/100 g), conductivity, and pH in ammonium acetate and water extracts of earth; and total S and $CaCO_3$ equivalent of earth from licks created by spreading NaCl. Data from Indiana courtesy H. P. Weeks.

State or province	Sample no.	Acetate extract Ca	Mg	Na	K	Water extract Ca	Mg	Na	K	SO_4	Cl	Cond.	pH	Total S	$CaCO_3$
		me/100 g				me/100 g						mmho/cm		ppm	%
Pennsylvania	186	0.75	0.16	0.79	.34	.04	.02	0.87	.07	.03	0.67	2.98	4.7	108	. . .
Pennsylvania	187	1.00	0.27	0.61	.05	.05	.002	0.45	.008	.04	0.48	1.27	3.9	209	. . .
Manitoba	145	15.46	4.49	3.64	.62	.15[a]	.04	1.13	.05	.03	0.21	0.22	7.4	231	.26
Indiana		1.79	1.42	7.83	.27	n.d.	n.d.	n.d.	n.d.	.08	1.41	n.d.	4.5[b]	n.d.	n.d.
Indiana		9.42	2.81	4.35	.19	n.d.	n.d.	n.d.	n.d.	.08	1.27	n.d.	5.1[b]	n.d.	n.d.
Indiana		1.98	0.63	2.96	.18	n.d.	n.d.	n.d.	n.d.	.08	0.005	n.d.	6.0[b]	n.d.	n.d.
Indiana		16.45	0.76	0.43	.32	n.d.	n.d.	n.d.	n.d.	.14	0.89	n.d.	7.1[b]	n.d.	n.d.

[a]Water extract in ratio of 5 parts water to 1 part soil, by weight.
[b]Determined on 1 part water by volume to 1 part soil, by weight.

Table 7.25. Contents of selected elements, conductivity, and pH in NH_4OAc and water extracts of earth and total S and $CaCO_3$ equivalent of earth from animal licks associated with wells.

State	Type of well	Sample no.	Acetate extract Ca	Mg	Na	K	Water extract Ca	Mg	Na	K	SO_4	Cl	Cond.	pH	Total S	$CaCO_3$
			me/100 g				me/100 g						mmho/cm		ppm	%
Pennsylvania	oil	103	12.08	1.44	1.24	0.56	0.20	.02	0.25	.005	0.16	0.14	0.73	5.0	778	. . .
Michigan	oil	104	20.53	1.77	5.03	0.10	1.17	.37	4.22	.05	0.49	4.69	1.30	6.9	215	. . .
Michigan	oil	232	12.60	0.51	0.53	0.15	0.16	.01	0.33	.009	0.08	0.13	1.49	7.9	129	1.0
Tennessee	gas	233	2.64	0.60	0.57	0.07	0.01	.02	0.21	.004	0.01	0.10	0.98	7.7	32	0.1
Wyoming	water	346	4.74	0.60	10.53	1.20	0.009	.003	2.05	.02	1.38	0.05	8.62	9.5	364	0.3

montmorillonite rich; but since a routine preparation of a saturation extract did not yield sufficient extract for analysis, a 5-part water to 1-part soil extract was performed. Sodium was clearly more abundant than chloride in this extract and can be accounted for by accumulation of sodium over the 20-odd years that this site has been salted (see App. 1). A nearby natural lick in Manitoba (143) was found to contain substantial amounts of sulfate and chloride salts, including about three times more sodium extractable by water than that found in the artificial lick.

Comparison of sodium in the ammonium acetate extracts with sodium in soils of eastern United States indicates that the Pennsylvania samples and the least concentrated of the Indiana samples were five or six times that expected in the surface soil. Extractable sodium in the Manitoba sample and in the other three Indiana samples is substantially higher than that expected in soils of the region. Calcium levels were high in three of these samples.

LICKS AT WELL SITES. Data are presented in Table 7.25 for the composition of five samples from widely distributed well sites. All sites are associated with oil drilling except the Wyoming locality, which is an artesian well. The water-extract data suggest that substantial salt occurs in sample 104 from Missaukee County, Michigan. The data for the Wyoming sample suggest that water from the well is of the sodium sulfate type. The remaining samples are not unique when judged from salt levels obtained from water extractions. Ammonium acetate extracted substantial calcium from the Pennsylvania and Michigan samples and sodium was high in the Pennsylvania; the Missaukee County, Michigan; and the Wyoming extracts. In all except the Tennessee sample (Fig. 7.10), it can be speculated that sodium and/or calcium were the elements sought. The Tennessee sample is low in all bases, the pH is relatively high (7.7), and the sample contains a very small amount of carbonate mineral (0.1%); therefore calcium from extractable sources and from carbonate minerals is assumed to be the sought-after mineral.

The Wyoming sample is the only antelope lick represented in this collection, and as mentioned earlier, observations of geophagous activity by pronghorn antelope are rare. This site is evidently used for its sodium content. The high pH (9.5) suggests that carbonate salts of sodium are present.

ANIMAL CAFETERIAS. Within the category of artificial licks, it seems appropriate to discuss the so-called cafeterias or experimental setups that scientists have used to determine the factor or factors in soil sought in geophagous activity. Cafeterias consist of wooden pegs impregnated

Fig. 7.10. White-tailed deer lick (center foreground, sample 233) at an abandoned gas well in Cumberland County, Tenn. (Courtesy K. Garner)

with salt solutions of different composition, pans or pots of salt-impregnated soil, or plots of salt-drenched soil. Blair-West et al. (1968) used salt-soaked pegs in studying the nutrition of wild rabbits in southern Australia. Stockstad et al. (1953) used pots containing 23 different salts and plots of soils drenched with the same salts in a study of licks used by mountain goats, mule and white-tailed deer, and elk in western Montana. Weeks (1974) used a peg cafeteria to study mineral nutrition of rodents in southwestern Indiana. Fraser and Reardon (1980) placed pails with sodium, potassium, magnesium, and calcium salts near a lick used by white-tailed deer and moose on the Sibley Peninsula, southwestern Ontario.

By weighing the amount of soil consumed or wood gnawed from the peg, investigators have inferred which of the salts is sought. Blair-West et al. and Weeks concluded that sodium was sought in each of rabbit, squirrel, and woodchuck populations that they studied. Stockstad et al. also found that sodium salts were preferred among the ungulate populations they studied.

Both Blair-West and colleagues (1968) and Weeks (1974) combined studies of adrenal glands with their behavioral studies of salt appetite. In both studies hypertrophy of the glands was apparent and substantial increases in the zona glomerulosa occurred, which is evidence for sodium conservation. In their summary, Blair-West et al. (p. 926) speculated on the very interesting mechanisms that underlie the animal's discovery of the specific pegs. We may similarly wonder about the discovery of natural licks.

Fraser and Reardon (1980) found that practically all the 1000 ml of sodium chloride, sodium sulfate, and sodium bicarbonate solutions (all 0.1 M) were consumed by deer and moose during the 30 hours that pails bearing the solutions were available. In contrast, potassium, magnesium, and calcium salts were little used and the authors concluded that the sodium ion was sought in the four natural (wet) licks in the area that had an average of 7.1 me Na/l.

Licks in Africa

In an account of geophagia in east Africa, French (1945) drew together what must have been all the published analyses for licks of the region. French believed that cattle and goats did not eat the soils for their sodium content, but his contention is not borne out by the analyses. Evidently, natives of Africa have utilized licks as long as they have domesticated animals, and the use of licks by wild ruminants at specific times is documented. Many licks listed by French are rich in sulfur—Glauber's salt being present. Some of the earths from licks are leached of their salts by the natives and subsequently precipitated and sold for livestock use. French concluded that the factor or factors sought in the licks was not apparent and he suspected that some trace element might be important. Data for certain of the analytes reported by French have been averaged and are recorded in Table 7.26. The contents of acid-soluble ash (method of acid treatment not specified) are high for

Table 7.26. Contents (means and standard errors in me/100 g) of selected elements in earths eaten by livestock in equatorial Africa.

Country	*n*	Ca	Mg	Na	K	P	S
Nigeria	16	23 ± 7	12 ± 3	988 ± 175	17 ± 6	4.6 ± 1.1	414 ± 247
Kenya	25[a]	52 ± 12	n.d.	230 ± 72	17 ± 5	11.1 ± 3.4	n.d.
Tanganyika	16[b]	11 ± 2	n.d.	112 ± 47	2.7 ± 0.1	7.8 ± 5.3	43 ± 16

Note: Some of the data are for impure salts beneficiated from earth materials by native herders and entrepreneurs. Analyses for Nigerian and Kenyan materials are evidently total analyses, whereas those for Tanganyika are HCl soluble (data calculated from French 1945).

[a]For P, $n = 7$.

[b]For K, $n = 14$; for P, $n = 12$.

Nigeria and Kenya, suggesting the salty natures of these suites of samples. Unlike the interpretation of French, we (on the basis of data in Table 7.26 and other data by French) are impressed that sodium is rich in these earths relative to the low levels present in soils of the native pastures, situated in many instances, no doubt, in highly leached and weathered soils.

Boyd-Orr (1929:81–84) refers to pathology in African stock that is in large part nutritionally based. Regrettably, at that time no patterns of etiology were detectable, and nearly two decades later when French (1945) published his essay the picture was no clearer except for the use of bone meal as a source of calcium and phosphorus. In the perspective of Boyd-Orr's compilation of worldwide data for the nutritional quality of pastures, it is unfortunate that French did not relate any of his observations to pathology, production, or life history of the herds with which he had come into contact.

In their monograph regarding rangeland management in Africa, Pratt and Gwynne (1977) mention that livestock crave salt in most of the region; "few of these natural licks show to advantage when analyzed, and it has often been noticed that they are abandoned when animals have access to an introduced mineral mixture, well suited to the area" (p. 199). Unfortunately, these authors do not give any analyses for the natural licks to which they referred.

Langman (1978) reported geophagy in giraffe in eastern Transvaal; 93% of the animals observed eating soil were subadults, and termite mounds were favored earth-eating sites. Osteophagia, involving particularly carpal bones of giraffe, also was common; Langman concluded that the calcium and phosphorus nutrition of these giraffes was inadequate and by osteophagy they were attempting to compensate for these deficiencies. He observes that rickets often develops quickly in young giraffes. Evidence that calcium was sought comes from the findings of Hesse (1955) on the mineral composition of earth materials in termite mounds in east Africa, particularly the termitaria of the genus *Macrotermes*. Hesse found that mounds built on poorly drained soils often contained relatively high concentrations of calcium carbonate, which sometimes became concentrated in very large pieces in the base of the mound. Carbonate also was disseminated throughout the mound as a fine powder, and the average calcium carbonate content of the entire mound amounted to 1 to 2%. Carbonate concentrations may accumulate in termite mounds for two reasons: (1) free carbonates may occur in the soils surrounding the mounds, particularly in the subsoil from which the mound is constructed; and (2) the mounds, because of their more aggregated structure (compared to the soil substrates on which they rest) and their slightly higher fertility, support vigorous vascular plant growth that transports large amounts of bicarbonate-bearing water from around the mound. Langman's giraffes probably were selecting this calcium carbonate-bearing earth. The fungal colonies maintained by the termites also were found to contain about 1% calcium (Hesse) as well as about 20% water solubles—hence, these colonies would contain about the same amount of calcium as an equivalent amount of the earth making up the termite mound. Availabilities might be quite different. French (1945:156) also reported goats, sheep, and cattle eating "ant earth" in Tanganyika (present-day Tanzania) and speculated that these animals were relieving a calcium shortage. Recall that North American wild sheep eat calcium-rich earths.

Penzhorn (1982) reported results of analyses of 3 lick earths eaten by Cape Mountain zebras and antelope (species not specified) in South Africa. Nineteen of 21 observations of lick use were made during summer months. From the data presented for ammonium acetate extracts, Penzhorn concluded (p. 84): "It appears that calcium may have been the essential element." He also recorded an observation of a zebra (p. 86) "biting a chunk off a termite mound and eating it." The mean contents (B. L. Penzhorn, pers. commun.) for calcium, magnesium, sodium, and potassium were 20.3 (s = 8.5), 10.4 (s = 9.1), 1.3 (s = 0.2), and 2.9 (s = 1.5) me/100g, respectively.

CHAPTER EIGHT

White-tailed deer and their mineral environment

OUR EVALUATION of mineral licks used by ungulates and ingestion of soil-contaminated forages has led us to consider (1) the relationship of rock substrates, soils, and mineral content of the diet in respective ranges to physical size, fertility rates, and antler size; and (2) the relationship of minerals consumed to physiology and biochemistry. This chapter is concerned with the relationship between soil mineralogy and soil fertility to whole-animal responses.

Relations of soils to weights and fertility

Biologists and soil scientists long have sought to associate phenotypic development and population density with soil fertility (Albrecht 1946). In 1921, Hesse pointed out that roe deer collected on chalk terranes weighed 4 kg more than deer from sandstone and graywacke bedrock types. In an earlier literature survey, we found what appeared to be relationships between body size and population densities of diverse species of animals and sulfur levels in their environments (Hanson and Jones 1976: App. 2), and as we have pointed out, this hypothesis lead to our study of animal licks. From the analyses of lick samples, we concluded that lick sites could be explained by levels of calcium, magnesium, and sodium that were in most cases well above levels found in the overall environment of a respective range. Ruminants are attracted to sodium-bearing soil as are a great many other animals representing diverse phyletic origins—from butterflies to rodents. But in addition to this sodium attraction, the lick data also emphasize the overriding significance of magnesium, calcium, and alkalinity in the earth materials of licks and in range soils in general. These latter elements, placed in the perspective of well-developed endocrine systems that conserve sodium with extreme efficiency, are now considered in relation to the whole-animal response, using the white-tailed deer as the classic example.

Calcium soils in perspective

Agriculturists long have realized the importance of calcium-rich soils to soil productivity. An understanding of this relationship was noted early in Europe and carried to the North American colonies. Early in this century, the relationships of soil calcium to yield, morphology of the plant (ecotype), plant geography, and quality of the crop were noted by the soil chemist Hilgard (1906), who clearly espoused the benefits of calcium-rich soils to plant agriculture. Hilgard identified calciphile vegetation with sandy soils having as little as 0.1% CaO, whereas clay-rich soils to support calciphile vegetation must have 0.5% CaO (1906:524). These levels evidently were determined by strong acid digestion with heat, and the difference associated with texture implied a relation to a cation-exchange capacity. Contemporary European scientists placed the division between calcareous and noncalcareous soils at 2% CaO. At the turn of the century the concept persisted that "a lime country is a rich country." However, doubts were emerging concerning the permanence of monoculture, for example, corn in the Midwest (Hopkins 1910), where field experiments were begun to test the effects of long-term cultivation. At about this time Hilgard (1906) indicated that calcium-rich soils were being condemned in Europe ("injuriousness of an excess of lime is among the foremost themes of European [especially French and English] agricultural writers" [p. 252]). It could be that the "poor chalk lands" spoken of were, in reality, eroded fields in which limestone did create infertility and an impediment to establishment of crops. It is a logical step beyond Hilgard's admiration of calcium-rich soils to his speculation regarding the establishment of civilization on these soils. The Nile floodplain and delta, the Middle East centers of culture including the Tigris and Euphrates plains, and the Indus plain all share a common calcium richness. The importance of soils in culture development in these areas and the influence of soils in the development of civilization was the topic of Hyams's book published five decades later (1952).

Ruminant populations in eastern United States before settlement were generally low. Deer were confined to edge habitats occurring in the vast forests (Leopold 1950). Extensive open regions like the Bluegrass Prairie of Kentucky and the Nashville Basin were certainly unique and attracted both cervids and bison in unusual numbers. With respect to these concentrations, native Americans developed specific hunting regulations and cultural practices such as burning to ensure production of forage suitable for deer. The game populations on these ranges naturally attracted the attention of the white scouts who first crossed the Appalachians. By comparison with dwindling game populations in Pennsylvania, Virginia, and New England, the herds seen in Kentucky were indeed large.

In the south, deer populations were never very high in the Coastal Plains and Piedmont. Before settlement, as now, deer sought out the bottomland hardwoods (Stransky and Halls 1967), and in contrast the extensive pinelands on the uplands maintained populations of perhaps one-fourth the density of the bottomland hardwoods.

In southeastern United States the effects of soil fertility are most readily apparent on free-ranging game. The lower floodplain and delta of the Mississippi River represent an extensive area of fertile soils in a region of relatively infertile upland soils. These soils not only contain more nutrients, but also pH is higher and within a range that is more amenable for microbial activity and vascular plant growth. A similar situation is found on the upland in the so-called Black Belt region of Alabama and Mississippi and the Jackson Prairie of Mississippi where Miocene chalks and limestones give rise to the gum-oak-juniper association. These arcuate outcrops of calcareous rocks represent fertile areas that were exploited in the early agriculture of the region. Tombigbee River flows for some 160 km in the Black Belt and its bottoms represent some of the best deer habitat in Mississippi and Alabama. To the east in Georgia, riparian woodlands of the Ocmulgee and Oconee rivers yield some of the largest deer in the state. In the Carolinas the largest populations tend to occur in tidewater counties where Tertiary chalks and marls outcrop. These points are not meant to suggest that the intervening country is or was devoid of deer or other ruminants in earlier times but that deer populations flourish best in habitats that occur over calcium and magnesium-rich parent materials.

Mean weights of fawns

The mean weights of fawn does in northeastern United States were chosen to best reflect the capacity of ranges to produce large animals; ranges that produce large animals also produce animals that reproduce early in life and have high fertility rates. The map of weights is presented in Fig. 8.1. Characteristics and origins of the data used in constructing Fig. 8.1 are given in App. 8. In this exposition, genetic controls on body size cannot be considered, but there is a direct relationship between body weight and latitude as they relate to soils. The data in Fig. 8.1 indicate that soil fertility rather than climate per se is the primary major factor determining deer size. We are aware that in some interstate comparisons

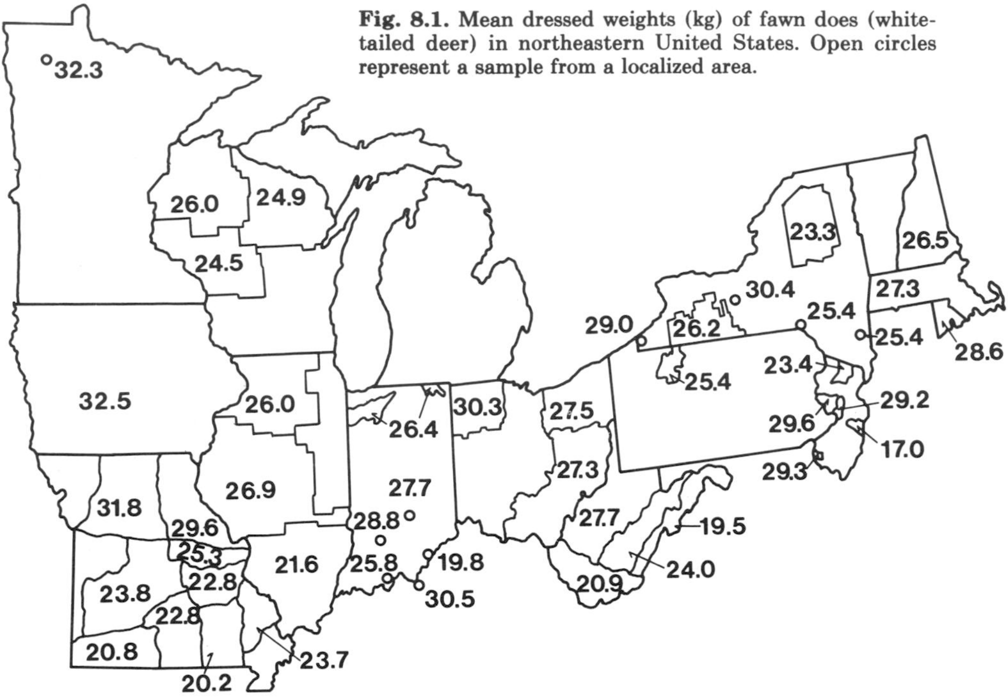

Fig. 8.1. Mean dressed weights (kg) of fawn does (white-tailed deer) in northeastern United States. Open circles represent a sample from a localized area.

ages of fawns examined at check stations vary between birth and hunting-season dates; nevertheless, these factors do not materially affect within-state comparisons. Given the uneven character of the data, both for animal and for soil, this attempt is obviously a first approximation. Eventually, we suspect that more regional data will accumulate to corroborate our findings and identify a mosaic of nutritional habitats within the more generalized regions. On a large geographic scale Olmstead (1970) studied white-tailed deer productivity data for Texas and states east of the Mississippi River, combining information for contiguous states into five regions. Olmstead considered the effects of great soil group and forest type on deer productivity. He concluded (p. 9): ". . . different forest types and soil types within these regions, except one, have only a nominal, if any, effect on deer productivity as measured by corpora lutea counts." Perhaps Olmstead's combination of data from diverse biogeochemical provinces obscured more localized and real differences. Later, Sileo evaluated similar data for number of embryos per fawn, yearling, and adult classes among animals from Minnesota and all states east of Mississippi River. Sileo's (1973) approach differed from Olmstead's in that he considered regions within 5 of the 18 states that he studied. Using the map of soil associations in the 1938 Yearbook of

Agriculture (USDA 1938) and a map of forest regions of United States (USDA Forest Service 1948), Sileo arbitrarily categorized great soil groups into low and high productivity classes and forest types into poor, fair, and good. Among the variables analyzed, including land in farms, physiography, and land resource unit (Austin 1965), the interaction of two variables—soil quality and forest quality—resulted in six "deer fertility regions." As would be expected, Sileo's map (his Fig. 6) has some affinities with our Fig. 8.1 and is another attempt at a broad geographic interpretation of soil fertility and the physiological responses of a native ruminant. Murphy (1970) presented a map of fawn weights, mostly male, for deer in midwestern states. Murphy said (p. 7) that regional differences in weight were a response to "amount of cultivated crops in the deer diet" and that soil fertility influenced condition to the extent that fertility affected land use.

In the following state-by-state accounts, an attempt was made to find associations between soils and size of female fawns; in cases where weight data are missing we have used fertility rates for New York and antler-beam diameters for Vermont.

EASTERN STATES

Vermont. McLaughlin et al. (1971) have analyzed antler-beam diameters and weights of yearling male deer in relation to the seven physiographic regions in the state (Fig. 8.2). Data for three of the region-deer groupings showed no significant overlap at the 5% level of probability (Student-Newman-Keuls test, Table 1, McLaughlin et al. 1971). Antlers having the largest beam diameters came from the Lake Plains in the northwestern corner of the state (16.46 mm) and from the northeastern corner (16.69 mm). The smallest antlers (15.14 mm) came from the foothills that form a north-south belt on the east flank of the Green Mountains. Soils of the Lake Plains are developed in calcareous lacustrine deposits (Pleistocene) in contrast with those of the foothills that are developed in drift derived from slates and schists. Large antlers from deer in the northeast likely reflect nutritional responses from areas of soils developed from calcareous sandstones (Northeastern Soil Research Committee 1954). McLaughlin et al. also presented weight data for yearling bucks in three game management zones (different from above) that they selected among the seven physiographic regions. Deer in the Lake Plains were heaviest, weighing an average of 53.0 kg (n = 187); and deer from their northeastern zone, corresponding largely to Caledonia, Essex, Lamolle, and Orleans counties, weighed 51.3 kg (n = 438). Deer from the remainder of the state weighed 46.8 kg (n = 2454). Antler-beam diameter tended to increase with weight, and the authors

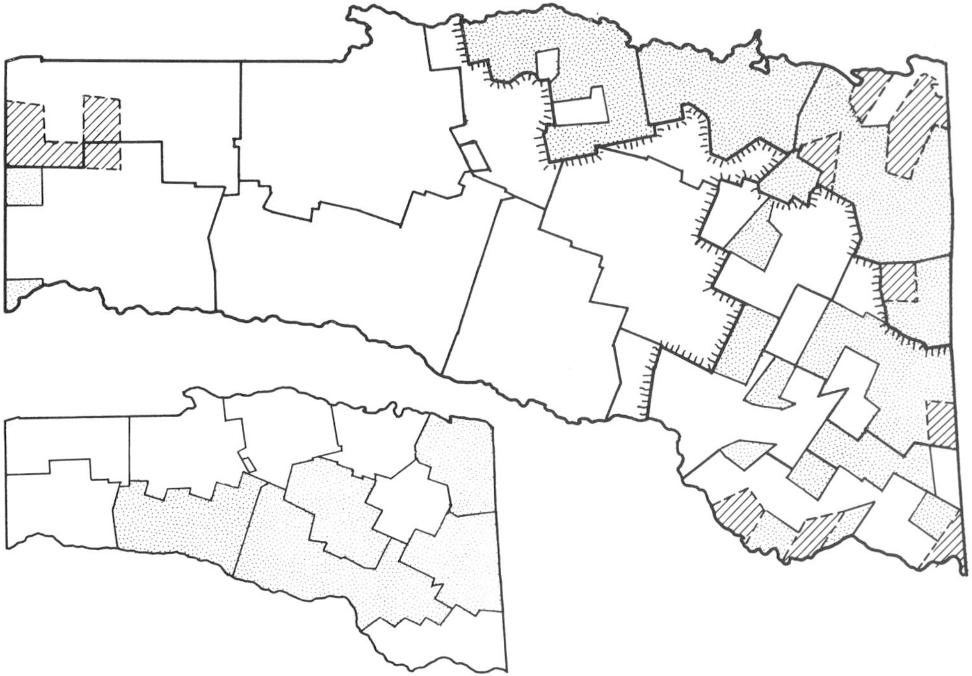

Fig. 8.2. Occurrence by towns (stippled) in Vermont of yearling deer with antler-beam diameters greater than 17 mm. Hachuring indicates no sample. Ticked lines divide Vermont into Lake Plains, northeastern, and southern zones (McLaughlin et al. 1971). Counties with soils of 10–20 me exchangeable Ca/100 g (from Lane and Sartor 1966) are stippled on the inset map.

Table 8.1. Values of GM (me/100 g soil) and GD of extractable bases in surface horizons of soils from New York.

Region	n	Ca		Mg		Na		K		Source
		GM	GD	GM	GD	GM	GD	GM	GD	
Adirondacks	18[a]	2.11	2.93	0.38	2.22	.1[b]	. . .	.35	2.53	R. W. Arnold, pers. commun. 1978
Adirondack periphery	8	6.7[c]	6.0[d]	1.5[c]	1.0[d]	n.d.	n.d.	.57[c]	0.15[d]	Cline and Lathwell 1963
Western	5	4.24	2.24	0.46	1.34	.1[e]	. . .	.23	2.35	Soil Survey 1971

[a] Except K, where $N = 16$.
[b] Nine samples were < 0.1, 8 were 0.1 and 1 was 0.3.
[c] Arithmetic mean.
[d] Standard deviation.
[e] 4 samples were 0.1.

speculated (p. 62) that indices of reproduction would have followed these condition indicators if sample sizes had been larger. Perhaps these findings should be interpreted in light of the lithology and soil conditions of the specific area representing the range of the animals in question and the data analyzed for differences among contrasting soil conditions. Indeed, in its publication regarding fertility of New England soils, the Northeastern Soil Research Committee commented: "In general this grouping [basis of carbonate content] also indicates the areas where the most productive soils occur."

New York. The fertility of white-tailed deer in the various regions of New York was reported in a seminal paper by Cheatum and Severinghaus (1950) who accumulated data on numbers of embryos and corpora lutea in about 900 does over a 10-year period. The regions corresponded to land use, native vegetation, and broad physiographic areas that are largely coextensive with contrasting rock types (Table 8.1), soil development, and native soil fertility. The data reflected substantial differences between the Adirondacks on the one hand and the northern extent of the Allegheny Plateau and the Catskills on the other hand. Soils of the latter regions are alike and markedly superior to the notoriously poor Podzols of the Adirondacks. As measured by productivity, the most marked differences occurred in the fawn age group in which the incidence of pregnancies increased in the order of inferred productivity potential of the soils (Table 8.2).

Later, Severinghaus (1955) reported weights of 23.3 kg for female fawns from the Adirondack Region compared with 26.2 kg from the western area (Fig. 8.1). Other data from Hesselton and Sauer (1973)

Table 8.2. Productivity of fawn and adult ($\geq$ 1.5 years) does among regions of contrasting soil fertility and geologic substrate in New York (Cheatum and Severinghaus 1950).

	Age class					
	Fawn			Adult		
Region	embryos	corpora lutea	pregnancies	embryos	corpora lutea	pregnancies
	no./animals		%	no./animals		%
Adirondacks	.03[a]	n.d.	3.4[a]	1.06[b]	1.11[b]	78.9[b]
Adirondack periphery	.08	n.d.	7.7	1.29	1.71	85.7
Central Catskills	.30	0.31[c]	20.0	1.37	1.72	86.8
Catskill periphery	.27	0.37	26.7	1.48	1.72	91.9
Western	.42	0.39	39.1	1.71	1.97	94.3

[a]Data for 1939–1943.
[b]Data for 1939–1943 and 1947–1949.
[c]Data for 1947–1949.

suggest that fawns in the Catskills would average about 25 kg (Fig. 8.1), and these authors give data for several localities in western New York that yielded fawns averaging nearly 30 kg.

Cline (1961) described the soils of the Adirondack Highlands (which are mostly derived from granitic rocks) and the Adirondack periphery (as defined by Severinghaus et al. 1950) as being "very highly acid and are low in all the major nutrients." Important soils of the Catskills are Walton, Wellsboro, and Oquaga, all derived from red sandstones and shales, and these soils are characterized by very strong acidity and low phosphorus content (Cline 1961). Potassium quickly becomes limiting when nitrogen- and phosphorus-bearing fertilizers are applied. Cline notes that legumes are difficult to maintain in pastures. Several soil associations that support the most fertile deer occur in the western area discussed by Severinghaus et al. (1950). Erie and Langford soils are important on the smooth slopes of the till-draped hills at the northern face of the Allegheny Plateau (Cline 1961). A combination of acidity, low phosphorus, and potassium limit production, as does wetness, which is caused by a subsurface pan horizon that is very slowly permeable. Generally to the south of Erie and Langford soils and farther into the Plateau, extensive areas of Volusia and Mardin soils occur on the long, gentle slopes of the till-mantled hills. Agriculturally, these soils demand management similar to that of the Erie-Langford association. West of the Genesee River in southern Cattaragus and Allegheny counties (more or less the physiographic extent of the Allegheny Hills), there are extensive areas of Lordstown soils that are situated on steep slopes and are characterized by thin sola. Agriculture is not practical in this area. To the east of the Genesee River the slopes in the Plateau are more moderate, and Lordstown soils that occur on steep areas are associated with Mardin and Volusia soils on these lesser slopes.

A few data for extractable bases were found for the Adirondacks and the western region (Table 8.1). Although far from sufficient in number, they are indicative of better soils of the western region, particularly as expressed by higher calcium content. The great variation in soils of the flanks of the Adirondacks shows the variation in rock types outcropping there. The heavy fawns reported by Hesselton and Sauer (1973) from the Seneca Army Depot in Seneca County and from Ripley near Lake Erie in Chautauqua County probably are attributable to development on fertile soils of high base status derived from moderately calcareous shales and from base-rich lacustrine materials from Pleistocene lake stands, respectively.

New Jersey. Perhaps no state can boast of such a variety of soils in so small an area as can New Jersey. From the sandy soils of dunes on the barrier islands through clay-rich soils developed on Cenozoic and

Paleozoic sediments in the Coastal Plain and Piedmont to crystalline rocks of both acidic and basic composition in the Piedmont, there is a striking array of soils of a marked productivity range. Limestone valleys of the folded Appalachians in the northwest add to the variety. These differences in lithology are expressed in productivity of soils and land use. A substantial body of data (Burke et al. 1975, 1976, 1977, 1978) for white-tailed deer weight by age and sex for the 40 deer-management zones provide a unique opportunity for comparison of deer size with soil conditions.

Five areas were chosen for comparison. The zone boundaries do not correspond to soil-area boundaries and mapping of soils at small scales has not been published for all of the areas; consequently boundaries in Fig. 8.3 have been generalized from the small scale maps of Quackenbush

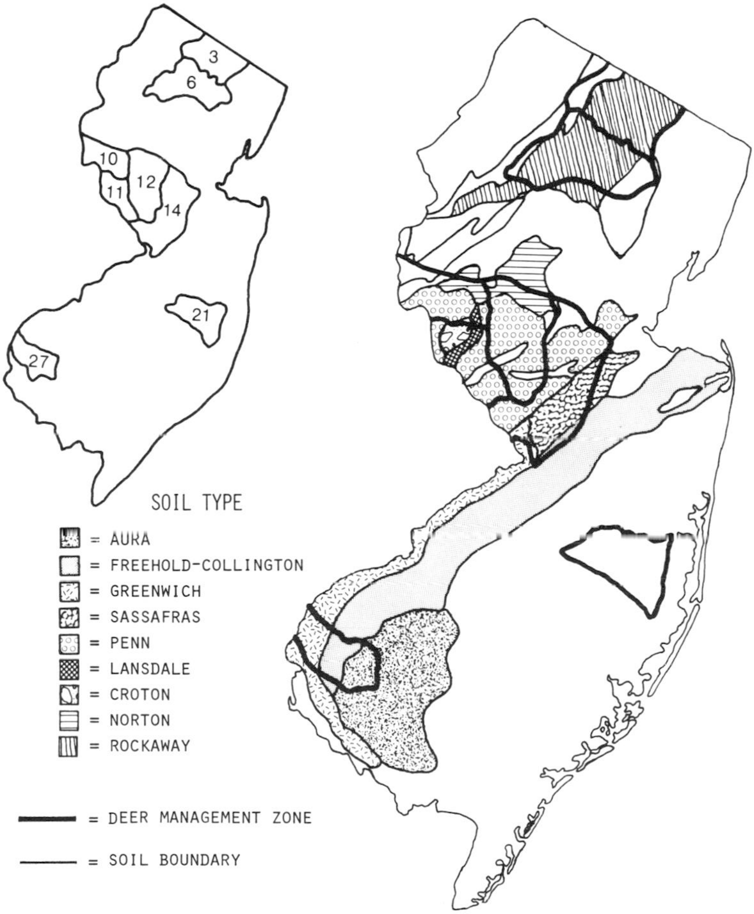

Fig. 8.3. Eight selected deer management zones and distributions of major soils (after Tedrow 1978) relative to zones in New Jersey. Data for soils and deer condition are found in Tables 8.3 and 8.4.

(1955) and Tedrow (1978). The four areas correspond to the Northern Highlands, the Hunterdon Plateau and adjacent Red Shale Country, the Greensand Belt, and the Outer Coastal Plain. These areas and the associated deer-management zones are presented in Fig. 8.3. Weights of female fawns for these areas for four successive years beginning in 1974 are presented in Table 8.3 together with deer density data and some relevant characteristics relating to deer production in these ecosystems.

The weight data are remarkable for the great range represented and are only approached by those for Missouri. The arrangement by physiographic region in Table 8.3 progresses from the poor sandy soils of the Outer Coastal Plain and Pine Barrens through intermediate environments (as identified by weight) to the relatively rich soils of the Inner Coastal Plain. The low agricultural potential of the Barrens is notorious; this lack of productivity is reflected in the lowest weights found for deer in the Northeast. Soils in the Northern Highlands are derived from thin glacial drift ultimately derived from underlying gneisses; the soils are acid, friable, and tend to be well-drained. Tedrow (1978:13) records that Rockaway soils (the most common soil series of the region) is potentially productive, but stoniness and steep slopes limit exploitation. We have combined the zones that are more or less coextensive with Hunterdon Plateau and the surrounding Red Shale Country to represent an area of sedimentary rocks ranging from feldspathic sandstone and shale to argillite that underlies the plateau. Soils developed from old glacial drift derived from red shales, crystalline rocks, and limestone occur in the northern part of the area. Deer from this region compare very favorably with deer from productive environments in northeastern United States. When any one of the four years is considered, it is noteworthy that Croton soils on the plateau support the largest deer of the three zones in the region. Tedrow (1978:15) has classed the agricultural potential of these soils as low compared to very good for Lansdale and fair for Penn soils in the Red Shale Country (Table 8.4). Norton soils in the northern part of these management zones are similar to Penn soils in potential. Sandy sediments form a band traversing northeast across the Inner Coastal Plain—the Greensand Belt of New Jersey, so-called because of the varying amounts of the green micaceous mineral glauconite contained in sediments. The Freehold-Collington soils in management zone 27 and Sassafras soils in zone 14 are the most productive in the state (Table 8.1). These areas support heavy deer and there tends to be little weight variation from year to year. Terraces bordering the Delaware River are composed of leached silts and fine sand and comprise the area of productive Greenwich soils. A substantial portion of zone 27 is in Greenwich soils. On the eastern third of the zone, about 0.67 m acid sands

Table 8.3. Physiography, bedrock geology, soils, land use, deer density, and weight of fawn doe white-tailed deer from selected deer management zones in New Jersey (Burke et al. 1975, 1976, 1977, 1978; Groff 1978). Zones represent a large range in bedrock geology and in productivity of soils.

Physiographic region	Rock type	Major soils	Deer manage. zone	Land use[a] (%) agri-culture	Land use[a] (%) wood-land	Deer density[b] (n/km^2)	Fawn doe weight, kg (*n*) 1974	1975	1976	1977
Outer Coastal Plain	quartzose sand	Lakewood	21	3.3	91.3	17.8	16.5 (20)[c]	16.7 (15)[c]	20.5 (5)[e]	16.8 (7)[c]
Northern Highlands	gneiss	Rockaway	3	4.2	76.9	43.5	25.2 (27)	19.6 (17)	24.3 (15)	21.7 (40)
			6	4.2	65.6	32.0	25.4 (13)	21.4 (51)	25.8 (19)	22.9 (20)
Hunterdon Plateau and Red Shale Country	argillite (plateau), sandstone, shale	Croton (plateau),	10	57.6	38.8	108.3	28.6 (87)	27.0 (83)	27.5 (75)	28.8 (45)
		Lansdale (sandstone),	11	59.3	33.8	85.8	31.2 (28)	27.8 (45)	28.9 (77)	30.2 (36)
		Penn (shale)	12	58.3	28.9	47.9	28.2 (8)	28.8 (41)	28.7 (50)	30.6 (48)[d]
Greensand Belt	glauconite-rich sand	Freehold-Collington	27	54.8	23.0	16.4	n.d.	29.2 (2)[d]	30.8 (2)[d]	28.6 (4)
Inner Coastal Plain	gravels, sands	Sassafras	14	46.3	23.3	23.3	28.8 (8)	30.1 (18)	29.2 (34)	28.6 (16)

[a]As percent of area in zone, derived from aerial photography of 1972.
[b]Average number of antlered deer harvested per square kilometer of range, 1974–1977.
[c]Lowest average weight among all zones in state.
[d]Highest average weight among all zones in state.
[e]Second lowest average weight among all zones in state.

Table 8.4. Classification, pH, and cation exchange capacity (CEC, me/100 g) of surface soil; productivity index; and natural fertility assessment of soils in New Jersey (Soil Conservation Service and Rutgers University 1974) associated with contrasting deer weights (see Table 8.3).

Soil series	Subgroup and family classification	Surface soil		Productivity index[b]	Natural fertility[c]
		pH[a]	CEC		
Lakewood	Spodic Quartzipsamment, mesic, coated	3.5–5.0	0–3	n.d.	very low
Croton	Typic Fragiaqualf, fine-silty, mixed mesic	4.5–5.5	10–15	3	high
Lansdale	Typic Hapludult, coarse, loamy, mixed, mesic	4.5–5.0	8–12	6–8	moderate
Penn	Typic Hapludult, fine-loamy, mixed, mesic	4.5–5.5	12–16	4–6	medium
Freehold	Typic Hapludult, fine-loamy, mixed, mesic	4.5–5.0	6–12	6–10	high
Sassafras	Typic Hapludult, fine-loamy, siliclous, mesic	4.0–5.0	4–12	6–10	moderate

[a]Natural, without limestone.
[b]Varies with slope and erosion; 10 is highest and is equivalent to 8.78 t corn/ha (140 bu/A).
[c]General expression of soil capacity to produce and sustain plants.

and silts overlie fine-textured sediments to form the area of Aura soils, which are productive for crops not limited by subsoil conditions that inhibit root development. Other evidence of the productivity of these areas is that 8 of the 18 deer weighing 91 kg (200 lb) have come from Mercer (6) and Hunterdon (2) counties (N.J. Div. Fish, Game, Shellfish. 1978). Perhaps these deer came from ranges with underlying glauconitic sands. Sussex County (far northwest corner of New Jersey in the folded Appalachians) is represented disproportionately. Limestones form prominent valleys in this area, and on the basis of the association of deer weights with calcareous terranes, it is likely that large deer are found in this county.

Pennsylvania. We have chosen three groups of counties to represent three distinct nutritional environments (Table 8.5). Berks, Bucks, Chester, and Montgomery counties in the southeast were selected to represent the Piedmont and an area of moderate- to high-soil productivity. Extensive areas of Triassic sedimentary rocks, Precambian igneous and metamorphic rocks, and Ordovician sedimentary rocks occur here. Beaver, Mercer, and Washington counties occur along the western border of the state in the Appalachian Plateau and represent an area of

Table 8.5. Distribution (%) of productivity groups among upland soils in selected Pennsylvania counties. Totals do not equal 100 because only areal proportions of principal soils in the soil association map (USDA Soil Conservation Service 1972) were used. Inclusion of all soil areas probably would not change distribution markedly. Soils were assigned to productivity groups using data supplied by L. J. Johnson (pers. commun.).

Physiographic region and county	Proportion of county sampled	Soil productivity group[b] 1	2	3	4	5	total	Region productivity
Piedmont								
Berks	100	19	12	28	7	1	67	
Bucks	97	5	2		33		40	
Chester	97	35	28	1			64	
Montgomery	97	12	10	15	21	5	63	
Region mean								2.37
Eastern Plateau								
Beaver	99		18	30	18		66	
Mercer	62[a]			10	32		42	
Washington	83		19	40	3		62	
Region mean								2.90
Northern Plateau								
Erie	54[a]			4	35		39	
McKean	95	12	6	34	3		55	
Potter	97	9	4	42	4		59	
Tioga	100		2	54	25		81	
Warren	89	19		16	20		55	
Region mean								2.98

[a]Balance of area in alluvial or lake-terrace soils.
[b]Group 1 is most productive.

outcrop of Pennsylvanian-age cyclothem deposits. A row of counties along the northern border (including Erie, McKean, Potter, Tioga, and Warren) represent an area of lower Paleozoic shales and sandstones. Approximately the northwest half of Erie County and the northeast third of Potter County were glaciated. Except in Erie County, calcareous rocks are rare. Data are available from these counties for elemental content of red clover at full bloom (French et al. 1957) and for corn silage (Stout et al. 1977). The Ridge and Valley Province is not represented among the divisions because the interaction of the complex pattern of rocks in the counties of the Province and the sampling of reproductive tracts results in a substantial range for the variables within the region. This is seen in Fig. 8.4 where embryo incidence in yearlings ranges from 0 to 72%.

Inspection of the data in Table 8.6 indicates that (1) conception rates in young does are substantially higher in the Piedmont and the western plateaus than in the northern Plateau, (2) there is a tendency for the number of embryos per adult to be higher in the Piedmont, and (3) the geological origins of these relationships are reflected in the mineral contents of the domestic forages of the regions. Thus, the content of magnesium in red clover from the Piedmont and of phosphorus, iron, and

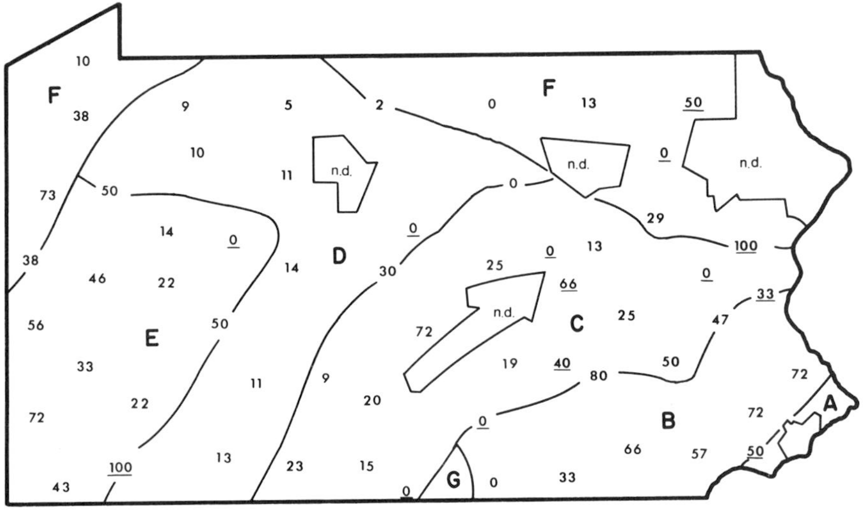

Fig. 8.4. Distribution of pregnant yearling does in Pennsylvania by county (autumn 1977). Counties with no sample (n.d.) are outlined. Percentages underlined indicate five or fewer samples. Land Resource Units of Austen (1965) are: (superimposed on the map) B, northern Piedmont; C, northern Appalachian Ridge and Valley; D, eastern Allegheny plateaus and mountains; E, central Allegheny Plateau; F, glaciated Allegheny Plateau and Catskill Mountains. (Data courtesy W. K. Shope)

Table 8.6. Proportion of pregnant fawn does, embryos per adult doe, and selected chemical properties of red clover and corn silage among areas of contrasting bedrock geology in Pennsylvania. Data for deer are from W. K. Shope (pers. commun. 1979), for red clover from French et al. (1957), and for corn silage from Stout et al. (1977).

Physiographic region and county group	Bedrock geology	Pregnant yearling does (*n*)	Embryos/adult doe (*n*)	Red clover (*n*)	Ca	Mg	K	P	Fe	Zn	Co	Corn silage, Mg
		%			-------------	%	-------------		-------	ppm	-------	%
Piedmont												
Berks	shales, sandstones, argillites,	58 (24)	1.85 (13)	4	1.53	.45	1.84	.27	141	29	.09	.18–.21
Bucks	local calcareous formations, basic	72 (25)	1.96 (23)	5	1.26	.47	1.44	.24	119	38	.09	.21–.24
Chester	and igneous rocks (metamorphosed in part)	57 (7)	2.00 (7)	7	1.12	.40	1.52	.22	100	29	.11	.18–.21
Montgomery	in west and south	72 (7)	1.67 (5)	4	1.24	.52	1.24	.24	134	34	.09	.21–.24
Eastern Plateau												
Beaver	cyclic coal-measure sequences	56 (16)	1.76 (17)	2	1.16	.30	1.18	.18	321	44	.13	.15–.18
Mercer		73 (15)	1.80 (15)	9	1.46	.37	1.26	.20	186	32	.08	.12–.15
Washington		72 (7)	1.73 (11)	5	1.32	.31	1.88	.22	200	35	.11	.15–.18
Northern Plateau												
Erie	shales, sandstones	10 (10)	1.71 (14)	10	1.43	.43	1.16	.19	155	34	.07	.15–.18
McKean		5 (19)	1.95 (49)	5	1.34	.33	1.34	.13	95	28	.08	.15–.18
Potter		2 (41)	1.56 (57)	2	1.19	.29	1.76	.22	65	24	.05	.15–.18
Tioga		0 (12)	1.65 (25)	4	1.29	.32	1.67	.17	101	24	.05	.15–.18

zinc in red clover from the Piedmont and western Plateau were higher than in forage from the other area(s). Cobalt tended to be slightly higher in red clover from the Piedmont and western Plateau. As in the case of red clover, magnesium was higher in corn silage from the Piedmont (Table 8.6).

It is reasonable to assume that native as well as domestic forage would be relatively more nutritious. The condition of deer on these ranges is reflected in their relatively high productivity. Additional evidence is found in the observations of Shope (pers. commun. 1979) that tooth annuli of the deer in this region of the state are relatively indistinct. Year-round availability without limitations in winter of more than adequate mineral content in forage may explain the poorly defined annuli.

West Virginia. On the basis of weight data of 17,000 deer, Gill (1956) divided the state into four regions (Fig. 8.5). Size and productivity decreased in the following regional order: west, Allegheny, south, and east. The average weights for female fawns in the respective regions were

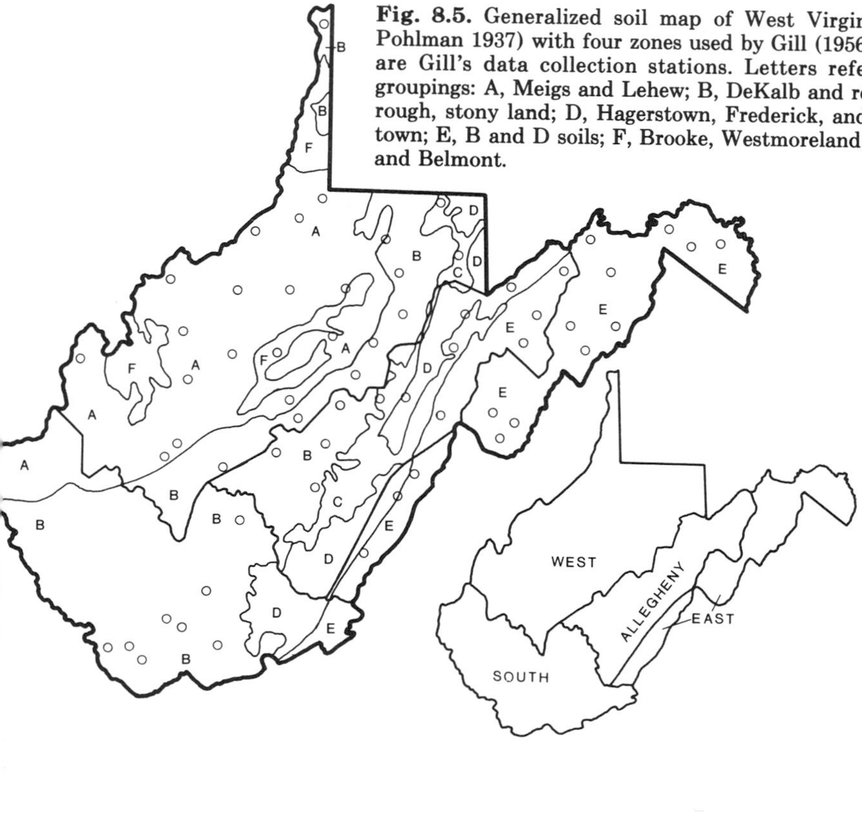

Fig. 8.5. Generalized soil map of West Virginia (from Pohlman 1937) with four zones used by Gill (1956). Circles are Gill's data collection stations. Letters refer to soil groupings: A, Meigs and Lehew; B, DeKalb and related; C, rough, stony land; D, Hagerstown, Frederick, and Frankstown; E, B and D soils; F, Brooke, Westmoreland, Upshur, and Belmont.

27.7 (n = 568), 24.0 (n = 707), 20.9 ($n \doteq$ 136), and 19.5 kg (n = 282). Similar differences occurred for male fawns and for both sexes in the 1.5 year age class. The number of fawns per 100 does (crude data) for the same respective regions were 114, 82, 78, and 72. Gill was careful to make the point that *quantity* of forage was not likely to be important in affecting productivity or size. He also concluded that growing season, precipitation, soil parent material, known mineral deficiencies, calcareous deposits, slope, land capabilities, and forest type did not account for differences. He said that only through study of confined areas could he sort out factors.

The regions Gill (1956) defined for collating his data correspond to the Allegheny Plateau in the west region, the Cumberland Mountains in the south region, the interior Allegheny Plateau in the Allegheny region, and the Ridge and Valley Province of the folded Appalachians in the east region. No recent soil association map for the state is available; hence the following discussion is based on the map of Pohlman published in 1937. The soils of the west are dominantly Meig's and Lehew soils, which are developed from Paleozoic sandstones and shales. Several extensive areas of Brooke, Westmorland, Upshur, and Belmont soils, derived from calcareous shales or limestones, occur. The Allegheny region consists of an extensive area of Dekalb and related soils and rocky land. Soils of the east region reflect the complex lithologic character of folded strata of the Appalachian Mountains, with northeasterly aligned belts of acid soils of low fertility in the mountains bordering fertile valleys eroded in limestones and base-rich shales. Soils developed in lightly metamorphosed rocks also occur in this region. Soils of the south region, an area largely in forest, are primarily DeKalb and related soils. Jencks (1969) has compiled a substantial body of data for extractable bases from surface horizons of virgin soils in the state. The geometric means and deviations for soil groupings derived from acid and calcareous parent materials are presented in Table 8.7. The contrasting calcium and magnesium concentrations are substantial. We speculate that the lower weights of deer from the south and east regions reflect the poorer nature of their soils, which developed from acid shales and sandstones.

Table 8.7. Contents (me/100 g) of bases extractable with NH_4OAc in surface horizons of West Virginia soils. Data are presented as values of GM and GD derived from data in Jencks (1969).

Parent material	n	Ca		Mg		Na		K	
		GM	GD	GM	GD	GM	GD	GM	GD
Shale, sandstone, acid	47	0.77	4.2	0.47	2.2	.06	2.4	.17	1.7
Limestone, calcareous shale	36	6.9	3.6	2.0	2.4	.08	2.4	.29	1.9

SOUTHERN STATES

Mississippi. Jacobson et al. (1977) sought to find relationships between three factors—weight of yearling male white-tailed deer, antler development, and reproductive fertility (measured by incidence of corpora lutea)—and selected soil fertility factors. Eight study areas, ranging from the fertile soils of the Mississippi River floodplain to the poor soils of the coastal flatwoods, were represented. The 8 to 36 soil samples from the study areas were analyzed for organic matter, pH, calcium, magnesium, phosphorus, and potassium, evidently by methods designed to indicate plant-available levels. Data for calcium, magnesium, and phosphorus and for weight of female fawns and yearling males are given in Table 8.8.

The authors chose calcium and phosphorus as the soil variables most closely associated with physical development of the deer. Doe fertility data were too few to give clear distinctions between areas. A plot of weight as a function of logarithm of calcium or phosphorus indicates that size for males increases with soil phosphorus to about 40 ppm phosphorus beyond which there is no increase; however, weight increases with calcium throughout the range of observed levels. The authors attached substantial significance to phosphorus even though the highest weights were clearly associated with the highest calcium levels. These findings are of particular interest with regard to the later discussion of deer weight and soil and mineralogy relationships in Missouri. Weight data for deer in Mississippi are of interest for comparison with weights recorded for deer in northern states.

MIDWESTERN STATES

Ohio. Nixon et al. (1970) summarized weight data and food habits by season for three regions (Table 8.9 and Fig. 8.6) that conform closely to extensive physiographic features. About 80% of their northwestern area corresponds to the black, poorly drained, clay-rich soils developed on sediments of the broad, nearly level plain of glacial Lake Maumee. The northeastern area is a region of rolling glaciated Appalachian Plateau in which the soils are derived from drift eroded from Paleozoic shales and sandstones. About 90% of the soils in the southeastern area or hill counties are derived from sandstones and shales, many of which are in the cyclical sequences of Pennsylvanian-age rocks common to the Plateau. Considered from the standpoint of native fertility, soils in the northeast do not differ appreciably from those in the southeast—the weighted average yield of corn under a low level of fertilizer input is 3.01 t/ha (48 bu/A) in each area (North Central Regional Technical

Table 8.8. Extractable Ca, Mg, and P in surface soils and mean weights of deer from Mississippi (Jacobsen et al. 1977).

Land resource unit[a] and study area	Soil fertility				Deer weight, kg (*n*)		
					females	males	
	n	Ca	Mg	P	0.5 yr	1.5 yr	> 2.5 yr
		------me/100 g------		ppm	--kg (*n*)--		
Southern Coastal Plain							
Leaf River W.M.A.	36	1.2	0.81	14	12.9 (2)	28.6 (9)	40.8 (10)
Meridian Naval Air Station	10	2.2	0.56	20	15.4 (3)	33.8 (22)	48.8 (26)
Choctaw W.M.A.	34	2.6	2.2	24	16.8 (3)	31.0 (4)	43.5 (7)
Noxubee Wildlife Refuge	18	7.5	2.9	35	n.d.	43.3 (39)	55.5 (15)
Alabama and Mississippi Blackland Prairies							
Talahala W.M.A.	36	7.1	3.0	17	15.3 (4)	32.6 (29)	44.8 (13)
Chickasaw W.M.A.	36	3.1	1.5	27	18.0 (8)	39.9 (54)	51.4 (24)
Southern Mississippi Silty Uplands							
Sandy Creek W.M.A.	18	4.1	0.67	44	15.8 (4)	38.1 (27)	50.1 (13)
Southern Mississippi Valley Alluvium							
Issaquena W.M.A.	8	21.3	9.3	239	n.d.	43.1 (14)	58.5 (5)
Benoit Hunt Club	12	50.4	24.6	229	21.5 (8)	n.d.	n.d.

[a]After Austin 1965.

Table 8.9. Weights of deer (kg) (field-dressed with heart and liver), P (ppm) extracted by mild acid fluoride, and pH of soils from Ohio. Deer weights followed by same letter in column are not significantly ($P \leq .05$) different. Soil fertility data are modes of numerous analyses that each represent a major soil series. Data for deer from Nixon et al. (1970) and for soil from Jones and Musgrave (1963).

Area	Fawn (*n*)		Yearling (*n*)		Adult (*n*)		Soil fertility		
	male	female	male	female	male	female	*n*	P (mode)	pH (mode)
Northwest	34.3 (98)a	32.1 (83)a	58.6 (51)a	47.1 (47)a	73.2 (45)a	52.4 (46)a	9,933	9, 9, 11, 12, 16, 19	6.0, 6.0, 6.0 6.2, 6.2, 6.6
Northeast	32.0 (65)b	29.3 (71)b	50.3 (50)b	44.0 (47)b	65.3 (70)b	50.6 (39)a	11,877	7, 9, 9, 9, 9, 10, 11, 14, 23	4.8, 5.5, 5.6, 5.8, 6.2, 6.2, 6.4, 6.4, 6.4
Hill	31.2 (58)b	29.1 (62)b	50.9 (34)b	44.3 (23)a,b	66.8 (80)b	45.6 (46)b	11,331	1.5, 2.5, 2.5, 2.5, 3.6	5.2, 5.4, 5.4,5.9, 6.2, 6.2

Table 8.10. Selected chemical properties of surface soils from Ohio. Boundaries are those in Fig. 8.6. Soil data are pH of a water paste and extractable cations.

Region	pH			Ca (me/100 g)			Mg (me/100 g)			K (me/100 g)		
	n	AM	SD	*n*	GM	GD	*n*	GM	GD	*n*	GM	GD
Northwest[a]	7	6.29	.58	7	12.0	1.8	7	3.7	1.8	7	.43	1.7
Northeast[b]	19	5.29	.94	19	2.4	3.4	18	0.8	2.0	19	.29	1.6
Hill[c]	12	5.55	.63	12	5.1	1.6	12	1.6	1.5	12	.35	1.5

[a]Data from Putnam County (Brock et al. 1974).
[b]Data from Ashtabula (Reeder et al. 1973), Richland (Redmond et al. 1975), and Portage (Ritchie et al. 1978) counties.

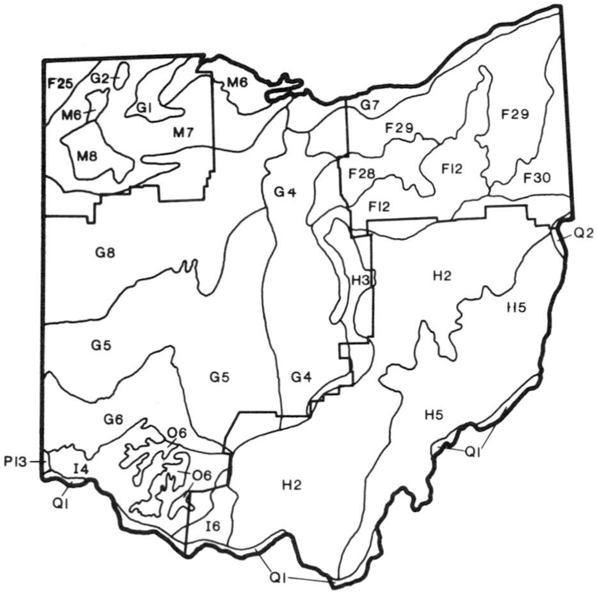

Fig. 8.6. Major soils of Ohio (North Central Technical Committee 3 on Soil Surveys 1965) and three sampling regions of Nixon et al. (1970) for deer condition. Soil series in the groupings are identified in App. 9.

Committee 3 on Soil Surveys 1965, calculations by methods discussed under Missouri), whereas yield in the northwest amounts to 3.70 t/ha (59 bu/A). Available phosphorus tends to be lower in the hill counties. Central tendencies and deviations were calculated for pH and extractable cations for the surface horizons of fine-textured soils in the three regions (Table 8.10). The soils originally were chosen for analytical work during soils mapping and do not represent an unbiased sampling and representation. Most of the data are for the plow zone of soils in agricultural production and consequently the levels of these analytes tend to be increased by amendments; some variability in the northwest is due to this source. Nevertheless, the averages follow a sequence similar to that expected from the fawn weight data; that is, the soils in the northwest are appreciably richer in bases, which also is implied by higher pH in the larger sample in Table 8.9. This contrast is expressed in the calcareous nature of the parent materials. The geometric mean of calcium-carbonate equivalent for parent materials from northwest Ohio taken from soil survey reports is 16.0% (n = 7; GD, 1.29), whereas the tills in northeast have a geometric mean content of 5.8% (n = 17; GD, 2.01).

Indiana. Data on female fawn weights from 10 collection points in Indiana were provided by H. P. Weeks, Jr. (pers. commun. 1979). The data for 6 of the sites in the drift plain of northern Indiana were combined to provide a weighted average (Fig. 8.1); this grouping was in accordance with the widespread occurrence of similar coarse-textured and productive soils in this region (USDA Soil Conserv. Serv. and Purdue Univ. Agric. Exp. Stn. 1977). The northwest group of 5 collection sites are largely in areas of Maumee, Gilford, and Sebawa soil associations; and Plainfield, Maumee, Fox and Oshtemo, and Fox associations occur on the Pigeon River State Fish and Wildlife Area in the northeast. The other sites are widely distributed in south central Indiana; the 2 northern localities in this group are Naval Weapons Support Center–Crane ($\bar{x}$ = 28.8 kg) and Camp Atterbury ($\bar{x}$ = 27.7 kg). Soils on the Crane area are predominantly the Wellston-Zanesville-Berks association, which also occurs in parts of southeastern Ohio where fawn doe weights are similar. The soils are developed in thin loess overlying shales, siltstones, and sandstones and are medium to strongly acid. At Atterbury, soils are predominantly the Cincinnati-Rossmoyne association developed in deeply leached Illinoian-age till. In the northern part of Atterbury, calcareous, loam-texture, Wisconsin-age till prevails; but less strongly developed soils of the Miami-Crosby-Brookston association also are found. South, near the Ohio River, 2 sites are represented. The western site is Mogan Ridge ($\bar{x}$ = 25.8 kg) where soils of the Wellston-Zanesville-Berks association occur; the association also is prominent at Naval Weapons Support Center–Crane where female fawns were 3 kg heavier. Whether this difference represents some slight difference in soil chemistry but significant nutritional influence or represents other environmental influences is unclear. The weights of deer at Mogan Ridge are similar to those from loess-covered terrains adjacent to the Missouri and Mississippi rivers (south of St. Louis in Missouri). The lightest deer recorded by Weeks in Indiana were weighed at the Charleston Depot. The Crider-Hagerstown-Bedford association is mapped for the area. These soils are strongly developed and occur in residuum derived from limestones, but at the Charleston Depot they are overlain by a thin layer of loess. Deer weights from this locality contrast sharply with the 30.5 kg found nearby on Fort Knox in Kentucky on Crider-Vertress and Garmon-Frederick associations (USDA Soil Conserv. Serv. and Kentucky Agric. Exp. Stn. 1975), both of which represent soils derived from limestone. No obvious, soil-related explanation exists for the low weights at Charleston Depot. However, Weeks (pers. commun. 1982) provided the most likely explanation—deer densities on this area have been as high as 38.5/km^2; this weight contrast can be considered a prime example of how overcrowding and its effect on nutrition is expressed in a deer herd.

Wisconsin. Dahlberg and Guettinger (1956) quoted data for deer weights of animals from good and poor ranges in each of three areas in the state (Table 8.11, Fig. 8.1). Their designation of poor range conformed to habitats associated with winter starvation. The Driftless Area of southwest Wisconsin occupies two-thirds of the central area; otherwise the soils for the three areas are derived from Wisconsinan-age glacial deposits. Upper Cambrian sandstones, with some shales and dolomites, underlie much of the northwest and most of the remainder of the central area. The northeast area and the eastern portion of the northwest area are underlain by Precambrian-age igneous rocks that are for the most part acidic. Significant areas of gabbro and basalt occur in the northwest area. These diverse kinds of bedrock have given rise to glacial sediments that are mostly coarse-textured and acidic in reaction (Hole 1976). Loess, thinning to the east but up to 3 m thick near the Mississippi River, overlies the glacial drift and bedrock over much of the northwest region. The range of differences in deer weights among areas rated as good range is small—equivalent to 1.5 kg—but the variation in average weights also is small for areas rated as poor range—only 0.9 kg. Because both range categories were rated on the basis of winter-carrying capacity, the lower weights of fawns in poor soil-range areas where winter is not a major stress factor may reflect the lower native fertility of the soils of these areas and/or the poorer physical condition of does during pregnancy and lactation.

The heavier weights of fawns from the northwestern sector may be traceable to higher calcium levels. For example, Lane and Sartor (1966) mapped the soils of Douglas and Ashland counties (Hibbing, Superior, and associated soils), finding an average 10 to 20 me Ca/100 g soil, levels that may be accounted for by the red-colored, calcareous tills of the region (Hole 1976:112). Other counties having calcium levels in this range are Clark and Pierce in the central area and Shawano in the northeast area. The small differences in deer weights reflect the essentially identical productivity indexes calculated from Hole's map (1976:Fig. 2-53). For the northeast, northwest, and central areas, the average

Table 8.11. Dressed weights of fawn does from good and poor ranges in Wisconsin. Poor range has insufficient food for overwintering deer. Data are for 1938–1947 from deer obtained by confiscation in October, November, and December (from Dahlberg and Guettinger 1956, Table 17). Soil productivity indexes are calculated from Fig. 2-53 in Hole (1976).

Area	Area soil productivity indexes	Good range, kg (*n*)	Poor range, kg (*n*)
Northeast	34	24.9 (186)	24.1 (235)
Northwest	35	26.0 (152)	24.8 (176)
Central	36	24.5 (45)	23.9 (226)

indexes (weighted for areal extent) are 34, 35, and 36, respectively. Hole assigned these dimensionless indexes on the basis of a number of soil properties and length of growing season but without consideration of drainage or irrigation.

Female fawn weights from good ranges of Wisconsin conform with those from other northern states. However, because the large area of fertile soils developed from calcareous glacial drift in the southeast quarter of the state is not open to deer hunting, weight data of potentially large deer are lacking for this area.

Illinois. There are substantial differences in the productivity of Illinois soils (Mausel et al. 1975); the most productive soils are located in a wide belt through east central Illinois and the least productive soils are generally found in the southernmost portions of the state. Calhoun and Loomis (1975) presented data for deer weights for three zones that included most of the state. The southern zone is nearly coextensive with the region of soils having below-average productivities (Mausel et al. 1975:Fig. 4); this southern zone yields the lightest deer—the average weight of female fawns is 21.6 kg. Weights in the central and northern zones are 26.9 and 26.0 kg, respectively. The basic productivity indexes of these zones (used earlier), calculated from county averages, are 55.0, 74.6, and 77.7, respectively. These productivity indexes reflect the capacity of the soils for grain production and are indexed in relation to the average of yields from a group of very productive soils in the state under a similar system, including fertilization inputs, of management. The basic productivity of a soil series is assumed to reflect better the native productivity of a soil series—an assumption that appears to be borne out by deer weight data for the three zones. Richie (1970) compared weight and antler-beam diameter data for Jo Daviess and Carroll counties in extreme northwest Illinois with similar data for deer from six counties in the extreme south. Differences in weight of female fawns—29.9 kg in the north versus 24.0 kg in the south—were significant ($P \leq .05$) as were areal differences for older age classes (through 4.5+ years); significant differences also were found for male deer through the 3.5-year class for these two areas. Antler-beam diameter was significantly smaller in the southern counties for all age classes. Richie calculated indexes comprised chiefly of subjective estimates of levels of 8 soil physical properties for the two-county groups (Richie 1970:Table 5), and the results tend to separate the northern counties into a higher group that produced larger, healthier deer. If the county productivity indexes of Mausel et al. (1975) are used as a base of reference, the northern counties have an average productivity index of 64.0 and the southern counties a productivity index of 51.0.

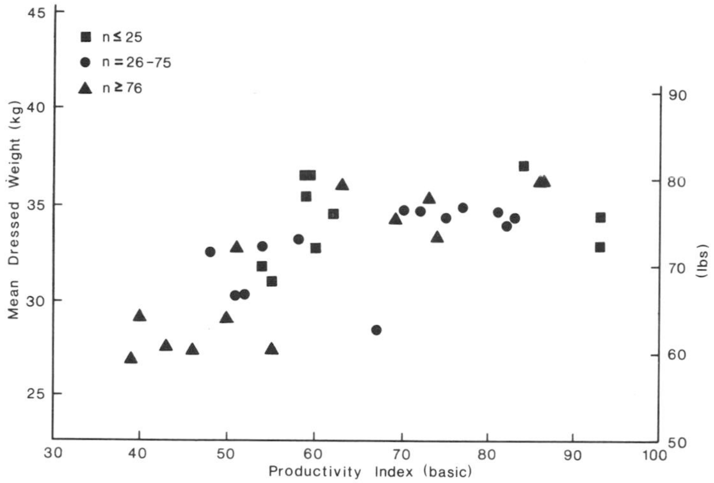

Fig. 8.7. Scatter diagram of mean live weight of fawn does plotted against mean productivity index for the respective county of collection in Illinois. Weight data are for collections in autumns of 1968, 1969, 1972, and 1973. Squares = sample sizes ≤ 25, circles from 26–76, and triangles ≥ 76. (Weight data from F. Loomis; indexes from Mausel et al. 1975)

Forrest Loomis of the Illinois Department of Conservation provided weight data collected at hunter check stations. Data for 1968, 1969, 1972, and 1973 were selected to illustrate in more detail the relation of fawn weights to soil productivity and to explore the state's conditions more closely. In Fig. 8.7, the mean weights of female fawns are plotted as a function of the average productivity indexes for the 36 counties represented in the data collection. The resulting scatter diagram suggests that weights of fawn deer are associated positively with incremental increases of productivity index, but weights do not vary above index level 60. Two averages of low weights above index 90 recorded from DeKalb and Piatt counties represent samples of three and six deer, respectively. These low weights are from counties generally characterized as having very productive soils and may reflect a factor that has much to do with the large variances in the diagram—whether the deer in the countywide samples actually represent prevailing soil types (especially pertinent in the case of counties containing an intimate intermingling of soils of different productivities). The use of woodland habitats and marginal nonagricultural lands, which more often are coextensive with soils of lower productivity than adjacent lands in row-crop production, may

Table 8.12. Values of GM and GD (me/100 gm) of extractable bases in surface horizons of soils from Illinois corresponding more or less to the regions for which deer-weight data were calculated.

	Ca			Mg			Na			K			
Region	*n*	GM	GD	*n*	GM	GD	*n*	GM	GD	*n*	GM	GD	Source
Northeast	31	13.9	1.4	31	5.9	1.7	27	.18	1.5	31	.44	1.8	Wascher et al. 1960
Northwest	36	9.8	1.9	36	3.0	1.8	31	.26	1.8	32	.36	1.8	Wascher et al. 1971
Central	25	8.9	1.6	25	1.9	1.5	16	.10	-[b]	23	.27	1.6	Soil Survey 1968
South	17	3.5	2.4	17	1.2	1.9	15	.11[a]	0.3	17	.32	1.9	Soil Survey 1968

[a]Arithmetic mean and standard deviation.
[b]Data invariant.

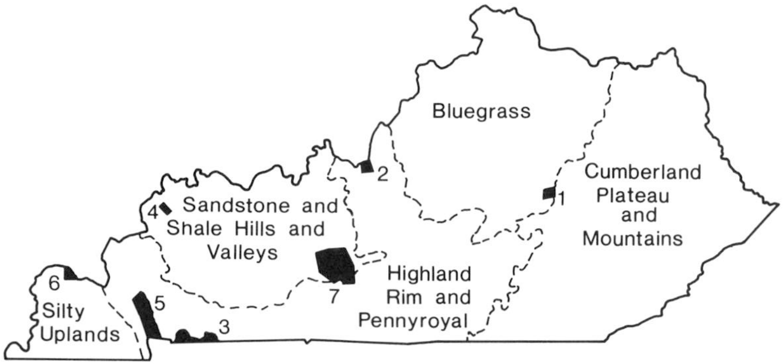

Fig. 8.8. Locations of yearling-doe collections and agriculture resource areas in Kentucky. Area numbers are identified in Table 57.

explain some of the variance found in relating deer weights to indexes of productivity.

Data for extractable bases in Illinois soils (Table 8.12) reflect the higher base status of areas corresponding to central and northern zones for female fawn weights. The northeast and northwest areas of the state, representing approximately the east and west halves of the northern zone, consist of soils largely derived from Wisconsinan-age till and loess, respectively. The till-derived soils tend to be richer in alkaline earths than the loess soils. The content of extractable bases in soils in the southern zone of the state is generally low, a deficiency related to the large proportion of Aquafs and Udolls (Planosols); but soils bordering the Mississippi, Ohio, and Wabash rivers have moderately thick layers of loess near the bluff and tend to be relatively high in bases.

If fawn weights representative of a single major soil type or of geographic units smaller than counties could be obtained, a closer correlation between soil productivity and fawn weight than shown in Fig. 8.7 probably would be found.

Kentucky. Weight data for yearling does representing seven areas of Kentucky were made available by J. Phillips and H. Barber. These data and the principal soils occurring in the ranges of the deer represented are given in Table 8.13; data collection sites are shown in Fig. 8.8. Although the weighing stations are widely separated, weights generally averaged 40 kg except at Fort Campbell where the mean weight was 34.1 kg. We have no explanation for the latter low weight. The general statewide uniformity of deer weights is presumably related to the rather uniform nature of soils and their parent materials, notably the prevalence of limestone parent materials. In comparison, yearling does from Naval Weapons Support Center–Crane (discussed earlier) in Martin County, Indiana, weighed an average of 44.6 kg for 1966–1969 ($n = 449$, Weeks 1974:274). Note that the heaviest Kentucky deer come from the Richmond Bluegrass Army Depot in the Outer Bluegrass region where limestone has been important in soil development. No doubt female fawn weights, if available, would follow the pattern displayed by yearling animals.

Missouri. The effects of differences in soil fertility on wild animal populations has been the object of study in Missouri for more than four decades. Denney (1944) reviewed the work done in the state and voiced strong support for continuing research directed toward establishing relationships between soil fertility, land use, and native animal populations. Denney's contemporary, Albrecht, also was a strong advocate of the effects of soil fertility on crop quality and its

Table 8.13. Major soil series of soil associations, parent materials, suitability for agriculture, and mean weights for yearling white-tailed does from areas in Kentucky. Map number refers to Fig. 8.8. Soils data from USDA Soil Conservation Service and Kentucky Agricultural Experiment Station (1975). Deer weights courtesy of J. Phillips and H. Barber (pers. commun. 1979).

Map no.	Area	Major soil series	Areal % of series	Parent material	Suitability for agriculture[a]	Doe weight, kg	Remarks
1	Richmond Bluegrass Army Depot	Lowell	39	clay-rich residuum from limestone	good	42.2	
		Fairmount	15	clay-rich residuum from limestone	poor		
		Shelbyville	13	loess overlying limestone residuum	good		
		Faywood	10	clay-rich residuum from limestone	fair		
2	Fort Knox	(1)[b] Garmon	38	loam-rich residuum from shaly limestone	poor	39.9	limestone at 0.5 to 1.0 m
		Frederick	14	clay-rich residuum from limestone	fair		
		(2) Crider	26	loess overlying limestone residuum	good		
		Vertrees	18	clay-rich residuum from limestone and shale	fair		
		Nicholson	12	loess overlying limestone residuum	good		
3	Fort Campbell	Crider	52	loess overlying limestone residuum	good	34.1	
		Baxter	22	clay-rich residuum from cherty limestone	fair		
4	Higginson-Henry Wildlife Management Area	Memphis	38	loess	good	39.3	
		Loring	33	loess	good		
5	Land-between-the Lakes	(1) Baxter	36	clay-rich residuum from cherty limestone	fair	39.7	
		Crider	18	loess overlying limestone residuum	good		
		Nicholson	18	loess overlying limestone residuum	good		
		(2) Brandon	28	from 0.6 to 1.2 m of loess overlying coarse-textured Coastal Plain	fair		

			Loring	16	loess	good		
			Saffell	16	coarse-textured Coastal Plain sediments	poor		
6	Western Kentucky Wildlife Management Area		Grenada	34	loess	good	41.7	
			Loring	19	loess	good		
			Calloway	10	loess	good		
7	Edmonson County	(1)	Caneyville	24	clay-rich residuum from limestone	fair	40.0	limestone at 0.5 to 1.0 m
			Zanesville	23	loess overlying residuum from sandstone, siltstone, and shale	good		
			Frondorf	20	loam textured residuum from interbedded sandstone, siltstone, and shale	fair		
		(2)	Zanesville	27	see above	good		
			Frondorf	21	see above			
			Wellston	13	loess overlying residuum from sandstone, siltstone, and shale	good		

[a]Poor, fair, and good scale.
[b]Indicates group of soils in area with two associations.

relationships to animal populations. In view of this historical tradition, it is particularly appropriate that the relationship between white-tailed deer and soil properties in Missouri be explored. For the present evaluation, weight data for female fawns killed during the 1975 and 1976 hunting seasons are used (Porath and Torgerson 1976, 1978). Averages for female fawns, weighted for sample size, have been plotted for 9 of the 11 deer management zones. The results define 2 extensive regions (Figs. 8.1, 8.9) that are separated by the Missouri River. The physiography and principal soil associations of each zone are described in Table 8.14. There is a range of about 11 kg between average fawn weights from the Ozark Plateau and from the till plains in northern Missouri. Some zones have

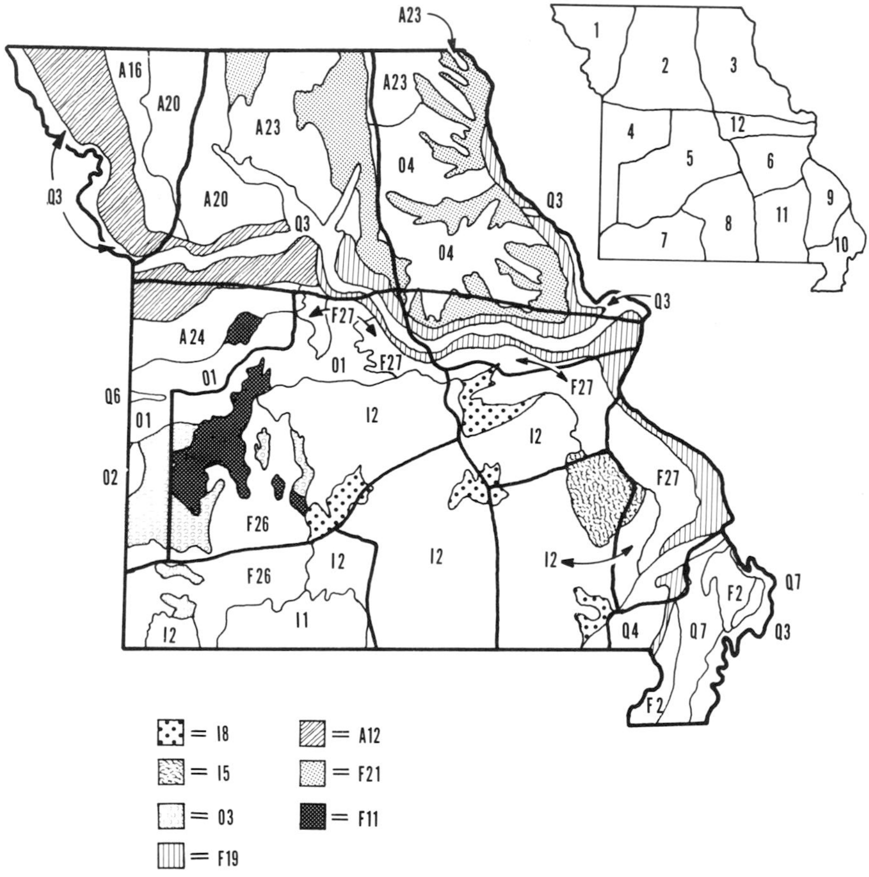

Fig. 8.9. Major soils (North Central Technical Committee 3 on Soil Surveys 1965) and 12 Deer Management Zones in Missouri (Porath and Torgerson 1976). Soil series in the soil groupings are identified in App. 10.

Table 8.14. Physiography and principal soil associations of zones in Missouri. Zones and soil associations are shown in Fig. 8.9. Corn yields for associations are arithmetic means for series within association (North Central Regional Technical Committee 3 on Soil Surveys 1965) and are not weighted for proportion of each series or factors affecting productivity. Yields calculated for average management for 1954–1963 and adjusted for secular increase in the late stages of the period.

Deer manage. zone	Physiography	Soil association: map no.	soil series	corn yield (mt/ha)	proportion of zone, %	remarks
2	rolling or dissected loess-covered Kansan till plain	A12	Marshall Knox Shelby	3.51	12	Marshall on gentle slopes
	rolling loess-covered Kansan till plain	A20	Shelby Grundy Haig	3.45	8	Shelby on steeper slopes
	same	A23	Shelby Seymour Edina	3.45	21	Shelby on steeper slopes
	dissected Kansan till plain, thin loess on ridge tops	F21	Lindley Weller Gara	2.70	20	Lindley and Gara on steeper slopes
3	see above	A23		3.45	13	
	see above	F21		2.70	26	
	nearly level Kansan till plain with moderately thick loess mantle	O2	Putnam Cowden	3.69	46	
5	dissected Ozark Plateau	I2	Clarksville Ozark	2.63	29	cherty limestone
	dissected Springfield Plateau	F11	Boone Bolwar	2.52	10	sandstone shale
	same	F26	Baxter Eldon Nixa	2.82	14	cherty limestone
	loess-mantled dissected upland	F27	Union Weldon	3.32	8	highly weathered loess over limestone residuum
	Springfield Plateau	O1	Oswego Woodson Bates	3.45	16	shale calcareous shale sandstone and shale
	same	O3	Cherokee Parsons	2.88	7	shale

Table 8.14 *(continued)*

Deer manage. zone	Physiography	Soil association: map no.	soil series	corn yield (mt/ha)	proportion of zone, %	remarks
6	Salem Upland	F27	see above	3.32	38	
	same	I2	see above	2.63	15	
	same	I8	Lebanan	2.76	40	loess overlying cherty limestone residuum
			Hanceville			sandstone
7	dissected Springfield Plateau	F26	see above	2.82	43	
	same	I1	Clarksville	2.63	29	cherty dolomite
			Taney			occurs in "gladelands"
	same	I2	see above	2.63	16	
	same	O3	see above	2.88	4	
8	dissected Ozark Plateau	I2	see above	2.63	96	
		I8	see above	2.76	4	
9	moderately thick to thick loess-mantled upland, dissected in part	F19	Menfro	3.51	20	
			Alford			
			Hosmer			
	same	F26	see above	3.32	43	
	same	I2	see above	2.63	15	
11	Salem Plateau, southern section	I2	see above	2.63	70	
	St. Francis Mountains, rounded hills	I5	Ashe	3.26	25	granite
			Tilsit			sandstone and shale
			Hagerstown			limestone
	Salem Plateau	I8	see above	2.76	5	
12	loess-mantled dissected upland	F19	see above	3.51	40	
	dissected Kansan till plain, moderately thick loess mantle	F21	Lindley	2.76	20	till
			Weller			loess overlying till
			Gara			
	loess-mantled dissected upland	F27	see above	3.32	17	

Table 8.15. Fawn doe weights, antler-beam circumference, and incidence of spike character in yearling white-tailed deer in Missouri. Data are weighted means for collections in autumns of 1975 and 1976 (Porath and Torgerson 1976, 1978).

Deer manage. zone	Soil productivity index[a]	Fawn doe weight (*n*)	Yearling buck (*n*) beam circumference	Yearling buck (*n*) spike incidence
		kg	mm	%
3	52	29.6(396)	70.1(327)	4(350)
9	52	23.7(24)	61.9(373)	11(365)
12	52	25.3(96)	65.9(450)	9(451)
2	51	31.8(293)	70.5(140)	4(179)
6	47	22.8(29)	64.3(385)	13(382)
5	46	23.8(163)	60.0(581)	16(606)
11	45	20.2(12)	54.6(13)[b]	54(13)
7	44	20.8(42)	59.3(102)	34(97)
8	42	22.8(138)	60.6(209)	17(207)

[a]Mean for soil association (see Table 8.14) and weighted for areal extent of association in zone (indexes calculated from North Central Regional Technical Committee 3 on Soil Surveys 1965).
[b]No 1975 sample.

a substantial range of soil properties and parent materials (e.g., Zone 5); hence, not surprisingly, the mean weight of female fawns in Zone 5 lies between averages recorded for the Ozark Plateau (Zone 7) and the loess-covered dissected country bordering the Missouri River (e.g., Zone 12). Using the mean of basic level productivity indexes for the soils in each association and taking into consideration the areal extent of the association in the zones, a productivity index for each zone was calculated. These indexes—female fawn weights, beam circumference, and spike incidence for yearlings—are given in Table 8.15. Imprecise as they are, these indexes tend to be positively related to weights and antler characteristics—the larger deer, both fawn and yearling bucks, are recorded from zones that contain the more productive soils. The correlation would be somewhat better if the weights and antler data could be paired with the soil associations representative of the primary ranges of the deer in each zone. A further indication of the relationship between fawn weight and the chemical characteristics of the soil is shown by reference to the comprehensive survey of Missouri surface soils carried out by Tidball (1974). Using data for calcium, magnesium, and sodium for soils from those counties in which weight and antler data were recorded in 1975 and 1976, the relationships are plotted against body weight in Figs. 8.10, 8.11, and 8.12, respectively. For calcium and magnesium, weight increases with increasing content of the alkalis and alkaline earths, but for sodium, an asymptote appears to have been attained. The relationships shown could be anticipated a priori on the basis of the nutritional importance of these three elements. At what level

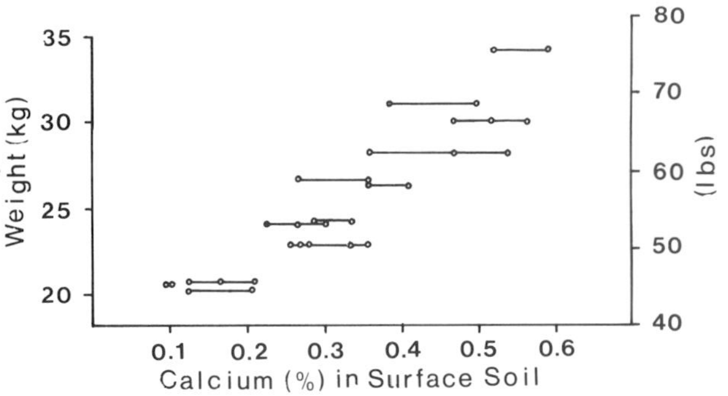

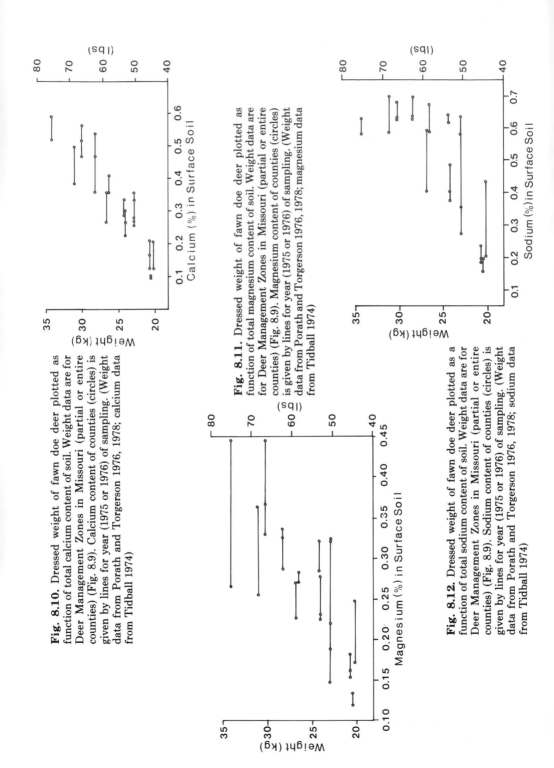

Fig. 8.10. Dressed weight of fawn doe deer plotted as function of total calcium content of soil. Weight data are for Deer Management Zones in Missouri (partial or entire counties) (Fig. 8.9). Calcium content of counties (circles) is given by lines for year (1975 or 1976) of sampling. (Weight data from Porath and Torgerson 1976, 1978; calcium data from Tidball 1974)

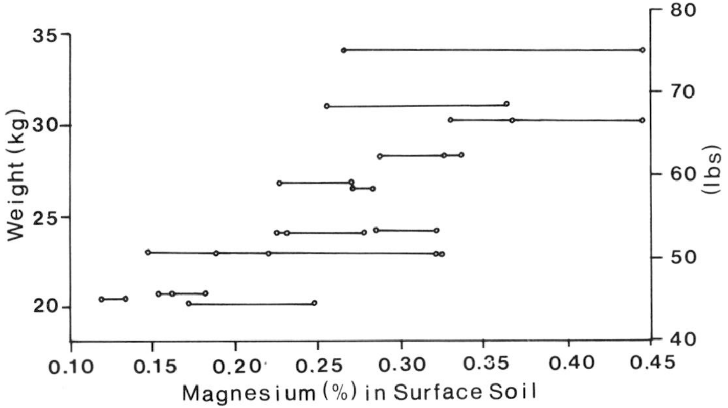

Fig. 8.11. Dressed weight of fawn doe deer plotted as function of total magnesium content of soil. Weight data are for Deer Management Zones in Missouri (partial or entire counties) (Fig. 8.9). Magnesium content of counties (circles) is given by lines for year (1975 or 1976) of sampling. (Weight data from Porath and Torgerson 1976, 1978; magnesium data from Tidball 1974)

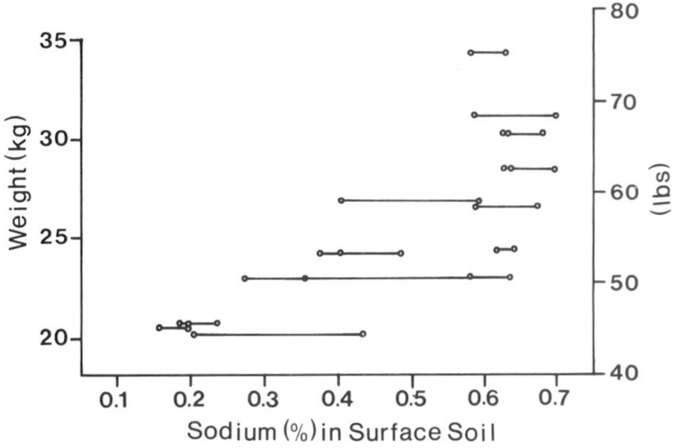

Fig. 8.12. Dressed weight of fawn doe deer plotted as a function of total sodium content of soil. Weight data are for Deer Management Zones in Missouri (partial or entire counties) (Fig. 8.9). Sodium content of counties (circles) is given by lines for year (1975 or 1976) of sampling. (Weight data from Porath and Torgerson 1976, 1978; sodium data from Tidball 1974)

of soil nutrients there would cease to be a body weight response cannot be answered. These relationships also suggest that under some circumstances the so-called total analyses of soil could be related in part, perhaps through geophagy, to certain conditions partially independent of a ruminant's foods. In general, levels of these cations in plants are related directly to their levels in soils except for some antagonism of potassium with calcium and magnesium (discussed elsewhere). From the compendium of analyses for a number of vascular species from vegetation-type areas of Missouri (Connor and Shacklette 1975), we have extracted data for phosphorus, calcium, magnesium, potassium, and sodium in sumac and cedar for three of the deer management areas (Table 8.16). Sumac and cedar are commonly used winter browse species in Missouri (Porath and Torgerson 1978: Tables 41–48). Magnesium is higher in both sumac and cedar from the glaciated prairies of the northern part of the state than elsewhere. Cedar, a calciphile, is notable for its high content of calcium and magnesium relative to sumac. Sodium in cedar growing on the thin soils of the cedar glade region of extreme southern Missouri is one-tenth that recorded for this tree in other vegetation areas. These plant tissue data suggest that small mineral differences are found among regions and they may account for the differences in weights found for the various zones. Note that both magnesium and sodium are highest in these plant species in the region of the state where the heaviest female fawns have been recorded and that these same elements distinguish the deer licks of eastern United States from surrounding soils.

STATEWIDE WEIGHT AVERAGES. Data for New Hampshire (Siegler 1968), Massachusetts (Shaw and McLaughlin 1951), Rhode Island (Rhode Island Department n.d.), and Iowa (Kline 1965) are summarized in Fig. 8.1. Weight data for Minnesota from Mud Lake (now, Agassiz) National Wildlife Refuge (Krefting and Erikson 1956) are included because of the unusual size of the deer and for comparative purposes. The mean size of deer in the New England states is near or slightly above the average for northeastern United States. A large percentage of the Massachusetts sample comes from the Berkshire Hills in western Massachusetts, an area noted for its productive soils. The relatively heavy deer from Rhode Island are difficult to explain because soils of the towns that evidently contributed most to the sample are strongly acid and sandy (e.g., data for Gloucester soil series in Smith and Gilbert 1945).

The weights of Iowa deer are the heaviest found; the average is derived from 288 animals taken between 1954 and 1962 (Kline 1965). Kline presented data for male fawns from four zones in the state. The heaviest deer (37.7 kg, $n = 53$) came from north central Iowa, the boundaries of which correspond closely to the Clarion-Nicollet-Webster

Table 8.16. Contents of selected elements in sumac and cedar from Missouri (Connor and Shacklette 1975).

Vegetation-type area	Deer manage. zone	Fawn doe weight	Smooth sumac[a] *n*	P	Ca	Mg	K	Na	Cedar[b] n	P	Ca	Mg	K	Na
		kg		%				ppm		%				ppm
Glaciated prairie	2, 3	30.5	50	.050	1.09	.092	.55	10	9	.14	1.48	.28	.52	160
Oak-Hickory-Pine forest	11	20.2	49	.047	0.92	.068	.41	10	6	.11	1.56	.22	.51	150
Cedar glade	7[c]	20.8	49	.043	1.04	.079	.36	8	50[d]	.078	1.86	.18	.37	13

[a]*Rhus glabra*, stems.
[b]*Juniperus virginiana*, stems and leaves.
[c]About one-half of the zone is in this vegetation-type area.
[d]*n* = 49 for Na.

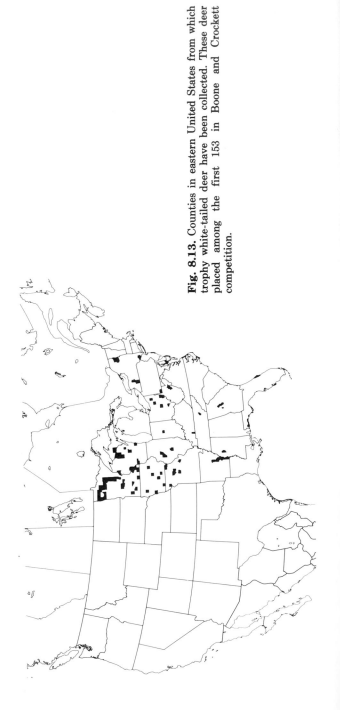

Fig. 8.13. Counties in eastern United States from which trophy white-tailed deer have been collected. These deer placed among the first 153 in Boone and Crockett competition.

and Kenyon-Floyd-Clyde soil association areas mapped by Oschwald et al. (1965). These are very fertile soils developed in calcareous, loam-textured till (Wisconsinan). The lightest male fawns (34.6 kg, n = 134) came from extreme northeastern Iowa, an area nearly coextensive with the driftless area. In this dissected region, the Fayette-Dubuque soil association occurs together with thin soils and exposed rock. The deer here are the lightest of the Iowa series but are nevertheless relatively heavy compared with mean weights from northeastern United States.

SOILS, DEER WEIGHTS, AND FERTILITY. Fairly clear-cut relationships between native soil fertility and mean weights of deer and fertility rates in deer populations are shown in the data above. These relationships are particularly apparent for Mississippi, Missouri, and Illinois, where especially relevant data are available. At the most elementary levels of soil chemistry, it appears that the amounts of available bases and the capacity to supply these elements are the cardinal parameters in the well-being of the deer. Calcareous rocks; limestones; dolomites and shales; and the more recent, thick deposits of loess—which tend to be base-rich in the upper Mississippi Valley—are the ranges of large animals of high reproductive capacity. The few outcrops of glauconitic sediments deserve special study to evaluate the unique soil fertility that seems to arise from their weathering and their related capacity to support populations of large phenotypes.

Forage growing on soils developed from all these rocks is highly nutritious. The contents of bases are higher and the nutritional needs for amino acids and energy are met at what appear to be optimal levels. These food quality relationships to free-ranging herbivores remain largely unstudied.

Geography of antler size in white-tailed deer

The distribution of 122 deer with scored antlers as measured by the Boone and Crockett system (Nesbitt and Parker 1977) that place them among the first 153 positions are shown in Figs. 8.13 and 8.14. The remaining 31 deer were taken in Canada. Several aspects of the geographic distribution appear to be clearly related to native soil fertility. The central sector of the country, between longitudes 84 and 102 degrees west, accounted for 62 large-antlered deer and was chosen to evaluate this relationship in detail (Fig. 8.15.) Austin (1965) delineated 156 land resource areas in the conterminous United States that have similar soils, climate, and land use and constitute valuable information for many biogeochemical interpretations—in this case, for judging the relationship between antler size and soil characteristics. In the following accounts we attempt to relate the occurrences of these trophy deer to several common

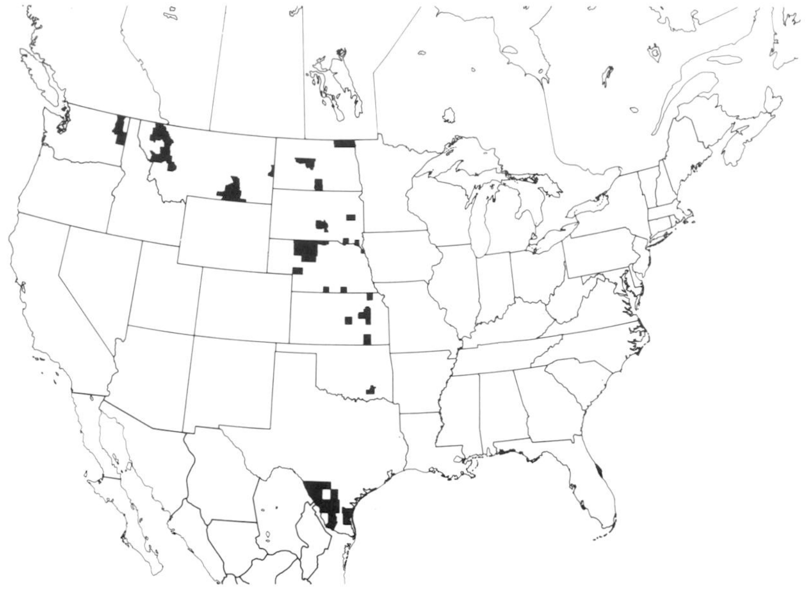

Fig. 8.14. Counties in western United States from which trophy white-tailed deer have been collected. These deer placed among the first 153 in Boone and Crockett competition.

geochemical factors in their ranges, realizing that the genetic background of some herds is not known and substantial differences for genetic effects between areas may result from restocking. To what degree the antler size of these introductions differs from antlers produced by remnant populations of native deer and the complex genetic interchange between these populations is not known.

Georgia. The concentration of large-antlered deer in central Georgia lies just to the east of the central section of the country, but it deserves specific attention. Jones, Jasper, and Newton counties, a south-to-north sequence, form a belt from which four record deer have been recorded. Lamar County, directly west, accounts for a fifth record deer. This group of counties lies within Austin's (1965) Southern Piedmont. About 60% of the area is in forest, about 20% in pasture, and 10% in cropland. The Oconee National Forest and the Piedmont National Wildlife Refuge occupy substantial portions of northwestern Jones and southwestern Jasper counties. The Ocmulgee River forms the western boundaries of Jones, Jasper, and Newton counties. Perhaps deer from these counties matured in the floodplain habitat, because the uplands are characterized

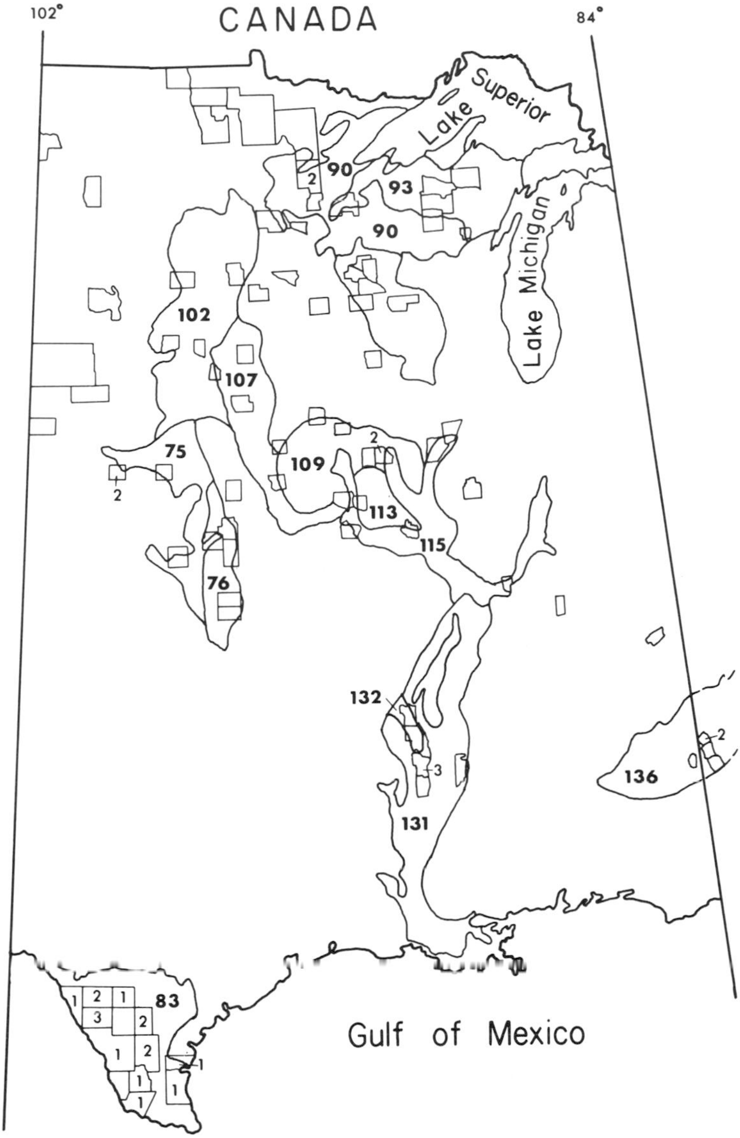

Fig. 8.15. Counties in central United States that have yielded trophy white-tailed deer. More than one deer are indicated by 2 or 3. Fourteen agriculture resource areas (Austen 1965) are depicted.

by strongly weathered, infertile soils derived from granites, biotite-granite gneisses, and amphibolite gneisses. No similar substantial alluvial plain occurs in Lamar County, but the bedrock and soils of the uplands are similar. W. I. Segars (pers. commun.) provided estimates (Table 8.17) of extractable bases in the alluvium of Ocmulgee River. Compared with data for other habitats supporting large-antlered deer, soil mineral levels in the Georgia environment seem very low. Lick samples 209 through 214 come from this area of Georgia and are substantially richer in calcium, magnesium, and sodium than the surrounding soils. From a biogeochemical standpoint, the relationship between antler size and soils in Georgia remains an engima, although large local outcrops of masses of ferromagnesian rocks may be the source of the available minerals for an especially nutritious forage.

Arkansas. The second significant cluster of large-antlered deer in the south is in southeast Arkansas and one deer from Leflore County, Mississippi. Desha County, Arkansas, alone accounts for three record animals—numbers 27, 64, and 109. This cluster is from the Mississippi River floodplain, although the deer from Arkansas and Prairie counties just to the north may have been using some upland habitat. Desha County is wholly within the floodplain where both the White and Arkansas rivers have contributed sediments to Mississippi River alluvium. An outstanding feature of the natural history of this region (the area of confluence of these rivers and north along White River) is the uniform development of floodplain forest (White River National Wildlife Refuge), which is so apparent on Landsat imagery (Fig. 8.16 and National Geographic Society 1976). The contents of extractable cations in soils of Desha County (Table 8.17) suggest the base-rich nature of the floodplain, particularly in comparison with the adjacent upland soils (red soils of the Coastal Plain in Table 8.18 and other upland soils in Arkansas). The alluvial soils developed by the White and Arkansas rivers appear to be particularly enriched with magnesium, judging from the limited data available (Table 8.18).

Minnesota-Wisconsin-Michigan. A large portion of these states is included in the USDA "Northern Lake States Forest and Forage Region," but counties peripheral to the Mississippi River are included in the northern Mississippi Valley Loess Hills (Unit 105) within the Central Feed Grains and Livestock Region. Sixteen record deer come from this combination of units. Unit 105 has many soil and geological affinities with Units 107 and 115 in the lower Missouri and middle Mississippi valleys. Its topography is characterized by almost complete dissection, yielding narrow ridges and valleys that eventually collect and empty onto the Mississippi River floodplain. Local relief is notable, but generally less than 70 m. Thick loess, especially on the east side of the Mississippi River, overlies limestone bedrock in the Driftless Area in southwest Wisconsin and extreme northwest Illinois and drift, mostly pre-

Table 8.17. Central tendencies and dispersion of data for pH and extractable bases (me/100 g) for soils in areas of deer with large antlers. Data were calculated for the darkened surface horizon. Wisconsin data are mostly for 0–17-cm interval. Resource areas are located in Fig. 8.15.

Locality	Agriculture area	*n*	pH		Ca		Mg		Na		K		Parent material
			AM	SD	GM	GD	GM	GD	GM	GD	GM	GD	
Desha County, Arkansas[a]	131	10	6.2	0.8	11.9	2.0	3.1	2.0	.2[b]	. . .	.7	2.1	alluvium
Georgia[c]	136	. . .			1.3		0.8		.04		.2		alluvium
Southwest Wisconsin, northeast Iowa, southeast Minnesota[d]	105	10	6.2	0.7	8.2	1.3	2.1	1.5					loess
Northern Wisconsin[e]	90, 93	4	6.1	0.4	13.1	1.9	4.1	1.4					alluvium
		4	5.3	0.3	6.2[f]	2.5	2.0[f]	0.7					sandy drift, outwash; well drained, moderately leached
		4	5.7	0.2	5.8[f]	0.4	1.6[f]	0.5					sandy drift, outwash; well drained, weakly leached
		5	5.4	0.2	18.3[f]	5.9	3.2[f]	0.6					sandy drift, outwash; wet site
		5	5.4	0.6	18.7	1.6	7.5	1.5					woody peat
		5	4.3	0.4	7.8	2.1	1.8	2.6					moss peat

[a]Gill et al. 1972.
[b]Data essentially invariant.
[c]Typical analysis of Ocmulgee River alluvium (W. I. Segars, pers. commun.).
[d]Muckenhirn et al. 1955.
[e]Wilde et al. 1949.
[f]Arithmetic statistics.

Table 8.18. Extractable cations and clay content of surface horizons of soils developed in alluvium of Mississippi River and extractable Mg and K in some upland areas of Arkansas. Data are arithmetic averages.

Area of alluvium or physiographic region	*n*	Extractable cations (me/100 g)				Clay, %	Source
		Ca	Mg	Na	K		
Mississippi River, north of White River	3	1.0	0.60	0.61	0.31	4.5	Brown et al. 1973
White River, west side of Crowley Ridge	3	2.8	1.0	1.16	0.42	8.4	Brown et al. 1973
White River, Mississippi River confluence	4	3.3	2.1	0.55	0.71	18.6	Brown et al. 1973
Mississippi River, west central Mississippi	3	7.7	3.5	0.11	0.42	17.6	Brown et al. 1973
Ozark Highlands	42	n.d.	0.58	n.d.	0.26	n.d.	Sabbe et al. 1972
Ouachita Mountains	13	n.d.	0.34	n.d.	0.16	n.d.	Sabbe et al. 1972
Coastal Plain	62	n.d.	0.77	n.d.	0.22	n.d.	Sabbe et al. 1972
Loessial Areas	79	n.d.	1.6	n.d.	0.24	n.d.	Sabbe et al. 1972

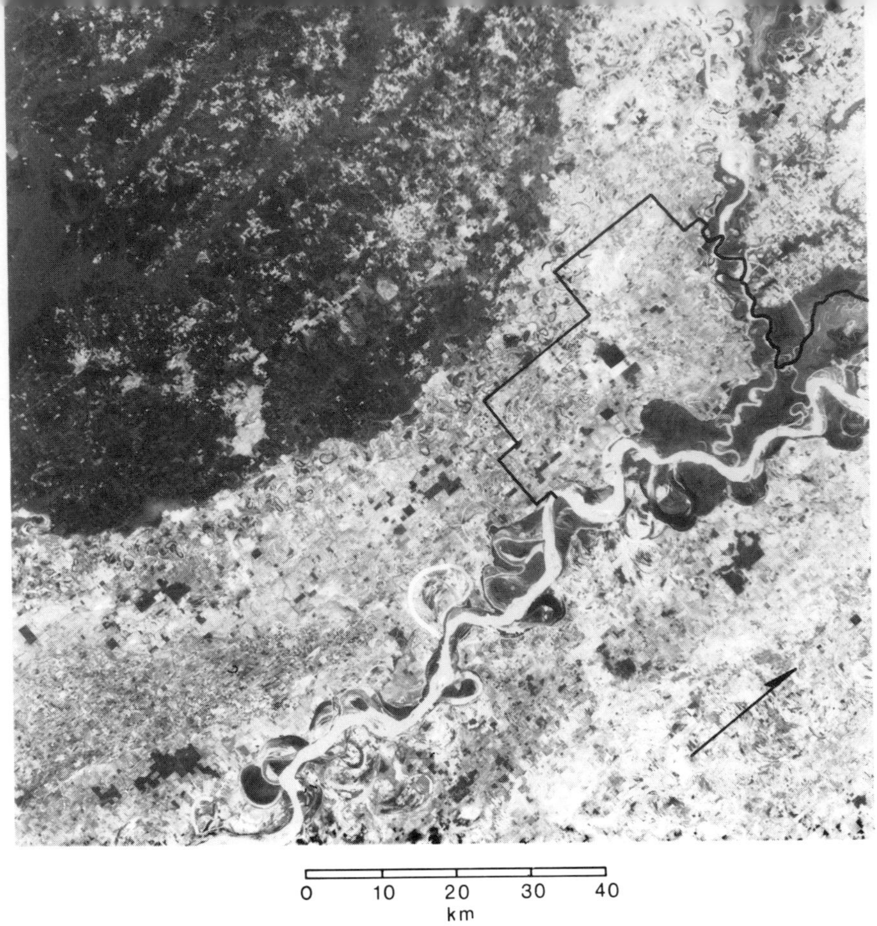

Fig. 8.16. Skylab 4 photograph (Jan. 31, 1974) of the Mississippi River alluvial plain and adjacent upland at the confluence of the Mississippi and Arkansas rivers (courtesy NASA and U.S. Geol. Surv., EROS Data Center). Desha County, Ark., from which three trophy deer have been taken, is outlined. Extensive tracts of floodplain forest are seen along the rivers. Light-toned areas in the alluvial plain are coarser textured, more highly developed soils; grayer areas next to the river are youthful and clay rich.

Wisconsin, elsewhere. About 40% of this unit is on cropland, another 20% is in pasture, and 33% is in forest. Only on rather broad and level highlands where prairie grasses could readily establish do prairie tracts and associated Brunizems occur; otherwise, Gray-Brown Podzolic soils dominate. Lithosols occur on the steep land adjacent to major streams. The geometric mean calcium and magnesium contents extractable by ammonium acetate are 8.2 and 2.1 me/100 g, respectively (Table 8.17).

To the northeast and northwest of the Driftless Area is the Central Wisconsin and Minnesota Thin Loess and Till Area (Unit 90), a region of level-to-rolling till plains with numerous lakes. Except for the forest area south and west of Lake Superior, practically all of the unit is in farms, of which about 25% is cropland in feed grains and dairy cattle forage and 16% of the area is in tame pasture. Potatoes and other specialty crops occupy other land. Gray-Brown Podzolic and Gray Wooded soils are the most common soils except in the northwest, where calcium carbonate is low or absent in the drift. Here Brown Forest soils and Sols Bruns Acides occur. Bog soils, Humic Gley, and Low-Humic Gley soils occupy the numerous wet sites in this region.

Northernmost Wisconsin and much of the Upper Peninsula of Michigan is included in Northern Michigan and Wisconsin Stony, Sandy, and Rocky Plains and Hills (Unit 93). Almost 80% of this area is in forest and only 10% of it is in cropland or pasture. It is an area of undulating till plains dotted with many lakes and ponds and relief is low, on the order of less than 10 m. The bedrock is crystalline rocks, ranging from acid to basic in composition, that give rise to Podzol soils. The drift is commonly very sandy and yields Regosols. Bog soils and Low-Humic Gley soils are characteristic of the wet sites. Areas of bare rock that have been scoured by glaciation are not uncommon.

Data for extractable calcium and magnesium in soils of Resource Units 90 and 93 in northern Wisconsin vary between 5.8 and 18.7 me Ca/100 g and 1.6 and 7.5 me Mg/100 g (Table 8.17). These relatively high concentrations are similar to levels recorded for the Mississippi Valley Loess Hills (Resource Unit 105) to the south and with the alluvial soils of Desha County, Arkansas.

The first-ranking trophy deer comes from Burnett County in northwestern Wisconsin. A map drawn by Twenhofel (1936:Fig. 1) shows an extensive area of glauconite-bearing Cambrian sandstone in western Burnett County. Although we do not know that this large-antlered deer was associated with the glauconite outcrop, an association of large-antlered deer with glauconitic sediments is discussed below in the case of Texas.

Early in this study, Richard A. Hunt (Wisconsin Department of Natural Resources) pointed out Schorger's unequivocal statement: "The Chippewa Valley was unquestionably the greatest game region of the state [Wisconsin]" (1937:121). Buffalo River and Buffalo County commemorate the former abundance of these animals, for (according to Schorger) "Carver ascended the river in June, and observed larger drives of buffaloes and elks than in any other part of his travels." Schorger (1954:56) cited Captain Goddard who accompanied Carver: "This is a fine river . . . , here is plenty of animals, such as stag, deer, bear, and buffaloes, of which, we killed every day one sort or other." Schorger (1954:6) reported that Schoolcraft found elk and deer plentiful on a prairie in Barron County and their tracks abundant at the mouth of the Chippewa. He cited Bunnell who came to La Crosse in 1842: "Elk were also abundant there [the mouth of the Chippewa River] in the Mississippi bottoms, on the prairie, and in the oak thickets below and east of Eau Claire, extending their range over the headwaters of all the streams south of the pipe-belt as far as Black River. The writer saw a band of sixty elk, in 1845, on a prairie about eight miles below Eau Claire." By the mid-1860s elk had been exterminated from Wisconsin—apparently the last in the area of Elk Mound. Elk Creek and the village of Elk Creek commemorate the former abundance of elk in this region of the state.

On the premise that there *had* to be a reason other than fertile prairies, which occur elsewhere in Wisconsin, for the former abundance of wild ungulates near the Chippewa River, a few stream waters were sampled on the assumption that they might be unusually high in sulfates (for reasons outlined). Analyses revealed that sulfur values were not unusual. Along several road cuts, samples of an unusual strata several inches thick of a greenish-colored sandstone were identified as being glauconitic. Associations between trophy records of white-tailed deer (in the Boone and Crockett system, see Chap. 9) and glauconite substrates occur in New Jersey and Texas and glauconite occurs in certain licks (148, 154, and 227). The known distribution of glauconite in Wisconsin is shown in Fig. 8.17. Whether this distribution is indicative of a true correlation with ungulate abundance and size remains to be determined. Some may argue that the dolomitic limestones of the region offer a more prevailing source of magnesium.

Missouri-Iowa-Illinois. Included in this geographic division are deer taken in western and southern Iowa, Missouri, and counties bordering the Illinois River in Illinois. The relevant units (107, 109, 113, and 115) occur in what Austin (1965) calls the Central Feed Grains and Livestock Region—also known commonly as the Corn Belt. Physiographically, there are two general types in this combination of units—the Iowa and Missouri Deep Loess Hills (Unit 107) and the Central Mississippi Valley Wooded Slopes (Unit 115), which have similarities but contrast with the Iowa and Missouri Heavy-Till Plain (Unit 109) and the adjacent Claypan Area (Unit 113) of Missouri. These four units account for 15 deer with record antlers. The Deep Loess Hills and Wooded Slopes units are characterized by 40 to 60% cultivated land and about 10 to 33% forested area, the latter occurring in the Wooded Slopes Unit. Much of the remaining land is in permanent pasture. On upland sites, relatively deep loess overlies drift. Narrow ridges with narrow, high gradient valleys that empty onto the broad alluvial plains of the Missouri, Mississippi, and

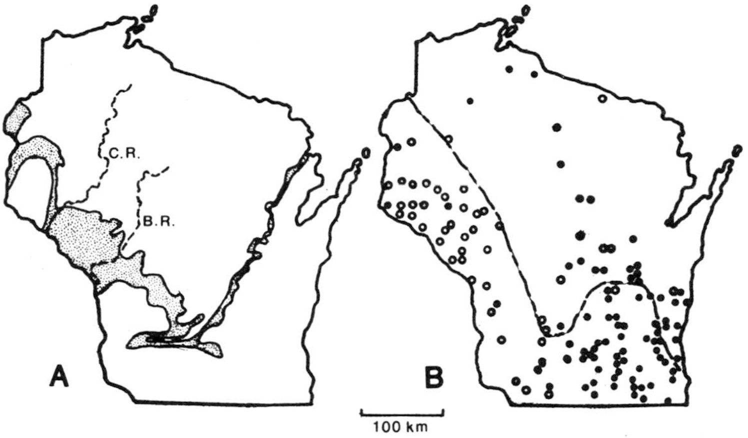

Fig. 8.17. (A): Distribution of greensand or glauconitic sandstone in Wisconsin (after Twenhofel 1936) (C.R. is Chippewa River and B.R. is Black River). (B): Distribution of bison (southwest of dashed line); elk observations in sixteenth and seventeenth centuries indicated by open circles and elk skeletal material by closed circles. Elk skeletal material is associated closely with major wetlands in Wisconsin (Hole 1976:Fig. 2-50). (Map B is redrawn from bison distribution of Schorger 1937 and elk occurrence map of Schorger 1954.)

Illinois rivers characterize the topography. In the Deep Loess Hills Unit, Brunizems have developed on the broad interfluves, Gray-Brown Podzolic soils occur on the forested sites, and Regosols can be found where erosion is substantial. Humic Gley and alluvial soils occur in the bottomlands and are important to row crop agriculture of the units.

Within the arc formed by the Missouri and Mississippi rivers are two units that contrast with the dissected topography and land-use character discussed above. The Claypan Area (Unit 13) may not be significant in our consideration of deer and soil relationships inasmuch as no county is wholly contained in it and the soils occurring in the area are of naturally low fertility. About 60% of the Claypan Area is in cropland and about 16% in forest. Another 10% is in permanent pasture. Planosols are the principal soils with Gray-Brown Podzols occupying sloping land. To the north and west in the Iowa and Missouri Heavy-Till Plain (Unit 109), land use is similar, and the topography is more rolling or sometimes hilly with strongly sloping land adjacent to valleys. Planosols occur on some broad, nearly flat upland sites; otherwise Brunizems occupy the interfluves. Humic Gley soils occur in poorly drained areas. The loessial soils are fertile and rich in calcium and magnesium.

South Dakota–Nebraska–Minnesota. The Loess, Till, and Sandy Prairies Unit (102) comprises easternmost South Dakota and Nebraska, extreme southwest Minnesota, and the northwest tip of Iowa. Practically all of this unit is in the farms; between 65 and 75% of the land is cropped to corn and small grains. About 20% is in pasture, most of which is native grasses. About 1% of the area is in forest confined to river valleys. Reference to Fig. 8.15 indicates that three of the five counties either wholly or partly within Unit 102 are bounded by the Missouri River. Lyon County, Minnesota (with more permanent streams) is, perhaps, more wooded than the average for the unit. The fifth county, Kingbury, South Dakota, is probably very close to average in the unit for land use. This unit as a whole has an unusual lack of wooded cover. The record deer that have been taken in this unit are from counties having greater than average forest acreages. The unit is mostly nearly level to very gently rolling prairie; along the streams erosion has produced valley floors that are seldom more than 10 m below the upland. However, slopes in this dissected terrain are steep and ribbons of trees follow the narrow valleys. Chernozems occupy the prairie with Humic Gley and Solonetz soils occupying the poorly drained sites. Regosols are associated with the steeply sloping, eroded land. All of these soils are base rich.

Kansas. A north-south lineament of six counties from which record deer have originated is found in eastern Kansas. Five of these counties lie within the Bluestem Hills resource unit (Austin 1965). About 60% of the land area is in native grasses, and beef cattle production is important. Narrow divides and valleys typify the landscape, which is underlain by Permian limestones and shales in the northern counties and by Pennsylvanian limestones and shales in the south. Nemaha County on the Nebraska border is largely underlain by Kansan-age drift. Lithosols are very extensive, Brunizems less so, and Planosols occur on more level upland sites (Austin 1965). Aandahl (1972) has mapped Argiustolls, Haplustolls, and Natrustolls in the area.

Immediately west in Resource Unit 75, which Austin (1965) calls the Central Loess Plains, three deer with notably large antlers have been recorded. Two of these probably were taken from riparian habitats along the Republican River in southernmost Nebraska, a region of low relief and rolling topography. About two-thirds of this area is in cash crops, practically all small grains, and the rest in range. The soils are similar to those in the Bluestem Hills, although Chernozems occur in northern parts of the unit. The soil parent materials in these counties consist of extensive thick loess in McPherson County, Kansas, and the Ogallala formation underlies the upland in Harlan County, Nebraska (Burchett 1969). The Republican River in Harlan County is exposing Pierre shale in its valley. To the east in Nuckolls County, the Niobrara formation covers about the western two-thirds of the county and the Carlile formation outcrops in the east. These rocks are either limestones or calcareous sediments.

Texas. A concentration of 16 record deer heads from the Rio Grande Plain of south Texas is one of the outstanding features of trophy deer distribution in the United States. About 80% of this region is in range. The soils are Grumusols, Reddish Chestnuts, Lithosols, Regosols, and Calcisols (Austin 1965). The distribution of soil orders (contemporary classification) in the area of the counties of interest is mapped in Fig. 8.18 (after Buol 1973). Vertisols, Mollisols, Alfisols, and a small area of Inceptisols are represented. Potential natural vegetation of the area is mesquite and acacia savanna (Küchler 1966).

The plain is underlain by Tertiary-age rocks except for the coastal counties, which are underlain by Quaternary-age sediments (American Association of Petroleum Geologists, Map 7). Shales and sandstones, many of them calcareous, are the most common rocks in the Tertiary sequence. The plain, viewed on the color photomosaic of the United States (National Geographic Society 1976) together with evidence from the four map sheets from the Geologic Atlas of Texas that cover this area

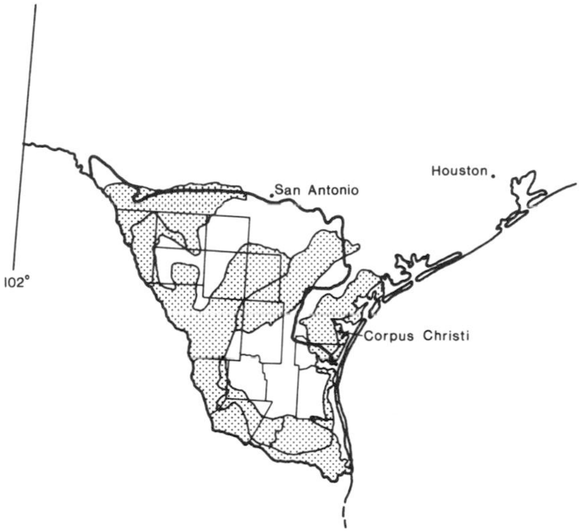

Fig. 8.18. Map of Rio Grande Plain with counties outlined in which deer with very large antlers have been collected. Stippled area represents Mollisol and Vertisol and a small area in the south represents Inceptisol soil orders (boundaries of great groups are located but not identified). Unstippled area represents Alfisols. (Soil distributions from Buol 1973)

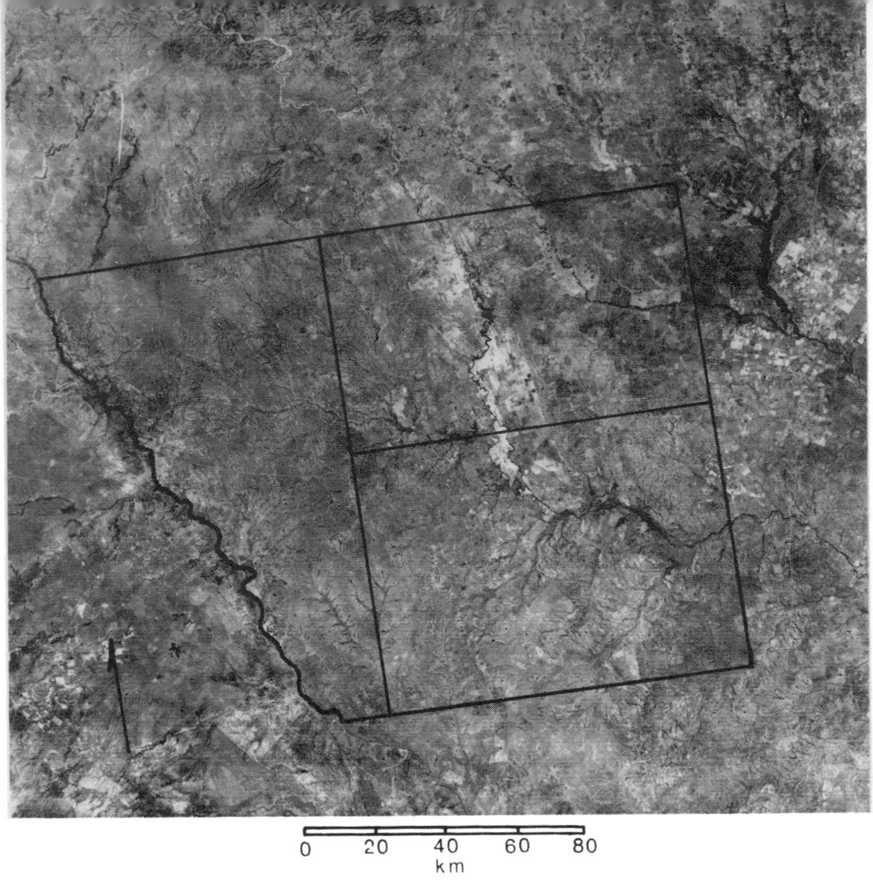

Fig. 8.19. Landsat image (Band 5) of the northern Rio Grande Plain taken Nov. 8, 1978. Maverick, Zavala, and Dimmit counties are shown (courtesy NASA and U.S. Geol. Surv., EROS Data Center). Dark tones are more densely vegetated and appear to be underlain by glauconite-bearing rocks.

(Corpus Christi Sheet 1975; Crystal City–Eagle Pass Sheet 1976; Laredo Sheet 1976; McAllen-Brownsville Sheet 1976; all Bureau of Economic Geology, University of Texas at Austin) afford clues to the unusual productivity of the plain. Seen on the mosaic (Fig. 8.19), two broad and one narrow green belts, presumably heavily vegetated, trend north and south on the plain. They appear to correspond to the Escondido formation (upper Cretaceous) in Maverick County, the Kinkaid formation (Eocene) in Maverick and Zavala counties, the Laredo formation (Eocene) in Zavala and Starr counties, and the Yegua formation (Eocene) in Frio and Duval counties; an outcrop of the Jackson group (Eocene) forms the narrow band in Duval, Jim Hogg, and Zapata counties. Comparison of the Soil Survey Reports for Maverick (Smith et al. 1942), Zavala (Smith et al. 1940), and Dimmit (Smith and Huckabee 1943) counties with the extent of the darker tones of the mosaic indicates that Duval, Maverick, Monteola, and Webb soils are common in the dark-tone areas. Webb and Duval soils are reddish soils, whereas Monteola soils are calcareous, black soils.

The heavily vegetated areas appear to be coextensive with the outcrop of glauconite-rich rocks. Often they are sandstones; in weathering the green iron-bearing glauconite gives rise to the red hues. Some gypsum is also present, for example, in the upper part of the Laredo formation, in which Turner et al. (1960:49) noted that vegetation of the outcrop "generally is dense, consisting of guajillo, mesquite, low shrubs, and grass."

Four decades ago, Fraps and Fudge (1937) presented data for several thousand samples of Texas soils. We have redrawn two of their summary maps that show acid-soluble calcium and magnesium in surface horizons of Texas soil (Fig. 8.20). Comparison of the levels of calcium and magnesium in soils of the Rio Grande Plain with the Edwards Plateau to the north and the prairies to the east show that calcium and magnesium contents are intermediate in concentration. The geometric mean calcium and magnesium concentrations (both n = 10) in light- or red-colored upland soils on the Plain—Duvall, Maverick, and Webb series—as measured by strong, hot-acid extractions are 20.4 and 14.8 me/100 g soil,

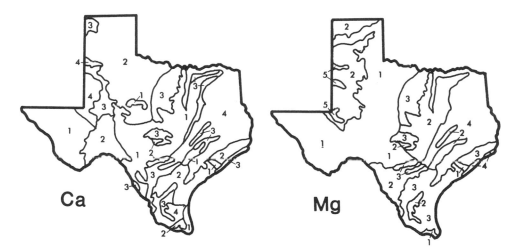

Fig. 8.20. Contents of calcium and magnesium in surface horizons of Texas soils. For calcium, the numerical categories are 1: $\geq$ 1.43%; 2: 0.29–1.43%; 3: 0.15–0.29%; 4: 0.079–0.15%; 5: $<$ 0.079%. For magnesium, 1: $\geq$ 0.37%; 2: 0.19–0.37%; 3: 0.096–0.19%; 4: 0.048–0.096%; 5: $<$ 0.048%. These determinations were made on solutions obtained after boiling the soil with 7 N HCl for 8 hours. (After Fraps and Fudge 1937: Figs. 9, 12)

respectively (from Table 17 in Fraps and Fudge 1937). The Edwards Plateau, underlain by Cretaceous-age limestone, is richer in both elements but has not produced deer with large antlers. Fraps and Fudge (p. 77) mapped the soils of the plain as containing from 110 to 220 ppm of total phosphorus, which is slightly lower than the 230- to 440-ppm range mapped for the Edwards Plateau. Note that the levels of phosphorus in the East Texas Timber Country and West Cross Timbers area, which are noted for small deer, are in the range of the Rio Grande Plain soils. Fraps and Fudge (p. 76) concluded that active phosphorus, as measured by solubility in 0.2 N nitric acid, was higher in soils of the plain than in the timber areas and was as high as that in soils of Edwards Plateau. Nitrogen tends to be less than 600 ppm over much of the plain (Fraps and Fudge 1937:75). We suspect that deer on the plain experience a superior nutriture that is largely based upon the relative levels of calcium and magnesium and may be closely related to the occurrence of glauconite-bearing rocks.

RELATION OF ANTLER SIZE TO SOIL: A HYPOTHESIS. We have looked closely at the habitats of large-antlered deer from a midsection of the United States. Much is known about the bedrock, soils, and phytogeography of the region; nevertheless, specific data for nutrients in the particular habitats that we were interested in are almost universally lacking. It appears that deer with antlers of unusual size occur in unique circumstances, since a large proportion of these deer in the midsection of the country come from a relatively few agricultural resource units that correspond to areas of similar soils, land use, and climate. Because these units represent a wide range of latitude and climate as well as other factors, is there a unifying factor among these units? In general, it appears from the forgoing description of the association of large deer and alkaline earth–rich habitats that the relatively base-rich character—specifically the adequate levels of calcium and magnesium—explains the occurrence of deer with exceptional antlers in these environments. (The size of antlers would be expected to be associated with size of the deer bearing them by some allometric relationship between the variables.)

Unusually large tracts of country exist within the white-tail's range that, in general, are characterized by soils of low fertility and have not produced large deer. In many respects, the dearth of record deer from these areas is as important in explaining phenotypic development as is the presence of clusters of animals. The great swath of the Piedmont, the Ozarks, Michigan, the Adirondacks, and the White and the Green mountains are examples of large areas that have not yielded record deer. Certainly the genetic potential has always existed in many of these areas.

CHAPTER NINE

Herbivores and the mineral environment: the common biogeochemical challenge

Physiological responses to alkaline earth environments

Hofmann (1973) has described the gross anatomy and histology of the stomachs of African ruminants and has documented the great variations of morphology exhibited that are associated with feeding habits. But a common denominator of their functional efficiency is the maintenance of an optimum pH for the maintenance of their microbial and protozoan populations as well as for preservation of the integrity of the mucosal epithelium.

The preference of ungulates for the new flush of grass, whether a product of spring growth or the onset of the rainy season, is universal. However, new plant growth, as pointed out earlier in respect to tetany, contains high levels of potassium relative to calcium and magnesium; if this ratio in the diet is too heavily weighted on the side of potassium, herbivores, especially those on acid-rock and highly weathered substrates, risk hypomagnesemia. Domestic cattle are particularly susceptible. For example, cattle in some areas of Scotland are subject to grass tetany, but we are not aware that red deer on these terranes are physiologically challenged to the same degree. However, the avidity with which red deer devour bone and antler is expressive of the same nutritional need (Darling 1937).

The ratio of potassium to calcium plus magnesium—expressed as equivalents—is widely accepted as an index of the tetany-producing ability of forages (Mayland and Grunes 1979). The phenological progression of change in this mineral ratio can be seen in data for forages from two widely separated areas—western wheatgrass from North Dakota (Fig. 9.1) and greenbrier and honeysuckle from South Carolina (Fig. 9.2). There are no reasons to believe that these patterns of mineral

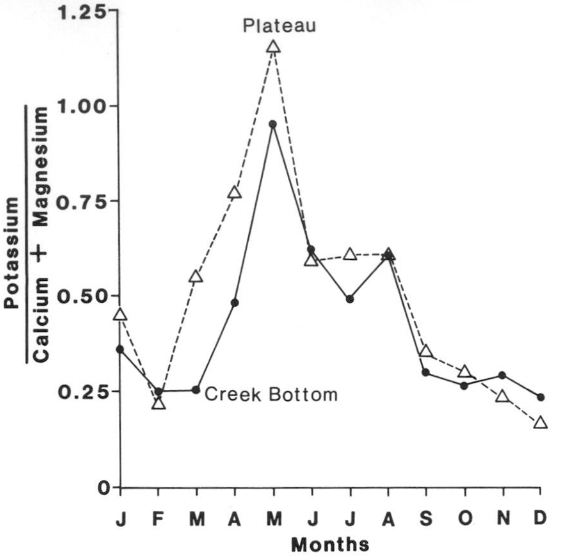

Fig. 9.1. Seasonal patterns for the ratios K/(Ca + Mg) on the equivalents basis, in *Agropyron smithii* for two sites in North Dakota. (Courtesy S. D. Fairaizl)

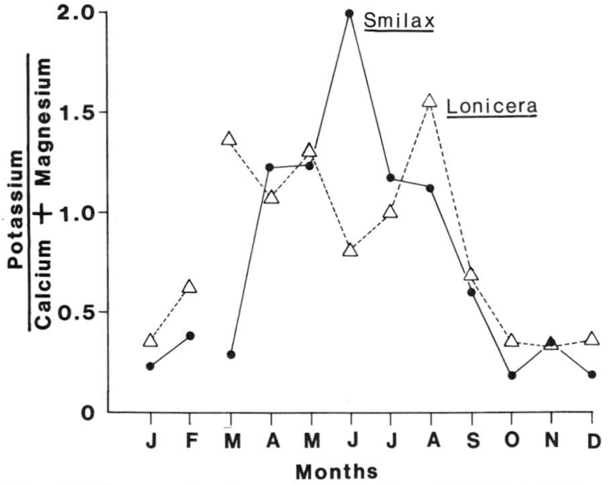

Fig. 9.2. Seasonal patterns for the ratios of K/(Ca + Mg) on the equivalents basis, for *Smilax* and *Lonicera* from South Carolina. (Calculated from Thorsland 1966)

content in forages have changed appreciably in late Cenozoic and Pleistocene times; correspondingly, it is equally unlikely that ungulate feeding behavior has changed over this time interval.

Clues to the physiological challenges faced by North American ungulates from spring dietary mineral imbalances were found in several blood chemistry studies from widely separated locales (Table 9.1). Weeks (1971) provided data for serum calcium, magnesium, sodium, and potassium of white-tailed deer in Martin County, southwestern Indiana. Karns and Kinsey (1972) have recorded data for serum calcium, magnesium, and phosphorus for captive white-tailed deer in Minnesota, and Seal et al. (1972) for serum calcium, magnesium, sodium, and potassium of captive white-tails in Michigan. Serum levels for calcium, magnesium, sodium, potassium, and phosphorus for mule deer in the Cache la Poudre River basin of Colorado have been reported by Anderson et al. (1972) (Table 9.1). Serum calcium levels changed significantly with season only in the Minnesota deer, declining in spring and summer; serum calcium levels in the Indiana deer were only marginally lower in spring than in winter and summer. Serum magnesium levels were significantly higher in spring in the Minnesota deer, but no changes were recorded in the other deer studied. Serum sodium tends to be lowest in winter. The reason for a significant difference for sodium content in sera taken in March from captive pregnant does maintained on white cedar is not apparent. In the case of potassium, the highest levels, which were significantly different from later seasons, were found in spring in deer from Cache la Poudre, Colorado. Potassium in the sera of deer fed white cedar declined significantly through the four-month trial (Seal et al. 1972). Note that red cedar in Missouri was rich in sodium, calcium, and magnesium (Table 8.16), factors that might affect serum potassium as discussed later.

Interpretations of blood chemistry often have been fraught with difficulty for all disciplines that have endeavored to correlate concentration of some blood analyte with another variable. Blood chemistry in relation to nutrition has been a particular enigma; only with the advent of radiochemistry was certain unequivocal evidence available. In this context it is not surprising that a significant phenological chain of events is not indicated in the data of Table 9.1; that is, the effects of changes in diet, substantial as they are from winter to spring, are not reflected in significant quantitative shifts in the major mineral elements of the plasma. However, the lack of response in blood levels in normal deer is to be expected as a priori an animal tries to achieve homeostasis in blood levels, if necessary, at the expense of other tissues. Furthermore, as Seelig (1980:93) points out, "magnesium is predominately an intracellular cation, and since levels in the blood are generally kept within narrow limits, relying on serum magnesium as the index of magnesium status of the body can give misleading information."

The potassium-rich vegetation of spring is perhaps reflected in the high values of the Colorado data as the authors speculated (Anderson et al. 1972). Although they are elevated, calcium and magnesium do not show significant compensatory change. The data for magnesium in captive deer in Minnesota (Karns and Kinsey 1972) reflect high circulating levels in the spring and suggest that magnesium may be

Table 9.1. Seasonal levels of Ca, Mg, Na, and K in sera of adult deer ($\geq$ 1.5 years).

Kind of deer	Locality	Months	Sex	Serum analyte (mg/100 ml) Ca	Mg	Na	K	Source
White-tailed deer	Indiana (Martin County)	Jan.–Mar.	M, F	10.6	2.7	374	47	Weeks 1974
		Apr.–Jun.	M, F	10.2	2.4	375	44	
		Jul.–Sept.	M, F	10.7	3.4	377	45	
		Oct.–Dec.	M, F	10.4	3.2	367	39	
White-tailed deer (captive)	Minnesota	Jan.–Mar.	F	$10.1^{a,b}$	$2.2^{a,b}$	...	...	Karns and Kinsey 1972
		Apr.–Jun.	F	9.4^{a}	$3.0^{a,c}$	...	...	
		Jul.–Sept.	F	9.5^{b}	$2.3^{c,d}$	...	...	
		Oct.–Dec.	F	9.6	$2.7^{b,d}$	...	...	
White-tailed deer (captive)	Michigan	Dec.	F	9.2	2.9	329^{a}	19.5^{a}	Seal et al. 1972
		Mar.	F	9.1	2.9	345^{a}	17.6	
		Apr.	F	9.0	3.3	334	16.0^{a}	
Mule deer	Colorado (Cache la Poudre River basin)	Dec.–Mar.	F	12.3	4.1	352	32	Anderson et al. 1972
		Mar.–Jun.	F	12.9	4.2	342	37^{a}	
		Jun.–Sept.	F	11.8	3.8	330	36^{b}	
		Sept.–Dec.	F	12.0	4.2	339	$28^{a,b}$	

Means with same superscript were found to be significantly different at $P \leq .05$ (refer to source for specific probability).

excreted in greater quantity at this time. We interpret the decline in serum magnesium that Weeks (1974) reported for spring samples of white-tailed deer as reflecting the substantial increase in potassium content of the diet in this population (Fig. 9.3).

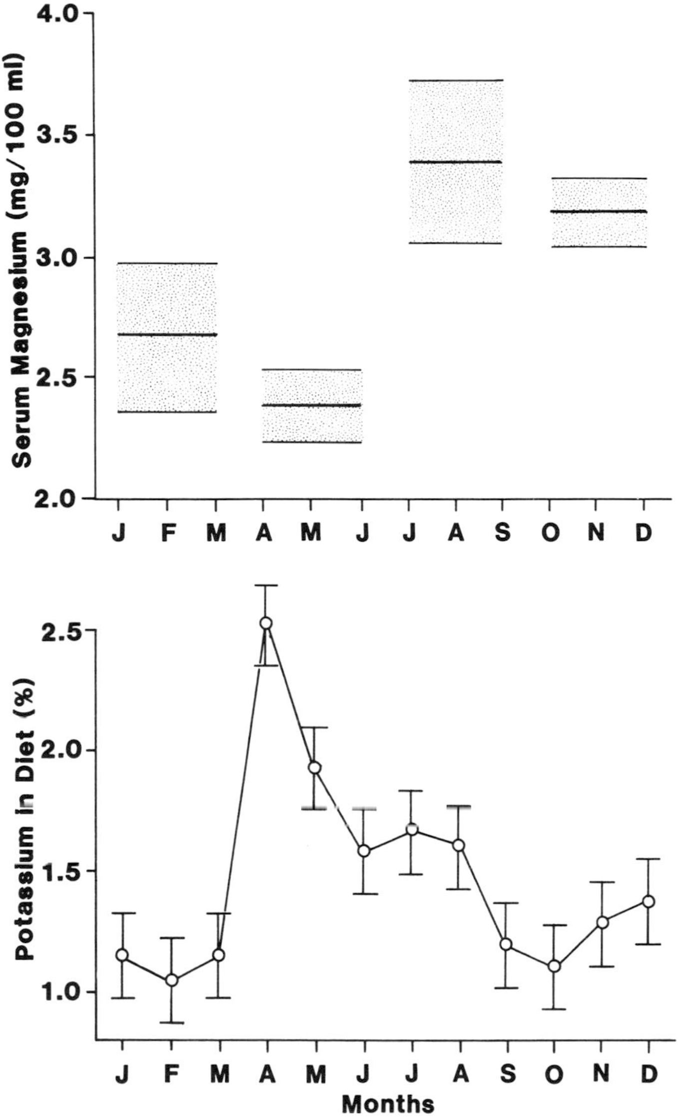

Fig. 9.3. Magnesium content (mean and standard error) of serum by season and potassium content of diet by season for white-tailed deer in southwestern Indiana. (Magnesium data from Weeks 1974; potassium data from Weeks and Kirkpatrick 1976)

The loss of magnesium is due partly to the effect of elevated potassium on the adrenal cortex and the subsequent production of aldosterone. Weeks (1974) has documented a substantial enlargement of the adrenal gland in the spring, which could be partially explained by a disproportionate increase in the cortical tissue. Weeks attributes this increase to the increased dietary potassium and water flux at this time of year, which he said acted to reduce sodium reserves of deer (assumed to be already in a depleted state). Aldosterone also increases fecal excretion of magnesium (Levin 1976) during a time when the animal can least afford it. The chronic diarrhea observed in spring also may contribute to magnesium losses, as has been found in humans (Hammarsten and Smith 1957). As a magnesium-depleted animal can call upon no substantial magnesium reserves (as in the case for calcium), a fast-acting, fatal tetany may be the outcome.

Two apparent cases of tetany in wild ruminants have been reported. Herbert and Cowan (1971) described tremors, which they attributed to selenium deficiency, in mountain goats during restraint. Shope (pers. commun. 1979) has described the death, attributed to white-muscle disease, of a white-tailed deer trapped in Centre County, Pennsylvania. No doubt exists among biologists who have attempted blood sampling of live free-ranging wild animals that the physiological effect of restraint is substantial. Barrett and Chalmers (1977) have demonstrated many effects of such stress on the concentration of different analytes in the blood of pronghorn antelope. The possibility that wild ruminants, particularly females, may be suffering borderline chronic magnesium deficiency over much of their range raises the question as to whether death seen in animals agitated by trapping, chase, etc., may be attributable to the morbid effects of magnesium deficiency. In further support of this thesis, note that Karns and Kinsey (1972) had evidence (not stated) of tetany in wild deer in Minnesota.

Another inference of the importance of maintaining normal magnesium metabolism is the avidity with which deer eat buckwheat when available. Hosley (1956) noted that culture of the plant in certain parts of Pennsylvania was impossible because of deer depredations. Evidently buckwheat is very sensitive to deficiency of magnesium in soil (Beaumont and Snell 1935). Beeson (1941) gives the magnesium and calcium contents of buckwheat as 0.60 (n = 3) and 0.27% and 2.62 (n = 3) and 2.59%, respectively. These data are for whole plant at anthesis or shortly after. Single determinations for sodium gave 0.28 and 0.01% for Massachusetts and French samples (Beeson 1941:156). Evidently there has been little recent research on buckwheat (the name, incidentally, is a derivation of German *buckweizen,* referring to resemblance of the seed to the beechnut), so these older analyses contain the only data on the mineral-rich character of its foliage and seeds.

Finally, Cowan and Brink (1949) wondered why licks in the mountains of southeast British Columbia, rich in calcium and magnesium, could be an attractant when they were located on calcareous terranes that presumably supported vegetation rich in these elements. As documented above, this relationship is particularly true for mountain goat licks. Perhaps further explanation occurs in research of Scott et al. (1950), who demonstrated the self-selection of rats of diets rich in alkaline earths. Their findings are relevant and raise a number of nutritional and behavioral questions; and in their words, "The results with calcium carbonate are unusual in one respect, in that both control and deficient animals preferred diets containing this substance in comparatively large amounts. This was presumably a flavor effect unrelated to deficiency . . . [there was] preference for diets containing as much as 3 to 6% calcium carbonate."

Inferred mineral patterns in moose forages

Further indication that magnesium is only marginally adequate in the forages of some northern ecosystems is found in the data of Franzman et al. (1975) for hair analyses of moose in the Kenai Peninsula of Alaska. Using 110 ppm magnesium in hair as their benchmark as reflecting adequate nutriture, these authors considered moose in their sample to be below normal, although they stress that the 110 ppm level is derived from studies on domestic livestock hair and nothing is known of values for moose hair that indicates the critical magnesium level. Copper and manganese also were found to be low in Kenai moose by domestic animal standards.

We have plotted the data for potassium and magnesium of Franzman et al. (1975; Table 1) in Fig. 9.4. The level observed in each sampling is considered to reflect hair grown over the preceding 2 months (Flynn et al. 1975). The rise in potassium precedes magnesium by about 2 months and is taken to reflect the increase in potassium in the early flush of annual growth. The subsequent rise in magnesium can result from mobilization of magnesium to offset loss through the kidney caused by mineral corticoid response to the potassium load. The increased magnesium also may be accounted for by use of licks and, certainly later in the season, by the natural phenological increase in forage magnesium. The data for these Kenai moose suggest very substantial changes in mineral loading and/or, less likely, electrochemical shifts in nature of transport in the plasma and hair follicle. An interesting and unaccounted-for aspect of these data is the increase of 53 to 125 me of total major-element cations—calcium, magnesium, potassium, and sodium—between May and October. Franzman and colleagues did not wash the hair in a polar solvent (rather, ethyl ether); therefore, some of

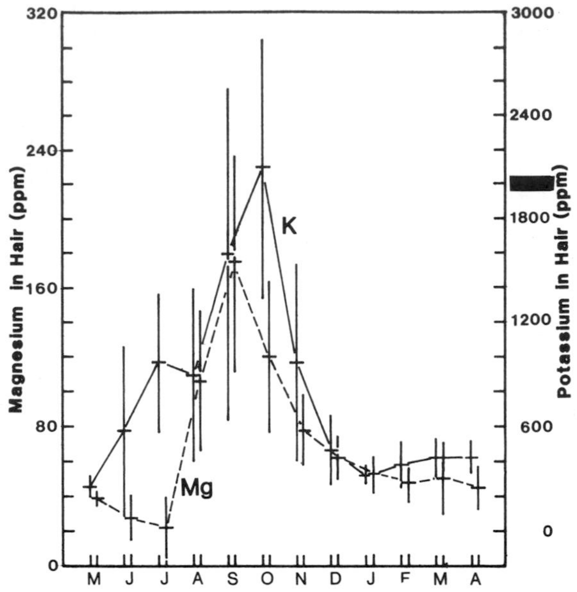

Fig. 9.4. Mean contents, and their standard errors by month, of potassium and magnesium in moose hair. (Data from Table 1 in Franzmann et al. 1975)

these cations may have been loosely held or actually were surface deposits on the hair that would have been lost in washing with water.

A unified theory for mineral lick use

Our study has shown that sodium and magnesium in eastern North American licks and calcium and magnesium in western North American licks are the major elements sought by ungulates. Sodium is the major and sought-after element in an important percentage of licks (Table 7.4). It also is apparent that native ruminants and other animals may express a similar desire for common salt, as humans do (Fig. 9.5). But if, as contended here, magnesium—found in concentrations above adjacent areas in most licks—is the sought-after element as an antidote to the high levels of potassium found in spring vegetation, surely sodium-rich licks could not possibly serve the same function. The physiological functions of divalent magnesium have relatively little in common with those of monovalent sodium and their roles need not be reviewed in any detail here. Their interactions in relation to absorption and excretion do, however, have an important bearing on the thesis presented here that herbivores, by utilizing either magnesium-rich licks or sodium-rich licks, achieve the same physiological goal—prevention of potential

Fig. 9.5. Free-ranging bighorn sheep eating salt out-of-hand. (Reprinted by permission. © University of Chicago Press, all rights reserved, and V. Geist)

physiological imbalances exemplified by tetany inherent in the consumption of potassium-rich spring growth of vegetation.

Rasmussen et al. (1976:226) have pointed out that homeostasis represents an organism striving to achieve physiological stability rather than maintaining absolute constancy of the internal environment. Thus, there are parallels with an animal's external environment; the "balance of nature" is achieved by oscillation of its components around a mean condition and this mean itself may shift or oscillate with time.

Homeostasis of essential mineral elements obviously is achieved in its grosser aspect only by control of rates of absorption and/or excretion. The main bulk minerals are regulated closely, although in the case of magnesium the exact mechanisms are not known. Trace elements, on the other hand—perhaps because their relative scarcity has not induced the evolution of control mechanisms—are not monitored closely. The greatly varying levels of trace elements found in the primary feathers of wild geese were considered eloquent evidence that neither absorption nor excretion of these elements was controlled (Hanson and Jones 1976).

The absorption of bulk minerals must be considered with respect not only to varying degrees of independent absorption mechanism but also to known antagonism between elements in respect to absorption processes.

Interactions of potassium, sodium, and magnesium in homeostasis

SODIUM CONSERVATION. Edmonds (1976:269) pointed out that although the effect of aldosterone on the kidney is transitory, "intestinal sodium conservation during periods of sodium depletion is largely if not entirely effected through the adrenal-aldosterone system. . . . Despite the emphasis that has been laid upon renal effects, therefore, it is quite conceivable that the primary role of aldosterone in sodium homeostasis may lie in the control of sodium transport across gut epithelia." The effect of aldosterone on the colon varies by section: the proximal colon is stimulated to increase sodium and chloride absorption but potassium absorption only slightly; the distal colon, however, is associated with increased potassium secretion. "Present evidence suggests that changes in permeability of the luminal membrane barrier are largely responsible for the effects of aldosterone on the colon" (Edmonds 1976:277).

SODIUM-LINKED POTASSIUM HOMEOSTASIS. Sodium-linked potassium homeostasis, subtitle from the *Handbook of Physiology,* sec. 8 (Laragh and Sealey 1973:886), describes with explicit brevity what we assume must be one of the key roles of sodium-rich salt licks. Further to the point: "The defense of plasma potassium levels within a relatively narrow range, despite wide variations in intake, is one of the most closely guarded homeostasis functions. In this defense, the direct effect of plasma potassium levels on aldosterone secretion can be viewed as part of a system that protects the organism from dangerous hyperkalemia via an aldosterone-induced kaluresis. . . . The maintenance of potassium balance also invokes coordinated changes in renal sodium transport," Actually, a double feedback system involving renin secretion by the kidney is involved in achieving sodium-potassium homeostasis (see comprehensive review by Laragh and Sealey).

A close linkage between sodium and potassium turnover in birds was perceived (Hanson and Jones 1976) in data for the feather mineral content of primaries of wild geese and, in effect, substantiated by data for two wild, localized populations of the ring-necked pheasant (Fig. 9.6).

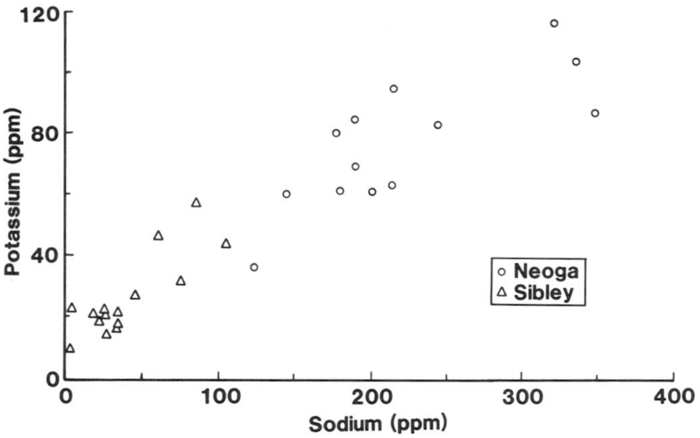

Fig. 9.6. Relationship of potassium and sodium contents in vane portions of primary feathers of ring-necked pheasants from two areas in Illinois. (From Fig. 262, Hanson and Jones 1976)

The close linkage in turnover of these monovalent ions was not anticipated to be expressed in the feather keratin. Note that feather keratin provides a permanent record of this metabolic relationship—one that also would be mirrored in the hair of herbivores.

DUAL ANTITETANY ROLE OF SODIUM. Data for mineral loading of feather keratin of wild geese indicate that sodium has a wider role than generally appreciated on the turnover of other minerals (Hanson and Jones 1976). It can be argued that it also serves a dual role in the prevention of grass tetany: (1) Because a stoichiometric relationship between sodium and potassium turnover appears to exist, an increased sodium intake would aid in lowering and reestablishing normal potassium levels (Fig. 9.6); and (2) by enhancing the rate of magnesium absorption, increased sodium intake would work toward reestablishing losses of magnesium resulting from dietary potassium loading. This latter relationship possibly can be inferred from Tansy's review (1971:195): "Herbage analysis from the pasture where grass tetany has occurred showed that the disease was not simply due to magnesium deficiency. The magnesium was in the pasture, but for some reason it could not be absorbed by the animal."

PHYSIOLOGICAL ROLES OF MAGNESIUM. The significance of magnesium in the ecosystems of herbivores cannot be fully appreciated without understanding some of its more salient physiological roles.

Although 65% of total body magnesium is incorporated in the mineral phase of the skeleton of mammals, it is not directly (and immediately) involved in the main pathway of metabolism; 34% of the total is intracellular and only 1% extracellular (Heaton 1980:43). The large number of biochemical reactions involving and depending on magnesium have been reviewed by Livingston and Wacker (1976) and Heaton (1980). Heaton pointed out that *in vitro* studies have shown that magnesium activates more than 100 enzymes. Of concern here is the requisite need of magnesium in the glycolytic pathway and the tricarboxylic acid cycle as well as for transphosphorylating reactions. Despite its widespread involvement in metabolic reactions, Heaton (1980:50) stated that "few disturbances in intermediary metabolism have been unequivocally established during magnesium deficiency in living organisms." We suggest here, however, that magnesium deficiency is the next proximate explanation beyond hypoglycemia for shock disease in snowshoe hares and, ostensibly, a similar syndrome in Norwegian lemmings prior to death during cyclic highs. In essence, as Tansy (1971:202) stated, "the role of magnesium lies predominately in regulating enzymatic activity and neuromuscular irritability." Although there is no specific hormonal control for magnesium homeostasis, intracellular ratios of magnesium to calcium ions in the normal mammal are closely maintained and directly influence cellular metabolism. Magnesium ions affect nerve impulses and are generally antagonistic to calcium ions at the neuromuscular junction. "An imbalance of the relationship of Ca^{2+} to Mg^{2+} secondary to magnesium depletion might result in heightened neuromuscular excitability" (Livingston and Wacker 1976:218). However, under normal conditions, the concentration of magnesium within a cell may vary twofold (Rassmussen et al. 1976:228). Although the relationship between magnesium insufficiency and tetany seemingly has been well established and understood, at times it has seemed almost paradoxical that mammals (in this case, the snowshoe hare) can seem physically competent one minute and be in a state of shock the next; as Rassmussen et al. (1976:232) emphasize, extremely small changes in ion concentrations alter the activity of many enzymes.

Although the onset of the clinical symptoms of grass tetany and hypomagnesemia apparently has been commonly assumed to be a manifestation of calcium-magnesium imbalance at the neuromuscular junction, studies of sheep (Meyer et al. 1980;801–6) implicate lowered magnesium levels in the brain as the primal cause of convulsions resulting from magnesium-deficient diets. Magnesium levels in the serum declined first, but clinical responses were found to be more closely related to declines in magnesium levels in the cerebrospinal fluid. They found age- and sex-related differences in the responses of sheep that have some parallels to the dependence of wild ungulates on licks. Lactating sheep responded rapidly to the deficient diet, nonlactating adults after some delay, and young lambs only very slowly.

If tetany in ungulates is an expression of brain dysfunction, it is likely that the same explanation would hold for similar manifestations in snowshoe hares and lemmings—the topic that concludes this chapter.

Tansy (1971:194) conceded that the report of hypomagnesemia in cattle did more to advance the knowledge of gastrointestinal absorption of magnesium than any other factor. Calves absorb dietary magnesium in both the small and large intestines; with increased age this ability becomes restricted to the small intestine, a factor associated with the onset of grass tetany. This reduction in absorption surface has resulted in a 25% to 4% reduction of magnesium absorption. Perhaps restriction of magnesium absorption to the small intestine in cattle past the calf stage may be an evolutionary development. However, susceptibility to magnesium deficiency in rats also has been found to be age related (Günther et al. 1980:57), whereby older rats are less susceptible. Findings for the rat also suggest partially why licks are used most heavily by female-calf groups in wild ungulates.

Livestock also first provided evidence for the development of cardiovascular lesions resulting from magnesium deficiencies in the environment. "The histologic arterial damages of magnesium deficiency were first characterized in cows, on spring forage on magnesium-poor soil, or where other factors interfered with the availability of magnesium. . . . The syndrome, leading to these cardiovascular lesions, was seen predominately in lactating cows in areas where 'grass tetany' or convulsions of magnesium deficiency (characterized by hypomagnesemia and hypocalcemia) occurred during late pregnancy or during lactation. Not only cows but ewes are susceptible to this disorder, and it has been noted that it is more prevalent in herds with a high incidence of toxemia of pregnancy" (Seelig 1980:169). Because of the high production of modern breeds of cattle, lactation places them in special jeopardy of grass tetany. Under less stress in this respect, and with the freedom to roam and balance their diet ecologically by geophagy or selection of mineral-rich plants, wild ungulates apparently seldom exhibit the more severe manifestations of magnesium deficiencies.

The development of hyperkalemia in herbivores as a result of high potassium intake in spring is a two-edged sword. By stimulating increased secretion of aldosterone, which increases not only potassium excretion but also magnesium excretion, magnesium is depleted in muscle tissue. According to Livingston and Wacker (1976:217), the most striking symptom of hypomagnesemia in human beings is tetany, which may mimic exactly hypocalcemic tetany.

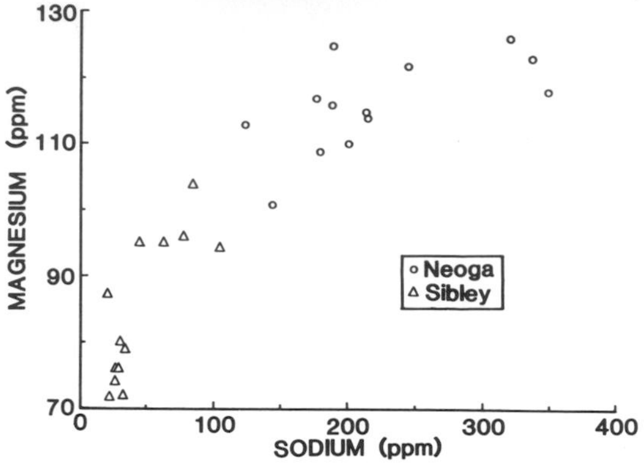

Fig. 9.7. Relationship of magnesium and sodium contents in vane portions of primary feathers of ring-necked pheasants from two areas in Illinois. (From Fig. 262, Hanson and Jones 1976)

An *in vitro* experiment with rat small intestine has shown that sodium ions increase the absorption of magnesium, and Tansy (1971:195) suggests that although such experiments may not be valid for the living animal, ". . .they suggest that attention should be given to the *influence of varying sodium intake on magnesium absorption in the ruminant*" (our italics). Of course, many licks that we studied were rich in both sodium and magnesium. A second line of evidence underlining a positive relationship of sodium to magnesium absorption is shown by levels of these two minerals in the primary feathers of ring-necked pheasants from two areas of Illinois, one (Neoga) having soils relatively high (mean, 0.82%) and the other (Sibley) having soils relatively low (mean, 0.68%) in total sodium (Fig. 9.7). In essence, mineral levels in feather keratin (probably more notably at higher levels) tend to reflect turnover rates of excess quantities of mineral elements, although not necessarily in proportion to their absorption rates.

POTASSIUM-LINKED MAGNESIUM HOMEOSTASIS. Linkage between potassium and magnesium in intermediary metabolism also has been demonstrated for cattle and sheep (Fontenot 1979) and rats (see Aikawa 1981:85–86). Reduction of both serum magnesium and absorption in the gastrointestinal tract have been observed in cattle and sheep that were fed high-potassium diets. Similarly, rats given a high potassium diet for four to six days had significantly lower plasma levels of magnesium than the controls (Duarte 1980:93), whereas potassium-depleted rats had elevated plasma levels of magnesium, presumably as a result of magnesium shifting from soft tissues to plasma (Aikawa 1981:86). Duarte's investigations showed that increased potassium load in the rat resulted in increased magnesium losses via the feces and increased deposition of magnesium in the tissues (in heart, bone, and kidney but not significantly in muscle and liver). Earlier, Suttle and Field

(1967) observed increased fecal output of magnesium with sheep on high potassium diets. Aleksandrowicz and Stachura (1980) speculated that heavy potassium fertilization in western Poland led to higher levels of cattle leukemia. They reasoned that empirical evidence has shown a relationship between magnesium deficiency, thymus atrophy and immune system derangement, and lymphatic pathology. Therefore, the cattle suffer because of the potassium-magnesium antagonism seen at soil, forage, and animal levels.

Mineral licks—the ecological link in mineral homeostasis

The main physiological threads involved in maintaining equilibrium between the major minerals, although mentioned only in outline, are sufficient to suggest one of the major reasons herbivores seek out mineral licks and engage in direct geophagy or osteophagy. Our emphasis here on achieving homeostasis in respect to high potassium in no way subrogates the significance of calcium and magnesium and their relationships to bone and antler formation, growth of the embryo and young, lactation, and a host of other biochemical-related functions (Fig. 9.8). Also, our emphasis is not unmindful of the paramount functions of sodium and the

Fig. 9.8. Diagram of annual phenology of lick use, diet, and some physiological demands generalized for nonyarding white-tailed deer of northeastern United States.

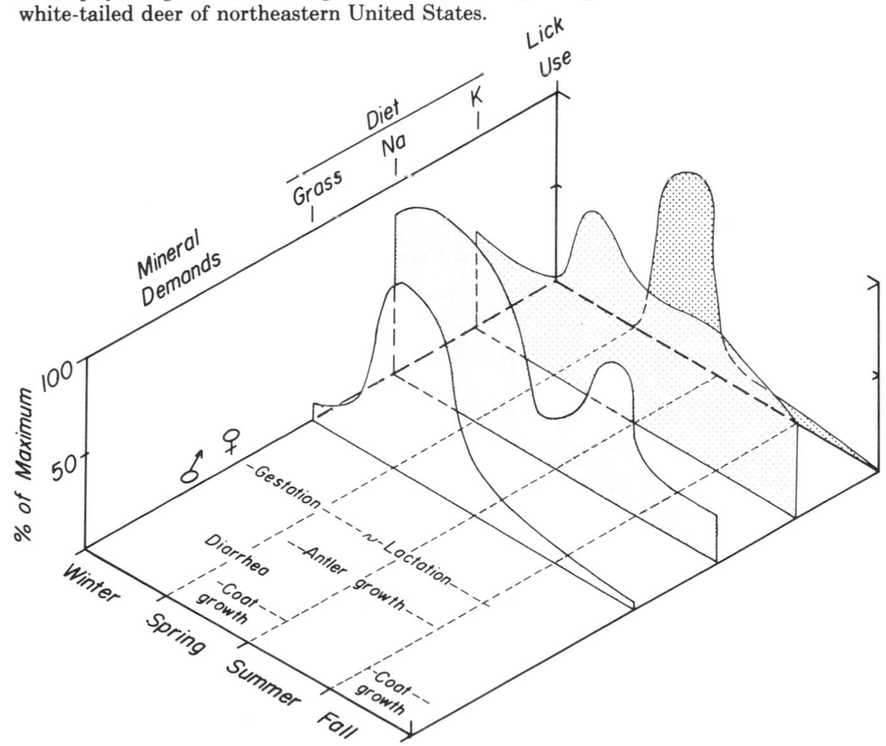

need for supplementation from localized sources in environments marginally adequate in respect to this element. And our findings by no means minimize the significance of varying levels of sulfur in the environment (Hanson and Jones 1976: App. 2). In many licks, sulfur also occurs in substantial amounts; a sulfate salt in these licks would contribute significantly to sulfur nutrition and, hence, protein nutrition in ungulates. Therefore its significance could not be readily evaluated independently of the levels of other major elements. However, note that ungulates use licks particularly in spring when they are replacing their hair coat (high in the sulfur amino acid cystine); and lactating females and young use licks in their major growth spurt, when protein formation is at a maximum.

The scheme in Table 9.2 sums up what is perhaps the major ecophysiological significance of mineral licks in the life of ungulates. To the extent that this specialized study (of the interrelationships between geology, soils, plants, mineral licks, and ungulate populations) may have raised more questions than it has answered is perhaps a measure of its worth.

Shock disease and alkaline earths in small wild herbivores

SHOCK DISEASE AND THE GENERAL ADAPTATION SYNDROME. Green and Larson (1937, 1938*a*, 1938*b*) describe a syndrome found in showshoe hares in Minnesota to which they applied the term *shock disease.* Live-trapped snowshoe hares (*Lepus americanus*) would develop tremors and die shortly after handling. Even unmolested hares in the wild were observed at times to "suddenly fall over" (Green and Larson 1938*a*: 192). Handling or other stress seemed to precipitate this condition. Not only is low blood sugar a biochemical disturbance characteristic of shock disease but also, on necropsy, affected animals display signs of pathology of the liver, spleen, adrenal, and thyroid glands (Green and Larson 1938*a*). However, the pancreas appears notably normal. Death was most common in late winter and spring and was most apparent during population highs associated with cyclical fluctuations common to the snowshoe hare (Keith 1963). Indeed, Green and Larson (p. 209) speculated that "this disease is truly the mechanism of the periodic decimation of hares." The condition was not sex-specific and hares 6 weeks old were dying with the symptoms. Green et al. (1939) induced the disease in a crowded population maintained in a fenced enclosure and observed that in the wild the disease occurred in patchwork distribution, implying that some *local* environmental factor or factors were significant.

Table 9.2. Characteristics of spring growth of forage in temperate and boreal lands, effects on herbivores ingesting forage, and beneficial effects of geophagy on animals eating earth material.

Forages	Effects on animals	Geophagy
K increases	mineral corticoids increase	
Mg is low	(1) K excretion increases	K may be preferentially absorbed and excreted
	(2) Mg excretion increases	Mg nutrition source
	(3) Na excretion decreases	Na nutrition source
Cell wall constituents are low	rumen pH decreases	pH increases due to $NaHCO_3$ ingested and/or buffering of colloidal aluminosilicates
	feed efficiency decreases	feed efficiency increases due to longer digesta transit time, may be particularly important in hindgut
Water increases	total water increases	effective water in gut decreases due to water associated with colloid
	diarrhea	decrease due to decreases in water activity and efficacious effect of adsorption of diarrhetic agents on mineral colloid
	gestation demand lactation demand antler demand	source of minerals, particularly Ca

About the time Green and Larson's observations were published, Selye began publication of a host of papers (about 1500 papers and 30 books) describing endocrine function and disorders and, particularly, the *general adaptation syndrome* (summarized in Selye 1946). The syndrome consists of "interrelated adaptive reactions to non-specific stress" (Selye 1950:6) and consists of three phases: an alarm reaction, a resistance phase, and an exhaustion phase. Selye (p. 10) defined an alarm reaction as "the sum of all non-specific phenomena elicited by sudden exposure to stimuli, which affect large portions of the body and to which the organism is quantitatively or qualitatively not adapted." The resistance phase represents "all non-specific systemic reactions elicited by prolonged exposure to stimuli" (p. 17), and the exhaustion phase is the constellation of nonspecific reactions developed due to chronic exposure to the stimuli and to which the adaptive mechanisms became incompetent (p. 12). Selye was careful to point out (p. 12) "that even a perfectly adapted organism cannot indefinitely maintain itself in a stage of resistance."

Christian (1950) reviewed the evidence for cyclic fluctuations in rodent populations and concluded (p. 250) that "Selye's adaptation syndrome (1946, 1947) provides us with an answer to the problem." Indeed, Selye had earlier (1946:194) called attention to the similarity of the symptoms of shock disease in the hare and the general adaptation syndrome. Christian cited the multiplicity of stress factors faced by the rodents: (1) food scarcity; (2) lack of suitable cover; (3) increased difficulty in finding food that also entailed exposure to predation, strife, and weather; (4) increased internecine and territorial strife, and (5) nutritional deficiencies. Later in a comprehensive review of endocrine physiology and population characteristics, Christian (1963) maintained the view that precipitous declines in rodent and lagomorph populations were due to the adverse effects of crowding as exemplified by stimulation of endocrine activity (pp. 322–23). He emphasized the interaction of sociopsychologic phenomena and the physiology of the endocrine system (p. 324).

Social stress as a result of high population densities per se has been questioned (Finerty 1980:154); nevertheless, adrenal hypertrophy has been associated with peak years in the cycles of both snowshoe hares and lemmings. It can be argued that adrenal hypertrophy at the time of population peaks is only a response to mineral imbalance.

MINERAL PATTERNS IN FORAGES. In the light of our findings on the relationships between mineral nutrition and ungulate populations, we decided to explore the possibility that deficits in the mineral nutrition of lagomorphs and microtine rodents may contribute a final, crucial stress to these animals—notably hares at a time of population highs and associated food shortages believed to be the ultimate cause of the die-off. Keith (1963:47) stated, "I believe the principal immediate causes of elevated mortality during the first year or two of the decline are starvation and starvation induced." Mortality in snowshoe hare populations is particularly common in late winter and spring, a time of substantial change in mineral patterns in forage—notably high potassium, the chief factor believed responsible for tetany in ungulates. The high levels of potassium and low levels of calcium and magnesium found in ungulate forages during the early spring flush of growth are of course also found in the plant foods of the smaller herbivores, which often utilize many of the same plant species. This mineral pattern places the same stress on adrenals of lagomorphs and microtines as on adrenals of ungulates. The imbalance in the mineral diet creates general as well as mineral stress on the adrenals beyond their capacity to respond.

Postulation of an association between shock disease and mineral nutrition partially depends on evidence of periodic or cyclic changes in the mineral content of the plant foods. These changes occur either independently of the herbivores or as an indirect result of their vegetation consumption. A limiting factor may be a time lag in the recycling of those minerals back into food plants. Evidently population crashes may occur regardless of food supplies (Christian 1963; Pease et al. 1979); however, high populations of small herbivores could subject some preferred plants to extreme pruning. Although a number of instances in the literature describe the local extirpation or near-extinction of highly palatable forages by ungulates, reduction of the more nutritionally desirable foods of small herbivores may not be so apparent. Furthermore, the interaction between small herbivores and the nutritional quality of their forages has not been clearly established. Schultz (1964:62), working at Point Barrow, Alaska, stated (no data) that nitrogen, phosphorus, and calcium are at their highest concentrations during peak brown lemming (*Lemmus*) populations and that magnesium (p. 66) does not show annual fluctuations. No data for potassium were presented. From a modeling of calcium and phosphorus nutrition in female lemmings based on data for forages at Barrow and, where necessary, physiological data for the rat, Barkley et al. (1980) concluded that the mineral nutrition of lemmings was at best marginal, and that the occurrence of moss in the diet reflected a need for minerals not supplied by grass species.

Evidence of pica as a response to mineral deficiencies is found in the review of Keith (1963:96), who recorded that during their peak populations hares had been observed eating flesh and bones of hares, moose, deer, fox, weasels, otters, ruffed grouse, and horse; articles handled by humans; sand; campfire ashes; and creosote-treated telephone poles. Keith speculated that these observations of pica might be related to other behavioral aberrations attendant with high densities.

Abnormal behavior by hares during population highs can likely be interpreted in terms of mineral—particularly magnesium—malnutrition. Note that Newson and Vos (1964) observed (during May and August) hares nibbling gravel on unsalted bush roads on Manitoulin Island. This gravel was probably limestone, given that the island is a glacially scoured limestone block with patchy thin soils.

The only long-term, retrospective data available for magnesium in food of a cyclical species are those for the snowshoe hare in areas near Rochester, Alberta (Pease et al. 1979). Six species of browse were analyzed for calcium; magnesium; potassium; and phosphorus and fat, carbohydrate, and protein. Collections were made in January of each year. Remarkable increases were found for magnesium and phosphorus in the terminal portions of woody vegetation in winters of 1969–1970 and 1970–1971, but a decline in the content of potassium occurred during these years. Calcium increased slightly in 1970–1971. In view of the importance attached here to magnesium in ungulate ecosystems, a further exploration of the Rochester, Alberta, data was desirable. Lloyd Keith generously provided the raw data summarized in Pease et al. (1979). The mean magnesium content of six browse species, together with hare populations (from Keith and Windberg 1978) and biomass of browse (from Pease et al. 1979) are plotted in Fig. 9.9 for the Landing Trail study area. The high magnesium content of browse produced in 1969 was undoubtedly significant in nutriture of the females producing the litters of the dramatic population increase in 1970. The browse produced during the 1970 growing season was again high in magnesium; and judging by analogy from the nutrient concentrations found by Grigal et al. (1976) in current and one-year twigs and leaves of shrubs sampled throughout the growing season in northern Minnesota, the magnesium content of the new growth of shrubs in 1970 at Rochester, Alberta, was greater than that in the following years.

Apropos of the above conjecture are the findings of Wolff (1977) near Fairbanks, Alaska, that smaller twigs (less than 3 mm in diameter) of willow and alder had much higher concentrations (significantly, it would appear in many cases) of 5 of 16 elements—zinc, calcium, magnesium, potassium, and nitrogen—than larger twigs (6- to 10-mm diameter). However, concentrations of these elements in the bark (Finerty 1980:124) of Alaska willow and alder sampled in spring either nearly equaled or exceeded levels in the small twigs. The bark category in this study presumably included the nutrient-rich cambium layer. Zinc concentration in the small twigs of willow exceeded concentrations in the larger twigs by five- to tenfold and in the bark by nearly threefold, notably exceeding comparable ratios for other elements. Studies by Prasad (1980) and other workers have revealed the crucial role that zinc plays in the normal development of the reproductive organs in humans.

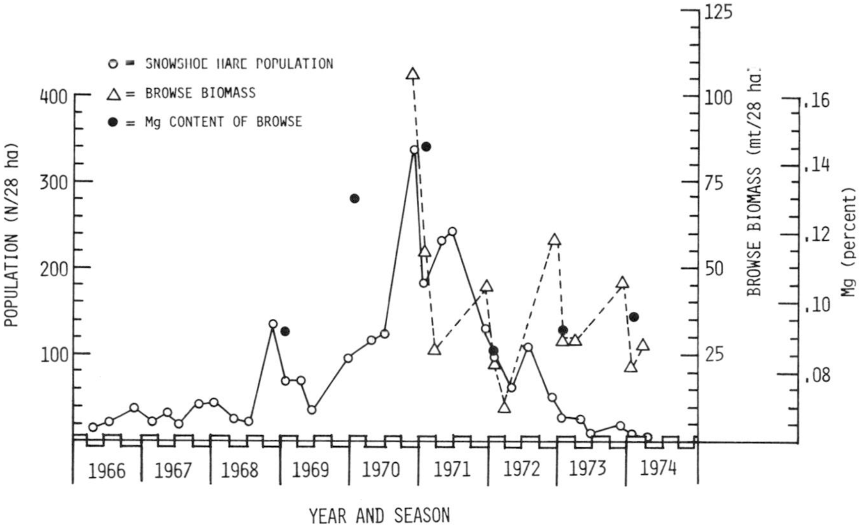

Fig. 9.9. Time-course of hare population, mean magnesium content of five browse species and of browse biomass on the Landing Trail study area, Alberta. (Population data from Keith and Windberg 1978; biomass data from Pease et al. 1979; magnesium data from L. B. Keith pers. commun. 1980.)

It is significant that the diameter of the browsed twigs in the Alaska study area was again in the 3-mm range when the hare population was reduced below carrying capacity. Thus, the population is again "fueled" with a key trace element and bulk minerals to assure maximum reproduction and survival.

Keith and Windberg (1978) concluded that early regression of hare testis weight appears to be closely linked to food shortage the previous winter. Was the food shortage more qualitative than quantitative in this case and was a zinc shortage the qualitative factor? If zinc concentrations in the terminal shoots of willow can vary tenfold at a locality in Alaska, it is not unreasonable that this element may vary at least several fold elsewhere.

These annual patterns raise questions regarding possible cycles in the levels of mineral elements in the forbs and grasses of the Rochester, Alberta, area. Phosphorus levels followed trends similar to magnesium, but potassium declined. If secretion levels of mineralocorticoids in small herbivores are high during cyclic highs (Ken Meyers of Guelph University studied the adrenals of snowshoe hares from the Rochester, Alberta, areas and found that they were hypertrophied in peak populations [pers. commun. 1980]), it would follow that increased release of the mineralocorticoid, aldosterone, would stimulate excretion of potassium and concomitantly, magnesium (Sharp and Leaf 1973). In many environments sodium levels in the ecosystem may be relatively but not critically low in supply as has been frequently suggested, or such areas would not support the animal populations that they do. Increased levels of aldosterone associated with increased intake of potassium would affect increased conservation of body sodium. Baestrup (1940) suggested that potassium nutrition was important in cyclical populations. In any event, it would appear that marked increases of both magnesium and phosphorus are associated with or followed immediately by cyclic highs in small herbivore populations.

It is equally noteworthy that sodium (Table 9.3) was low in browse produced at Rochester, Alberta, during 1969, 1970, and 1971—years that span the population peak. These low sodium levels in the food plants of snowshoe hares undoubtedly contributed to the adrenal hypertrophy observed by Myers; conversely, a rise in sodium content of the forages occurred during the growing seasons of 1972 and especially 1973, during which the hare population steadily declined. Hence, large increases in the hare population cannot be associated with a sodium increase in the browse. The relationships between forage sodium levels and hare population obviously deserve study.

There are hints in these data of an empirical verification of Pitelka's (1964) nutrient recovery hypothesis, which had its origins in Baestrup's (1940) speculations regarding the effects of mineralization of nutrients by soil microflora and subsequent mineral levels in vascular flora. Pitelka (1964:55) hypothesized that "the cycle is a result of interaction between herbivore and vegetation mediated by factors of nutrient recovery and availability in the soil." Data in Fig. 9.7 suggest that, in the Landing Trail area near Rochester, Alberta, there is a degree of "homeostasis" exhibited by the browse species studied; in January of 1969, 1972, 1973, and 1974, they contained 0.092% magnesium on the average. Regrettably biomass of browse was not measured in 1969 and 1970 because it is likely that browse production increased dramatically during the 1969 growing

Table 9.3. Mean contents (me/100 g) of Ca, Mg, Na, K, and P and P/(Mg + K) in duplicate samples of 5 browse species from the Landing Trail study area, Alberta (courtesy L. B. Keith; also see Pease et al. 1979).

January	Ca	Mg	Na	K	P	P/(Mg + K)
1969	6.6	7.6	0.75	11.8	17.5	0.91
1970	6.2	10.7	0.16[a]	10.5	24.6	1.16
1971	5.8	11.9	0.16[b]	10.5	25.4	1.13
1972	5.6	7.1	0.06[a]	9.2	17.9	1.09
1973	5.6	7.7	1.35[a]	9.0	19.5	1.17
1974	6.3	7.9	1.19[a]	13.3	24.6	1.16

[a]Calculated by best linear estimate for censored data (Gupta 1952). Data below detectable limit were 6 in 1970, 7 in 1971, and 1 each in 1973 and 1974.

[b]Calculated by maximum likelihood method of Cohen (1959) from censored data, which may be biased estimate; 6 analyses were below the detectable limit, the remaining 4 data were grouped (2 pairs).

season. Such an increase—if it occurred—could be attributed to the markedly high population of hares during the 1968–1969 winter that, from an energy standpoint, was cropping an adequate food resource. This pruning stimulated plant growth and thereby provided abundant, nutritious food for the wintering population in 1969–1970. This population, which had a favorable survival from December to April, further stimulated plant growth during 1970. Plant growth was measured for the first time late in 1970. The browse that year had about the same magnesium concentration as that of 1969 (Fig. 9.9).

The increase of magnesium level recorded for snowshoe hare browse in 1970 and 1971 is of special interest and raises the question as to whether it fluctuates from year to year in the arctic forage of the lemming. Shultz (1969) published a figure showing high phosphorus content during the peak lemming years of 1960 and 1965 at Barrow, Alaska. He mentioned (p. 87) that calcium, potassium, and nitrogen followed a similar trend but that sodium and magnesium levels in the vegetation were not associated with lemming density. These two studies at Barrow and Rochester of cyclical populations record nutrient increases in both monocotyledons and dicotyledons during years of high populations. Fertilizing the tundra at Barrow with nitrogen, phosphorus, potassium, and calcium raised forage production three to four times; protein levels a striking four to five times; and, important for this discussion, removed annual variation in element concentrations (Schultz 1969:89). Both calcium and phosphorus were high for the four years of observation. The data for mineral nutrients in forage at Barrow and Rochester are both high in phosphorus, calcium, and magnesium during peak years. A second increase in phosphorus in the forage produced during the 1973 growing season at Rochester was not accompanied by an increase in hare population. Using the data for Rochester, we have summed the equivalents of magnesium and potassium by season and compared them with the equivalents of phosphorus (Table 9.3). There is remarkably little difference among years, suggesting that in seasons when magnesium is high, an equivalent decrease occurs in potassium. Some of this relationship is undoubtedly due to the well-established reciprocity between magnesium and potassium in respect to absorption rates by plant roots (Metson 1974).

PHYSIOLOGY OF THE DIE-OFF IN SMALL HERBIVORES. Although Keith (1976) stated that he has not observed shock disease as described by Green and Evans in snowshoe hares, the description of shock disease syndrome of hares clearly implies that it is biochemically and physiologically homologous to grass tetany of cattle. Superficially examined, the above statement of commonality of origins is not supported by presently accepted explanations. It has long been clearly established that grass tetany in cattle is due to imbalances in the ratio of potassium to magnesium in spring forages; shock disease, on the other hand, has been attributed to hypoglycemia, a condition that also could arise from a food shortage per se. However, if starvation in a quantitative sense were the proximate factor of death in hares as sometimes suggested, why (apparently) have there been no documented reports of hares in a weakened state associated with starvation and low body weight prior to death?

We believe that a case can be made for qualitative rather than quantitative starvation as the principal cause of death of the individual hare as well as of the overall die-off initiating the sharp downward thrust of the cycle. Magnesium is required as a cofactor for 8 of the 13 enzymes involved in glycolysis (White et al. 1975:426). In addition, Rassmussen et al. (1976:258) pointed out that calcium, magnesium, and hydrogen ions are now realized to be the major regulators of glycogenolysis and glycolysis in muscle tissue. Thus it can be postulated that the reduced availability of magnesium in forage associated with population crashes in hares leads to reduced levels of readily available magnesium as a cofactor for the glycolytic cycle and, in turn, to a state of hypoglycemia in the hare.

Leite (no date) reviewed causes of death in herbivores using tall fescue grass to give perspective to observed mortality of cottontail rabbits eating this grass. Certain pathologies in ungulates were traceable to products of fungi living on the grass or to metabolites of the grass; and, of course, tetany in livestock eating the early growth of fescue is well documented. Of interest to us were Leite's citations of rabbit mortality. Cottontails, evidently thriving on a commercial ration in a study in Kentucky, were obligated to eat fescue (Kentucky 31) in the pen area when the commercial ration was withdrawn. After 7 to 10 days the rabbits began to die. Enteritis and diarrhea were observed. Leite also gives an account of a beagle club located near Elizabeth, extreme south central Indiana, that constructed a fenced area of 81 ha of fescue (also Kentucky 31). Soon the cottontail population declined to the extent that field trials were no longer feasible. Evidently other beagle clubs have had similar experiences and Leite concludes (p. 4), "In most instances the clubs had good rabbit populations until the fescue was planted." These observations lead us to conclude that the morbidity in these cottontail rabbits is qualitatively similar to that in the hares described earlier.

But what of the lemming, cycling in numbers on a 4- to 5-year basis rather than the 9- to 11-year cycle of the hare? Finerty's (1980) review of population cycles in small mammals has provided what we believe is an irrefutable base for postulating a unified concept of the factors that lead to physiological crises in herbivores on acid-rock substrates and/or northern ecosystems—namely, the surge of potassium in the spring vegetation combined with low concentrations of magnesium in the rock substrates, soils, and forage plants. Finerty (1980:117) cited a translation of Collett's (1911) now classic description of a lemming die-off in Norway. Finerty noted that "(1) migrating lemmings have a propensity to die of slight injuries; (2) [quoting Collett] 'sometimes individuals will suddenly fall, have convulsions in their hind legs and under convulsions they will die in a few minutes'; and (3) 'people say that they eat themselves to death on new grass.' " Reordering points 2 and 3, we have a classic, vernacular description of the principal factor leading to the development of grass tetany and the subsequent physiological state of tetany. More recent observations by Rausch (1950) of a cyclic die-off of lemmings (*Lemmus*) at Point Barrow, Alaska, add a convincing note to our thesis that tetany is probably the proximate cause of mortality. In Rausch's words (p. 176): "It was noted that dying lemmings observed on the surface of the snow [late March 1949] *usually manifested considerable spasmodic or convulsive activity just prior to death* [our italics]. Although the presence of more obscure conditions was not determined, it was evident from examination of numerous animals, secured both alive and after death, that there were no grossly visible lesions such as might be expected from bacterial infections. No effort was made to determine the presence of viruses or other organisms . . . there is nothing tangible to indicate that the decline in numbers resulted from starvation." Rausch, an authority on diseases and parasites of arctic and boreal wildlife, concluded (p. 176): "Helminth parasite infections could not be considered abnormal, nor detrimental to the condition of the animals infected. In fact, parasitism was very light when compared with that usually observed in microtine rodents. As concluded for other species (Rausch and Tiner 1949), there does not appear to have been any increase of helminth parasites under conditions of high population density." In this die-off, Rausch clearly discounted the commonly used explanations for cycles—starvation and/or parasitism. We believe that he observed the same symptoms that the Norwegian farmers saw, but he did not connect the symptoms with diet.

Several other elements—for example, calcium, sodium, and phosphorus—have been less convincingly suggested as the mineral elements involved in small herbivore cycles; but from reviews by Keith (1963) and Finerty (1980), it is clear that past studies have glaringly ignored two factors in the nutrient chain: (1) the geology of the rock substrate of ecosystems of the populations considered and (2) the interface of mineral metabolism with lower trophic levels of the nutrient chain.

Elton (1942:49) hinted at the importance of soil characteristics that might influence population levels in describing the vole and mouse outbreak of 1917 and 1918 on the northern plain of Germany: "On the southern edge of this plain . . . The outbreak was not of course universal in this huge area. For one thing the voles were limited by the type of soil, flourishing most on one of a medium, loamy kind, and less on sand or heavy clay. Also, they were thought to increase more on dry than on damp ground." Later, in describing the catastrophic population declines of snowshoe hare in Minnesota (as mentioned earlier), Green et al. (1939) remarked that something in the immediate environment of the animal was responsible for the outbreak of disease; indeed, these authors used the word *patchwork* to describe the distribution of the syndrome.

FACTORS AFFECTING NUTRIENT FLUCTUATIONS IN NORTHERN FORAGES: PAUCITY OF DATA. The empirical evidence to satisfy hypotheses relating the interplay of mechanisms underlying mineral nutrients and population cycles is still lacking. Observations on spring growth of apple trees (Burke and Morris 1933) suggest that about 60% of the magnesium in spring growth comes from root and soil and the remainder from magnesium stored near the shoot tip of the previous year, whereas most of the phosphorus and potassium come from root and soil. If there is dependence for spring growth on magnesium stored near the shoot tip, heavy browsing may remove critical amounts for the early flush of growth. The capacity of the soil to provide magnesium in spring is probably naturally low because of reasons described in the discussion of potassium-magnesium interactions at the soil-root interface in the production of tetany in cattle (Chap. 4). The high levels of magnesium in browse may represent the optimum conditions for nutrient recovery of environments in the Rochester, Alberta, area. Presumably this fluctuation in magnesium level would hold regionally as population fluctuations in showshoe hares occur (in general) simultaneously over large geographic areas. In his review of carbohydrate reserves and browse use, Garrison (1972) concluded that the standing crop in northern temperate regions could sustain from 40 to 60% use but that 75% pruning—the level of use during peak hare populations—made a severe demand on the resource.

Much remains to be learned about ecosystems characterized by cyclical populations. Foremost are the relationships between nutrient flows among the vascular flora, animal excretions, microflora, soil, and detritus. Some substantive evidence suggests large cyclic fluctuations in the overturn of macronutrients, but (considering their importance in nutrient chains) documentation of such rhythms is scanty.

Post hoc, ergo propter hoc

Recently, people have become acutely aware of the interconnected nature of the manifold processes in environments on earth. The science of biogeochemistry deals with some of these relationships, particularly those relating living systems to soil, water, air, and rocks. In this monograph we have related how we first endeavored to determine if ruminants visited earth licks to satisfy a deficiency of sulfur. This hypothesis was discarded. Instead, we found that magnesium and calcium and, to a lesser extent, sodium, were the most important constituents of lick earths. But apropos of our earlier emphasis on the role of sulfur in the environment and our present emphasis on the role of magnesium in sustaining high animal populations, the conclusion of Maianu (1981)—that magnesium sulfate characterized North Dakota surface waters and that they were unique in the world in that respect—can be construed as supporting evidence for both concepts. In respect to ungulate populations, it can be reasonably argued that North Dakota was indeed the "East Africa" of the North American continent, perhaps stretching back into the Cenozoic era.

The roles of alkaline earth elements in plant and animal nutrition have long been appreciated and intensively studied. Terranes underlain by calcareous rocks have been identified with better crop production and better animal and human health than nearby acid lands. From our study, we believe there is much circumstantial evidence for widespread marginal magnesium deficiency in wild animal populations. This conclusion parallels studies in recent decades of the relation of magnesium levels in human diets to many aspects of health.

The origin of these nutritional problems begins in the soil but is not separate from the phenology of certain nutrient absorption patterns. First in the sequence of events, potassium activity in soil water is highest in early spring. The rapid absorption of large amounts of potassium by plants is associated with lowered absorption levels of calcium and magnesium. The initial high level of available potassium may derive from the long period of high water content of soil during the winter (from wetting and drying cycles, some of which are due to freezing) and from the nature of potassium to have a higher activity than calcium and magnesium as volume of the double layer surrounding colloidal surfaces increases. The identity and content of various aluminosilicate colloids, particularly the micas, in controlling the extent of this nutritional stress are substantial and worthy of early study.

Cool season grasses seem well adapted to the cation complement available and undergo rapid growth. Note that the ammonium ion is important in the soil solution at this time and plays an important if not preeminent role in nitrogen nutrition.

Herbivores find this lush, first flush of growth especially palatable, but the physiological costs of the grazing may be high. The forage potassium is readily absorbed by the animal, elevating serum levels and eventually causing the adrenal gland to secrete increased amounts of aldosterone, which in turn serve to enhance potassium excretion by the kidney. However, increased aldosterone secretion also results in increased excretion of magnesium, which is particularly critical for herbivores. Body stores of magnesium are low and not quickly available; therefore the onset of pathologies due to loss and deficiencies can be rapid. Grass tetany is an acute form of magnesium insufficiency that can occur in domestic livestock. We suspect that in some situations a variety of wild, native herbivores also suffer tetany. No doubt other subacute pathologies present in wild herbivores are due not only to the magnesium deficiency seen in spring but also to chronic magnesium malnutrition endemic to certain geologic terranes.

Magnesium and sodium are the dominant ions of licks used by white-tailed deer in northeastern United States. A dual theory of mineral licks is postulated to account for the several roles that these ions assume in preventing grass tetany: (1) Magnesium intake compensates for losses due to high potassium intake, and (2) sodium intake has a subtractive role in effecting increased potassium losses. Calcium seems to be the sought-after constituent in elk, mountain sheep, mountain goat, and moose licks.

The quantitative geochemical cycle of magnesium in soils is essentially unknown, especially for boreal and arctic environments. The slow rate of organic matter decomposition and low levels of magnesium in parent rocks and low cation exchange capacity (coarse-textured soils) in some northern regions may be the proximate cause of cyclical population fluctuations of small herbivores.

Our conclusion that magnesium and, to a lesser extent, calcium nutrition are limiting or even marginal across large areas of North America is borne out by compilation of data for white-tailed deer weights and by size of antlers. Where calcium and magnesium contents are high, physical development tends to be optimum.

In searching for explanations for the effects of the abiotic on the biotic world it is easy to fall into the logic trap of this section's heading: After it, therefore due to it. We have used this logic to some degree in several arguments to explain the ordering of concatenations—as we comprehend them—in the natural histories of several herbivores.

Certain of our conclusions are not new in form or for the geographic regions to which we have applied them. Rather, this study provides biogeochemical data for large parts of the North American continent—thereby providing additional circumstantial evidence for hypotheses that have yet to undergo critical and costly empirical testing if we are to understand more fully the relationships of animals to their environments.

Appendixes

1. Lick localities and samples

Locations of licks, species using licks, lick-earth descriptions, and geologic interpretations of lick sites. Samples were described by R. L. Jones using procedures outlined in Chap. 2. Licks are arranged alphabetically by provinces and territories of Canada (from north to south within the province or territory) and states of the United States (alphabetically by county except from north to south in Alaska).

St./prov.;	Location			Species		
sample no.	twp.	rng.	sec.	using lick	Lick-earth description	Geologic interpretation
Alberta						
309	53	8W6	33	mountain goats, sheep, caribou, moose	Black (10YR 2/1) clay loam; weakly calcareous. Coarse fraction: subangular very dark grayish brown (10YR 3/2) shale.	No detailed account found; probably in Fernie group (Jurassic), which consists of soft-weathering fissile shales that are in part calcareous (Irish 1965).
329	55	7W6	5	mountain caribou, sheep, goats, moose, elk	Dark gray (10YR 4/1) sandy loam; strongly calcareous. Coarse fraction: rounded medium dark-gray (N 4/0) limestone; rare white (N 9/0) calcium carbonate aggregates and occasional thin discontinuous coating; very minor twigs and wood fragments.	Probably in Whitehorse formation (Triassic), which is composed of limestone, sandy limestone, sandy dolomite, argillaceous limestone, and dolomite (Irish 1965).
330	55	7W6	17	mountain sheep, goats	Brown to dark brown (7.5YR 4/4) loamy sand; strongly calcareous. Coarse fraction: rounded very dark-gray (N 4/0) asphanitic rock, pyrite on one fractured face; rare thin white (10YR 8/2) coatings.	Cadomin formation (Mesozoic), which is a prominent ledge and hogback forming stratum in the area. Irish (1965) does not mention calcareous members in the Cadomin formation, which he describes as chiefly quartz and chert pebbles set in a silicious matrix.
332	54	8W6	13	elk, moose, mule deer	Dark gray (N 3/0) silt loam; weakly calcareous; hydrophobic. Coarse fraction: mixed subrounded dark-gray (N 3/0) and light gray (N 7/0) fine-grained limestones; minor subrounded dark-gray (N 3/0) scoria and medium light-gray (N 6/0) pumice; minor subrounded brownish black (5YR 2/1) gabbro; rare continuous thin white (N 9/0) calcium carbonate coatings.	May be Whitehorse formation (see 329) judged from geology mapped about 0.5 km to east; igneous rock fragments do not conform to local bedrock (Irish 1965).
314	46	23W		mountain sheep	Very light grayish white (N 8/0) and white (N 9/0) with minor yellow (10YR 7/6) coating on medium dark gray (N 4/0, dry) coarsely bedded shale.	In surface coal mine; coating appears to be an alum (by microscopy, positive uniaxial crystals, columnar aggregates).

412	42	11W5	11	moose, mule deer	Black (10YR 2/1) silt; weak peaty aroma; smooth consistency.	Paskapoo formation (Paleocene and upper Cretaceous), which consists of gray to greenish gray calcareous and cherty sandstone, gray and green siltstone and mudstone, minor conglomerate, thin limestones and tuffs, and minor coal beds (Green 1970).
413	44	8W5	13	moose, mule deer	Very dark grayish brown (10YR 3/2) silt loam; strongly calcareous; smooth consistency; weak peaty aroma. Coarse fraction: dominantly rectangular wood fragments up to 1.5 cm long; minor subrounded pink (5YR 8/3) granite, pink (5YR 7/3), pinkish white (5YR 8/2) and very light gray (N 8/0) quartzites and olive-gray (5Y 5/2) gabbro (?); rare calcareous brown (7.5YR 5/2) platy siltstone.	In alluvium of Clearwater River.
237	39	13W5	8	elk, moose, mule deer	Dark grayish brown (2.5Y 4/2) loam. Coarse fraction: dominantly moderately platy subangular and subrounded light-gray (N 6/0) and olive-black (5Y 2/1) shale; minor subangular medium gray (N 5/0) chert; rare twigs.	Blackstone formation (upper Cretaceous) (Erdman 1950), chiefly fissile black shale with ironstone concretions and limey bands, formation poorly exposed. Compare with sample 316 about 8 km NNE.
310	40	10W5	30	elk, moose, mule deer	Very dark grayish brown (10YR 3/2) peat; hydrophobic. Coarse fraction: twigs, wood fragments.	No detailed report found. Erdman's study (1950) of Saunders quadrangle to west suggests that upper Cretaceous and Paleocene sandstones, shales, conglomerates, volcanic ash, and minor coal seams occur. No formal name is given to these rocks.
316	40	13W5	3	elk, moose, mountain sheep (?), deer	Mixed brown to dark brown (7.5YR 4/4) and very dark grayish brown (10YR 3/2) clay loam. Coarse fraction: strong, platy, angular very dark gray (N 3/0) shale; clay-rich, soft.	See 237, which is 8 km SSW in same formation (Blackstone).
344	39	11W5	35	deer, horse, moose	Dark reddish brown (5Y 2/2) silt loam; somewhat hydrophobic.	No detailed report found. See description for 310.
414	38	11W5	18	mule deer	Very dark (10YR 3/1) loam; weakly calcareous. Coarse fraction: dominantly angular medium dark gray (N 4/0) calcareous sandstone, with thin discontinuous very light gray (N 8/0) or white (N 9/0) efflorescence of sulfate salt (by $BaSO_4$ precipitation), dendritic in part; rare rounded medium light gray (N 6/0) and pinkish gray (7.5YR 7/2) limestones.	At slumped river bank located in area of Paskapoo formation (Paleocene and upper Cretaceous), which consists of gray to greenish gray calcareous and cherty sandstone, gray and green siltstone and mudstone, minor conglomerate, thin limestones and tuffs, and minor coal beds (Green 1970).

St./prov.; sample no.	Location twp.	rng.	sec.	Species using lick	Lick-earth description	Geologic interpretation
Alberta *(continued)*						
236	33	14W5	26	elk, deer, moose, mountain sheep	Dark brown (10YR 3/3) silt or silt loam; very strongly calcareous.	In alluvium of Clearwater River. Calcareous nature reflects large proportion of carbonate rocks of Paleozoic and Mesozoic ages in river basin (Price and Ollerenshaw 1971).
240	35	13W5	33	elk, moose, deer	Very dark grayish brown (10YR 3/2) loam; moderately calcareous. Coarse fraction; dominantly subangular light-gray (N 7/0) coarse-grained limestone and dark grayish brown (10YR 4/2) very fine-grained sandstone; minor dark reddish brown (5YR 3/2) ironstone; rare wood.	No detailed report found. Occurs in area mapped as Alberta group (Cretaceous) on provincial map (Green 1970). Group consists of dark-gray fissile silty shale, fine- to medium-grained sandstone (glauconitic, in part), siltstone, thin coals, and carbonaceous shale.
243	35	15W5	34	elk, moose, mule deer	Very dark grayish brown (10YR 3/2) loam; weakly calcareous.	See 241.
342	...[a]	...	...	moose, mountain sheep	Brown (10YR 5/3) silt loam with rare very light gray (N 8/0) grains; sand-size grains; very strongly calcareous.	In alluvium derived from complex lithology of Rocky Mountains (note rock types found at nearby lick sites in mountains).
343	...[a]	...	...		Brown (10YR 5/3) silt; very strongly calcareous.	See 342.
312	35	20W5	...	elk, mountain sheep	Dark brown (10YR 3/3) silt loam; very strongly calcareous. Coarse fraction: subangular very dark-gray (N 3/0) limestone coated with light-gray (5Y 7/1); very friable rounded white (10YR 8/2) calcium carbonate crumbs (tufa); minor twigs.	No detailed report found. In area of lower Paleozoic rocks on provincial map (Green 1970). Gray limestones and dolomites common in this sequence.
235	(51°56′ N, 116°10′ W)			elk	Very dark grayish brown (10YR 3/2) silt loam; very weak aroma of peat on rubbing. Coarse fraction: subangular grayish brown (10YR 5/2) and grayish black (N 2/0) fine-grained sandstone.	See 241.
238	32	12W5	18	mountain sheep	Very dark brown (10YR 3/2) loam; strongly calcareous. Coarse fraction: subangular medium light-gray (N 6/0) fine-grained limestone and medium dark-gray (N 4/0) coarse-grained limestone.	May be Triassic at contact of Sulphur Mountain formation (Price and Mountjoy 1971) with Rocky Mountain group (Pennsylvanian-Permian). Sulphur Mountain formation is composed of dark-gray and brown thin-bedded

[a]Along Saskatchewan River, Banff National Park.

						siltstones, silty mudstone, shale, and dolomitic siltstones. Rocky Mountain group consists of light-gray quartzose sandstone, dolomitic sandstone, silty dolomite, and chert.
241	34	16W5	24	elk, moose, mountain sheep	Grayish black (N 2/0) silt loam; abundant organic matter; one coarse-sand-size pyrite grain seen. Coarse fraction: dominantly moderate platy subangular and subrounded medium dark-gray (N 4/0) argillaceous limestone; accessory moderate platy subrounded dark-brown (7.5YR 3/2) shale; minor twigs and spruce leaves.	No detailed report found. In area mapped as lower Cretaceous, Jurassic, and Triassic on provincial map (Green 1970). Sequence includes dark-gray to black siltstone, dolomitic siltstone and limestone, silty dolomite and limestone, dark-gray carbonaceous and calcareous shale, and fine- to coarse-grained, cherty sandstone.
311	30	14W5	...	elk	Very dark grayish brown (10YR 3/2) peat: algal odor on rubbing; hydrophobic. Coarse fraction: dominantly moderate platy subangular and subrounded brownish yellow (10YR 6/6) siltstone, surface weathered to dark grayish brown (10YR 4/2); minor subangular very pale brown (10YR 7/3) fine-grained sandstone; rare angular light-gray (N 7/0) coarse-grained limestone, porous; rare twigs.	Fernie group (Jurassic) consisting of dark-gray to black shale, dark-gray siltstones and sandstones, dark-gray platy silty argillaceous limestone and brown limonitic quartzose sandstone (Price 1971*b*).
394	31	13W5	28?	elk, deer	Very dark grayish brown (10YR 3/2) silt loam; very strongly calcareous. Coarse fraction: weak platy subangular gray to light-gray (10YR 6/1) limestone; minor subangular dark gray (N 3/0) limestone; one gastropod shell.	Similar to that of 238. Lick along Tyrrill Creek at contact of Sulphur Mountain and Whitehorse formations, both Triassic (Price 1971*b*). Mountjoy (1963) reported gypsum in Whitehorse formation.
395	31	13W5	22?	elk, mountain sheep	Brown to dark-brown (7.5YR 4/4) loam; very strongly calcareous; odor of amine (fish origin?) on rubbing. Coarse fraction: subrounded and rounded dark grayish brown (10YR 4/2) limestone, weathers to very pale brown (10YR 7/3); minor pinkish gray (5R 7/2) coarse-grained limestone.	See 394.
239	32	10W5	16	deer, moose	Very dark grayish brown (10YR 3/2) loam or silt loam; moderately calcareous.	Just north of Burnt Timber map area (Ollerenshaw 1966), probably in Brazeau formation (see 315, 335).

St./prov.; sample no.	Location twp.	rng.	sec.	Species using lick	Lick-earth description	Geologic interpretation
Alberta *(continued)*						
242	34	10W5	35	elk, deer, moose	Grayish black (N 2/0) silt loam; hydrophobic.	See 241.
315	31	8W5	22	mule and white-tailed deer, feral horses, moose	Very dark grayish brown (10YR 3/2) peat; hydrophobic. Coarse fraction: twigs, wood fragments. From denuded position of lick.	Brazeau formation (upper Cretaceous) consisting of massive sandstone, gray siltstone, green and gray mudstone, conglomerate, minor coal, and bentonite (Ollerenshaw 1966).
335	31	8W5	22	mule and white-tailed deer, feral horses, moose	Dark reddish brown (5YR 3/2), dry peat. From bare hammock faces.	See 315.
322	31	11W5	32	mule deer, moose, elk	Dark reddish brown (5YR 2/2) peat, weakly calcareous.	Marsh or bog (Price 1971*a*); sediments of Fernie group (see 311) underlie site; a contact with Sulfur Mountain formation (see 238), which may contribute to weakly calcareous nature of lick earth, occurs within 300 m.
313	22	11W5	...	elk, moose	Water sample from Spray River, hydrogen sulfide odor at site.	None.
218	5	3W5	10	deer, moose	Very dark grayish brown (10YR 3/2) loam. Coarse fraction: dominantly subrounded and rounded medium dark-gray (N 4/0) and light-gray (N 7/0) fine-grained limestone; minor subrounded grayish brown (2.5Y 5/2) fine-grained sandstone; minor subangular dusky yellowish green (5GY 5/2) chert.	Belly River formation (upper Cretaceous) consisting of gray and greenish gray sandstones and mudstones (Norris 1958).
British Columbia						
397[a]	(59°50′ N, 127°45′ W)			Stone sheep	Brown (10YR 5/3) silt loam; strongly calcareous; slippery consistence when wet. Coarse fraction: moderate platy subangular dark grayish brown (10YR 7/2); minor moderate platy subangular white (10YR 8/2) coarse-grained limestone, weathers to yellow (10YR 7/6) and brownish yellow (10YR 6/6); rare twigs.	Similar to 396 (Gabrielse 1962*b*).

[a]NTS map 94M, grid reference WR6161.

396[a]	(58°50′ N, 127°50′ W)	Stone sheep, moose (?)	Olive-gray (5Y 5/2) silt loam; strongly calcareous; weak odor of amine (fish origin?) on rubbing. Coarse fraction: mixed angular white (N 9/0) massive calcium carbonate and very light-gray (N 8/0) quartzite; subangular strong platy medium light-gray (N 6/0) phyllite; minor medium dark-gray (N 4/0) limestone; rare moderate platy subangular weak-red (2.5YR 5/2) ferruginous limestone.	Sample suggests sequence of Cambrian and Ordovician limestone, phyllitic limestone, phyllite, argillite, sandstone, and limestone conglomerate (Gabrielse 1962*a*).
234	(53°43′ N, 121°13′ W)	mountain caribou, moose	Black (5Y 2/2) silt; strongly calcareous.	Probably in terrace of Slim Creek (Campbell et al. 1973). Rocks of uplands are Cambrian and Proterozoic shales, siltstones, artillites, limestones, and phyllites.
404	(50°54′10″ N, 116°01′10″ W)	elk	Grayish brown (2.5Y 5/2) fine sandy loam; very strongly calcareous; weak argillaceous aroma. Coarse fraction: light brownish gray (5Y 6/1) and medium gray (N 5/0) subrounded limestone.	See 400.
405	(50°54′50″ N, 116°00′30″ W)	mountain goat	Light brownish gray (2.5Y 6/2) silt; strongly calcareous. Coarse fraction: strong platy medium gray (N 5/0) and light gray (N 7/0) weathering to pink (5YR 7/3) limestone fragments.	See 400.
406	(50°55′20″ N, 115°59′50″ W)	mountain goat	Light grayish brown (2.5Y 6/4) silt; very strongly calcareous; weak argillaceous aroma. Coarse fraction: moderate platy medium-gray (N 5/0) and light-gray (N 7/0) limestones.	See 400.
400	(50°38′00″ N, 116°01′10″ W)	mountain sheep	Light brownish gray (2.5Y 6/2) silt; very strongly calcareous; weak argillaceous odor. Coarse fraction: dark gray (N 4/0) and medium gray (N 5/0) subrounded limestone.	McKay group (Cambrian and Ordovician) (Reesor 1973) consisting of gray argillite, argillaceous limestone, and dark shale. Reesor (p. 49) described interbedded buff gray and brown limestone on Mount Wardle (see 404, 405, 406). Ungulates seeking these rocks over at least 32 km (20 mi).

[a]NTS map 94L, grid reference WR7633.

St./prov.; sample no.	Location			Species using lick	Lick-earth description	Geologic interpretation
	twp.	rng.	sec.			
Manitoba						
126 (?)[a]	45	25	8	deer, moose	Yellowish brown (10YR 5/6) sandy loam; very strongly calcareous. Coarse fraction: subrounded light-gray (10YR 7/1 and 7/2) fine- to medium-grained limestone, weathers to light yellowish brown (10YR 6/6).	Salt flat along Red Deer River in area of Devonian-age limestone, dolomite, and red and buff shales (Wickenden 1945).
127	44	25	10	white-tailed deer, moose	Dark-brown (10YR 3/3) sandy loam; strongly calcareous.	A spring. See 126.
148	43	26	23	white-tailed deer	Olive (5Y 4/3) and minor grayish olive-green (5GY 3/2) loamy sand; moderately calcareous. Coarse fraction: dominantly subrounded very light-gray (N 8/0) and very pale brown (10YR 7/3) limestone; minor subrounded greenish gray (5GY 5/1) coarse-grained glauconitic sandstone; rare fine-grained salt and pepper sandstone.	At a spring in upper Swan group (lower Cretaceous) consisting of calcareous light greenish gray sandstone, glauconitic and quartzose sand, green plastic clay, and green clayey sand (Wickenden 1945).
150	40	26	27	white-tailed deer, moose	Very dark-brown (10YR 3/2) peat; strongly aggregated; abundant organic matter; fragmented and whole gastropod shells; aroma of dry dog food when rubbed.	Riding Mountain formation (upper Cretaceous) (Wickenden 1945) consisting of gray and greenish gray shales containing abundant ironstone concretions.
143	34	27	14	white-tailed deer, elk	Grayish black (N 2/0) silt loam; strongly calcareous; strong crumb structure; hydrophobic; abundant organic matter. Coarse fraction: dominantly subangular to subrounded white (10YR 8/2) limestone; common subrounded dark grayish brown (10YR 4/2) sandstone or quartzite.	A spring in basal portion of Riding Mountain formation that is either hard silicious shale or very soft greasy gray shale (Wickenden 1945, p. 49). Glacial erratics in sample. Lick 149 8 km south in similar geologic setting.
145	34	27	10	white-tailed deer, elk	Dark yellowish brown (10YR 3/4) and minor brown to dark-brown (10YR 4/3) loam; sticky; very weak sweet nondescript aroma. Coarse fraction: dominantly subangular and subrounded white (10YR 8/2) and medium-gray (N 5/0) quartzite; minor rounded red (10YR 4/8) and dark-red (10YR 3/6) quartzite, polished.	Artificial lick. Basal units of Riding Mountain formation underlie lick. Stickyness probably due to sodium saturation of montmorillonite in formation (Wickenden 1945, p. 49). Unlike nearby natural licks 143 and 149, this material is noncalcareous.

149	33	27	22	elk, moose	Very dark grayish brown (10YR 3/2) silt or silt loam; strongly calcareous.	See 143.
144	31	23	12	white-tailed deer, elk	Very dark grayish brown (10YR 3/2) loam; very strongly calcareous; abundant organic matter; weak aroma of menthol. Coarse fraction: dominantly subrounded white (N 8/0) fine-grained limestone; minor subangular pale red (10YR 6/4) granite.	Lower Riding Mountain formation (upper Cretaceous, see 150). Limestone and granite are glacial erratics derived from Paleozoic and older rocks to northeast.
Northwest Territories						
100	(63°44′30″ N, 127°10′00″ W)			woodland caribou, moose	Very dark-gray (5Y 3/1) sandy loam; strongly calcareous; abundant plant remains. Coarse fraction: well rounded medium-gray (N 5/0) to medium light-gray (N 6/0) dolomite.	No detailed report found. Possibly Sekwi formation (see 101).
101	(63°12′12″ N, 127°56′06″ W)			Dall sheep	Black (5Y 2/2) loamy very fine sand; strongly calcareous.	Probably Sekwi formation (lower Cambrian) consisting of dolomite and limestone with sandy dolomite and sandstones (Handfield 1968).
112	(63°51′30″ N, 128°53′00″ W)			woodland caribou, moose	Dark olive-gray (5Y 3/2) silt loam.	Sekwi formation (see 101).
339	(61°10′ N, 124°05′00 W)			moose	Dark reddish brown (5YR 3/2, dry) peat.	Thermal spring area at edge of Tlogotsho Plateau, here underlain by dark-gray shale with rare limestone and sandstone beds of probable upper Devonian age (Douglas and Norris 1960). No mention of hot springs by Douglas and Norris; their map does not show nearby faults or other structural features that might explain spring development. Rock outcrops evidently rare in this region.
340	(61°10′ N,			moose	Dark-brown (10YR 3/3, dry) peat.	Thermal spring area (see 339).
341 (?)[a]	(61°30′ N, 125°45′ W)			woodland caribou, moose	Dark-gray (N 3/0) silt loam; very strongly calcareous.	River terrace of sediments probably derived from Sunblood formation, a limestone (Devonian) (Douglas and Norris 1960).
347	(61°15′36″ N, 124°20′42″ W)			Dall sheep	Strongly calcareous; insufficient sample for other characterization. Preferred site in lick complex.	Dolomite unit (Devonian) (Douglas and Norris 1960).

[a]Questionable lick

St./prov.; sample no.	Location twp.	rng.	sec.	Species using lick	Lick-earth description	Geologic interpretation
Northwest Territories *(continued)*						
354	(61°15′36″ N, 124°23′18″ W)			Dall sheep	Strongly calcareous, insufficient sample for other characterization. Preferred site in lick complex.	See 347.
272	...[b]	...	...		Very light-gray (N 8/0) silt; weakly calcareous; chalk odor; friable.	Exact location not specified.
Nova Scotia						
151 (?)[a]	(45°14′30″ N, 63°04′00″ W)			white-tailed deer	Reddish brown (5YR 4/3) silt loam.	Lower Windsor group (Mississippian), which consists of red and green shales, marine limestones, gypsum, and anhydrite (Stevenson 1958).
Ontario						
199	(see map)			moose	Very dark grayish brown (10YR 3/2) loamy coarse sand. Coarse fraction: angular grayish black (N 2/0) aphanitic rock (basalt?).	Licks 199–205 located along lineaments of rocks that trend generally east-west (Ontario Division of Mines 1973). Lineament probably controls groundwater flow in this area of largely basic igneous rocks. All licks located in diabase terranes except lick 199 in area of conglomerate and sandstone of Osler group and lick 203 in area of sandstone of Sibley group. Latter sandstone is somewhat calcareous and lick material is calcareous. Also, the sandstone represents an aquifer.
200	(see map)			moose	Like 199 except plant fragments apparent. Coarse fraction: angular to subangular dark-gray (5Y 5/1) and black (N 2/0) igneous rock (gabbro?); partly weathered to reddish yellow (7.5YR 6/6); friable.	See 199.
201	(see map)			moose	Grayish black (N 2/0) peat; conifer leaves identifiable; hydrophobic.	See 199.
202	(see map)			moose	Dark yellowish brown (10YR 3/4) loamy sand. Coarse fraction: subangular dark reddish brown (5YR 3/3) and dusky red (10YR 3/3) rhyolite; rare twigs and wood fragments.	See 199.

[a]Questionable lick.
[b]Fortress Mountain, Keele River.

203	(see map)		Dark grayish brown (10YR 4/2) loam or sandy loam; moderately calcareous; moderately hydrophobic. Coarse fraction: mixed subrounded weak red (10R 4/3) rhyolite, medium dark-gray (N 4/0) basalt and weathered red (10R 5/6) porphyritic rhyolite; light-gray (N 7/0) granite; rare somewhat friable saprolite ("corned beef").	See 199.
204	(see map)	moose	Dominantly very dark grayish brown (10YR 3/2) sand with various colored individual grains; strongly calcareous. Coarse fraction: mixed subrounded and rounded medium dark-gray (N 4/0) basalt (?); white (N 9/0) rhyolite (?); and pale red (10YR 6/4), medium light-gray (N 6/0) granite.	See 199.
205	(see map)	moose	Brown to dark-brown (10YR 4/3) sand; moderately hydrophobic.	See 199.
384	(see map)	moose	Dark reddish brown (5YR 2/2) silt loam; tobacco aroma on rubbing. Coarse fraction: bark and wood fragments, twigs. See 389 for control samples.	Kelvin Island (Ontario Department of Mines 1965, Map 2102) wholly underlain by diabase (Keweenawan) covered by ground moraine ranging from silt to sand texture; local varve deposits (Ontario Department of Lands and Forests 1965, Map S265).
385	(see map)	moose	Very dark grayish brown (10YR 3/2) silt loam; strongly calcareous. Coarse fraction: dominantly light olive-gray (5Y 6/1) wood fragments that apper to have weathered subaqueously; very minor subrounded very light-gray (N 8/0) fine-grained sandstone and subangular white (10YR 8/2) chert. See 390 for control sample.	Similar to Kelvin Island; see 384.
386	(see map)	moose	Very dark grayish brown (10YR 3/2) silt; moderately calcareous; somewhat hydrophobic; tobacco aroma on rubbing. See 391 for control sample.	Like Kelvin Island (bedrock geology from Nipigon-Schreiber Map, Ontario Department of Mines 1972). Calcareous nature of sample suggests calcareous rock (Sibley group) outcrop nearby, perhaps beneath water surface (drumline lineation indicates ice moved from just north of east).

St./prov.; sample no.	Location			Species using lick	Lick-earth description	Geologic interpretation
	twp.	rng.	sec.			
Ontario *(continued)*						
387	(see map)			moose	Very dark grayish brown (10YR 3/2) silt loam; tobacco aroma on rubbing. See 392 for control sample.	Similar to Kelvin Island; see 384.
388	(see map)			moose	Olive-gray (5Y 5/2) silty clay loam; strongly calcareous; slippery consistence when wet; earthy aroma. Coarse fraction: mixed subangular very light-gray (N 8/0), light olive-gray (5Y 6/1), and medium light-gray (N 6/0) limestones; minor subangular dark-gray (N 3/0) igneous rock; rare subangular pale red (10R 6/4) potash feldspar; minor twigs and leaves. See 393 for control sample.	At or very near contact of Sibley group rocks with diabases (Nipigon-Schreiber Geology Map Sheet, Ont. Dep. Mines 1972). Calcareous sediments described in Sibley group (Giguere 1975) on islands in Nipigon Bay (see 203). Bedrock is overlain by ground moraine.
389	...	...	...	control for 384	Very dark grayish brown (10YR 3/2) silt loam; faint tobacco aroma on rubbing.	See 284.
390	...	...	...	control for 385	Dark grayish brown (2.5Y 4/2) silty clay loam; strongly calcareous; slippery consistence when wet; weak tobacco aroma on rubbing. Coarse fraction: dominantly twig and wood fragments; very minor mixed subrounded dark-gray (N 3/0) and white (10YR 8/2) plagioclases and light olive-gray (5Y 6/1) fine-grained igneous rock.	See 285.
391	...	...	...	control for 386	Very dark grayish brown (10YR 3/2) silt loam; weakly calcareous; somewhat hydrophobic; weak tobacco aroma on rubbing.	See 386.
392	...	...	...	control for 387	Very dark grayish brown (10YR 3/2) loam; very weak tobacco aroma on rubbing. Coarse fraction: dominantly subangular and subrounded medium dark-gray (N 4/0) and dark-gray (N 3/0) fine-grained and aphanitic igneous rocks; minor subangular grayish red (5R 4/2) fine-grained rock and subangular grayish orange (10YR 7/4) plagioclase; minor twigs.	See 387.

393	...	...	...	control for 388	Very dark grayish brown (10YR 3/2) silt loam; weakly calcareous; weak tobacco aroma on rubbing.	See 388.
Quebec						
190	(48°40′16″ N, 66°45′56″ W)			moose	Very dark grayish brown (10YR 3/2) loamy coarse sand; weakly calcareous. Coarse fraction: somewhat platy angular to subangular gray (10YR 5/1) siltstone.	At or near contact of Cap Bon-Ami limestone that overlies sequence of shale, sandstone, limestone, and siltstone (McGerrigle and Skidmore 1967).
191	(48°44′29″ N, 66°30′12″ W)			moose, deer	Dark grayish brown (10YR 4/2) loamy coarse sand; hydrophobic. Coarse fraction: dominantly subrounded and rounded light-gray to gray (10YR 6/1) siltstone; minor subangular light-gray (10YR 7/2) very fine-grained sandstone; rare spruce leaves.	Grand Greve formation—a sequence of impure gray limestone, siltstone, and shale with sandstone and conglomerate (McGerrigle and Skidmore 1967).
192	(48°39′26″ N, 67°01′16″ W)			moose, deer	Very dark grayish brown (10YR 3/2) loamy coarse sand; weakly calcareous; hydrophobic. Coarse fraction: subrounded light-gray (10YR 7/1) fine-grained limestone.	See 190.
193	(48°39′05″ N, 66°55′36″ W)			moose	Very dark grayish brown (2.5Y 3/2) loamy coarse sand. Coarse fraction: moderate platy subrounded greenish gray (N 3/0) shale.	See 190.
Alabama						
227	7N	1E	21	white-tailed deer	Strong brown (7.5YR 5/6) with minor reddish yellow (5YR 6/8) silt loam.	Salt water seep. Tallahatta formation (Claiborne group, Eocene) consists of white to very light greenish gray thin-bedded to massive silicious and aluminous claystone interbedded with thin layer of clay, sandy clay, glauconitic sand, and sandstone (Causey and Newton 1972). Gosport and Lisbon formations (Eocene) composed mostly of glauconitic sandstones overlie Tallahatta formation. Salt water arises along flank of Hatchitigbee anticline and Jackson fault in area. (In lick vicinity and immediately north is Upper Salt Works, one of three factories in county that produced salt for Confederacy; Barksdale 1929).

St./prov.; sample no.	Location twp.	rng.	sec.	Species using lick	Lick-earth description	Geologic interpretation
Alaska						
	4, 5, 7, 8S	24, 25 26E	. . .	mountain sheep	CANNING RIVER LICK COMPLEX	Leffingwell (1919) produced reconnaisance report for Canning River region. Licks are in an area he mapped as Lisburne limestone (Mississippian), which is gray limestone, black near base and about 1000 m thick. Lisburne overlies carboniferous shales and underlies Sadlerochit sandstone. Leffingwell found two springs coming from Lisburne limestone that flowed all winter. The most recent U.S. Geological Survey map (Beikman and Lathram 1976) for northern Alaska includes this area (undifferentiated) in Endicott group, which is described as composed of Kekiktuk conglomerate and Kayak shale (both Mississippian) and Lisburne group, which is composed of Alapah and Wahoo limestones (Mississippian and Pennsylvanian). From nature of rocks found and described in lick samples, licks are almost certainly in Lisburne group.
287					Dark-gray (N 2/0) loam (macerated rock); strongly calcareous. Coarse fraction: moderate platy angular medium dark-gray (N 4/0) and medium-gray (N 5/0) silty limestone; rare thin discontinuous white (N 9/0) calcium carbonate coatings on face; rare white (N 9/0) calcium carbonate.	
317					Medium dark-gray (N 4/0) silt; very strongly calcareous. Coarse fraction: subangular very dark-gray (N 3/0) limestone coated with light gray (N 7/0); very friable rounded light-gray (N 7/0) calcium carbonate crumbs (tufa).	
288					Medium dark-gray (N 4/0) silt loam; very strongly calcareous. Coarse fraction: angular and subangular medium-gray (N 5/0) fine-grained limestone; rare subangular yellow (10YR 7/6) quartzite.	
289					Dark-gray (N 2/0) loam (shale); very strongly calcareous. Coarse fraction: strong platy angular medium-gray (N 5/0) limestone.	
301					Dark-gray (N 3/0) silt or very fine sand (macerated rock); very strongly calcareous. Coarse fraction: angular to rounded medium-gray (N 5/0) to light medium-gray (N 6/0) fine-grained limestone.	
293					Dark-gray (N 2/0) loam or silt loam (shale ?); very strongly calcareous. Coarse fraction: angular very dark-gray (N 3/0) bituminous limestone; rare very light-gray (N 8/0) calcite; [illegible]	

294	Dark gray (10YR 4/1); very strongly calcareous. Insufficient sample for texture determination. Coarse fraction: angular medium-gray (N 5/0) bituminous limestone; thin very light-gray (N 8/0) coatings.
295	Light gray (10YR 7/1); very strongly calcareous. Insufficient sample for texture determination.
296	Dark-gray (N 2/0) loam or silt loam. Coarse fraction: moderate platy angular medium-gray (N 5/0) shaly limestone; very thin patchy, very light-gray (N 8/0) coatings.
297	Dark-gray (N 2/0) loam; weakly calcareous. Coarse fraction: like 296 without coatings.
299	White (10YR 8/1); moderately calcareous. Insufficient sample for texture determination.
300	Very pale brown (10YR 8/3) silt or very fine sand; strongly calcareous. Coarse fraction: angular very light-gray (N 8/0) limestone; continuous 0.5 to 2.5 mm white (10YR 8/2) calcium carbonate coatings, sometimes with botryoidal surface.
280	Gray (10YR 5/1); weakly calcareous. Insufficient sample for texture determination.
281	Dark gray (N 3/0) sandy loam; weak pineapple aroma on rubbing. Coarse fraction: mixed moderate and strong platy light-gray (5Y 6/1) and medium-gray (N 5/0) shale; strongly calcareous.
282	Grayish black (N 2/0) loam; abundant organic matter. Coarse fraction: strong platy angular medium-gray (N 5/0) shale; minor roots and areal grass parts; medium-gray (N 5/0) mineral-coated fecal pellets (sheep).
283	Grayish black (N 2/0) loam; moderately calcareous.

St./prov.; sample no.	Location			Species using lick	Lick-earth description	Geologic interpretation
	twp.	rng.	sec.			
Alaska *(continued)*						
					CANNING RIVER LICK COMPLEX *(continued))*	
284					Medium gray (N 5/0). Insufficient sample for texture determination.	
285					Medium light-gray (N 6/0). Coarse fraction: moderate platy angular medium-gray (N 6/0) limestone; rare thin continuous white (N 9/0) coating on face. Insufficient sample for texture determination.	
298					Dark grayish brown (10YR 4/2) sandy loam; moderately calcareous. Coarse fraction: moderate platy subrounded light-gray (5Y 6/1) silty limestone.	
286					Olive-yellow (2.5Y 6/6) with minor grayish black (N 2/0) clay loam (macerated rock); earthy odor.	
290					Very dark grayish brown (10YR 3/2) silt loam or loam; very strongly calcareous. Coarse fraction: angular and subangular medium light-gray (N 6/0) and dark-gray (N 3/0) limestone, thin continuous white (10YR 6/2) calcium carbonate coating; broken mineral-coated fecal pellets (sheep).	
291					Very dark grayish brown (10YR 3/2) loam; strongly calcareous. Coarse fraction: moderate platy angular and subangular light brownish gray (10YR 6/2) siltstone (?), occasional thin continuous white (10YR 6/2) calcium carbonate coating.	
292					Dark-gray (N 2/0) loam (shale?); strongly calcareous. Coarse fraction: moderate platy angular and subangular medium dark-gray (N 4/0) silty shale, rare thin patchy light-gray (5Y 7/1) coating (not calcium carbonate).	

	(67°23′ N, 150°39′ W)	mountain sheep	LASALLE CREEK LICK	Based on preliminary geologic map of Beikman and Lathram (1976); rocks of LaSalle Creek-Emma Dome-Bluecloud Mountain area are rocks of Skajit limestone (Devonian/Silurian). The formation in this area appears to be confined to this geographic extent. Skajit limestone in this area of the Brooks Range consists of limestone, dolomite, marble, and interbedded shale. A few samples contained phyllite, suggesting that some licks are close to the more extensive outcrops of lower-upper, and/or upper-middle (?) Devonian rocks including conglomerate, graywacke, chloritic phyllite, calcareous shale and sandstone, siltstone, and minor limestone (Beikman and Lathram 1976). Lower Paleozoic and/or Precambrian rocks made up mostly of schists and quartzite occur immediately south of the Dome in contact with the Devonian sequence. Tailleur et al. (1977) report a thermal sulfurous spring located 170 km west of Wiseman at the contact of Skajit limestone with Cretaceous-age acid-igneous rocks. This spring is used as a lick.
355			Dark-gray (N 3/0) friable acicular laths breaking to loam or silt loam; strongly calcareous; earthy odor. Coarse fraction: moderate platy or lath forming medium light-gray (N 6/0) phyllite; very minor white (N 9/0) crust, 0.5 mm thick.	
369		control sample for 355	Dark-gray (N 3/0) friable rock breaking to loam or silt loam; very strongly calcareous. Coarse fraction: moderate platy angular (forming splintery laths) medium-gray (N 5/0, sheen) limestone, very low density; thin continuous medium light-gray (N 6/0) calcium carbonate coating, botryoidal.	
357			Dark-gray (N 3/0) friable acicular laths breaking to loam; very strongly calcareous; earthy odor. Coarse fraction: like 369 above.	
358		control sample for 357	Like 357. Coarse fraction: like 369 except coatings more apparent and common.	
361			Medium dark-gray (N 4/0) friable rock breaking to loam; very strongly calcareous. Coarse fraction: strong platy subrounded medium-gray (N 5/0, sheen) graphitic limestone (cf. 360).	

St./prov.; sample no.	Location twp.	rng.	sec.	Species using lick	Lick-earth description	Geologic interpretation
Alaska *(continued)*						
					LASALLE CREEK LICK *(continued)*)	
367				control sample for 361	Dark-gray (N 3/0) friable rock breaking to loam or silt loam; very strongly calcareous; slippery consistence when wet. Coarse fraction: strong platy angular medium light-gray (N 6/0) phyllite, very fissile (cf. 365).	
	(67°25′ N, 150°40′ W)			Dall sheep	BLUECLOUD LICK	See 355 (LaSalle Creek Lick).
356					Olive-gray (5Y 4/2) loam; strongly calcareous; aroma of carrots. Coarse fraction: strong platy angular light-gray (N 7/0) and medium-gray (N 6/0) phyllite; minor light-gray (N 7/0) fine-grained limestone weathering to light yellowish brown (10YR 6/4).	
362				control sample for 356	Light brownish gray (10YR 6/3, dry); strongly calcareous. Coarse fraction: moderate platy angular medium-gray (N 5/0) limestone, thinly bedded, somewhat friable; minor moderate platy medium dark-gray (N 4/0) silty limestone, graphitic and bituminous in part.	
363					Olive-gray (5Y 4/2) with rare dark-gray (N 3/0) silt loam; very strongly calcareous. Coarse fraction: strong platy angular light-gray (N 7/0) phyllite, highly calcareous.	
359					Gray (5Y 5/1) with rare reddish brown (2.5Y 4/4) and medium light-gray (N 4/0) friable rock breaking to loamy sand; very strongly calcareous. Coarse fraction: like 357 but less lath forming, weathers to light gray (N 7/0) and less abundant reddish yellow (7.5YR 6/8).	

368			Olive-gray (5Y 4/2) silt loam; very strongly calcareous; micaceous. Coarse fraction: strong platy subangular light-gray (N 7/0) phyllite, thin discontinuous white (N 9/0) calcium carbonate coatings on reddish yellow (5YR 6/6) weathered faces.	
370			Gray (5Y 5/1) silt loam; very strongly calcareous. Coarse fraction: moderate platy subangular fragments bearing very thin lamellae of medium-gray (N 5/0) mica and very light-gray (N 8/0) calcite, some faces weathered to a thin coat of reddish yellow (5YR 6/6).	
	(67°25′ N, 150°40′ W)	Dall sheep	LICKS BETWEEN BLUECLOUD AND LASALLE CREEK	See 355 (LaSalle Creek lick).
366			Dark-gray (N 3/0) friable rock breaking to loam or silt loam. Coarse fraction: strong platy angular medium light-gray (N 6/0) phyllite, very fissile; some medium dark-gray (N 4/0) phyllite, slightly calcareous.	
364			Mixed dark-gray (N 3/0) and dark grayish brown (10YR 4/2) friable rock breaking to loam or silt loam; weakly calcareous. Coarse fraction: strong platy medium dark-gray (N 4/0) phyllite, thin discontinuous white (N 9/0) calcium carbonate coatings.	
371			Very dark grayish brown (10YR 3/2) friable rock breaking to loam; strongly calcareous; micaceous. Coarse fraction: strong platy subrounded light olive-gray (5Y 6/1) phyllite, calcareous; thin continuous botryoidal light olive-gray (5Y 6/1) calcium carbonate coatings on exposed faces.	
365			Dark-gray (N 3/0) loam; moderately calcareous. Coarse fraction: strong platy angular medium light-gray (N 6/0) phyllite, very fissile.	
360			Dark-gray (N 3/0) friable acicular laths breaking to loam; very strongly calcareous; earthy odor. Coarse fraction: strong platy subrounded medium-gray (N 5/0, sheen) graphitic limestone.	

St./prov.; sample no.	Location twp.	rng.	sec.	Species using lick	Lick-earth description	Geologic interpretation
Alaska *(continued)*						
					LICKS BETWEEN BLUECLOUD AND LASALLE CREEK *(continued)*	
270 (?)[a]	(63°55′ N, 147°35′ W)			mountain sheep	White (5Y 8/2, dry). Insufficient sample for texture determination. Coarse fraction: strong platy light greenish yellow (5GY 7/1) and greenish gray (5GY 5/1) shale, weathers to reddish yellow (7.5YR 6/8).	Probably in Tertiary-age sequence of rocks mapped by Capps (1940). See also Dry Creek lick complex (250).
271	(63°55′ N, 147°30′ W)			mountain sheep	White (10YR 8/2, dry). Insufficient sample for texture determination. Coarse fraction: subangular brownish yellow (10YR 6/6) ferruginous chert.	Probably in Tertiary-age sequence of rocks mapped by Capps (1940). See also Dry Creek lick complex (250).
273	(63°50′ N, 147°25′ W)			mountain sheep	Pinkish gray (7.5YR 7/2) with minor very light-gray (N 8/0) very coarse sandy loam (macerated rock); strongly calcareous. Coarse fraction: angular very light-gray (N 8/0) and white (N 9/0) quartz.	Totatlanika schist (Capps 1940), near Dry Creek lick complex (see 250).
176	(63°55′ N, 147°28′ W)			mountain sheep	Pale olive (5Y 6/3) and minor medium light-gray (N 6/0) clay; rock that slakes on analysis; strongly calcareous; argillaceous odor.	See 250.
177				control sample for 176	Reddish gray (5YR 5/2) and minor strong brown (7.5YR 5/6) loam, moderately calcareous, argillaceous aroma.	
178	(63°50′ N, 147°30′ W)			mountain sheep	Same as 177 but strongly calcareous.	See 250.
179				control sample for 178	Grayish brown (2.5Y 5/2) loam; micaceous; argillaceous aroma.	
180	(63°53′ N, 147°25′ W)			mountain sheep	Olive (5Y 5/3) loam; hydrophobic.	See 250.
181				control sample for 180	Very dark grayish brown (10YR 3/2) coarse sand.	

[a]Suspected lick.

	(63°55′ N, 147°25′ W)	mountain sheep	DRY CREEK LICK COMPLEX	Dry Creek flows more or less at right angles across the contacts of several formations (Capps 1940). Samples 178 and 179 taken in headwaters are probably Birch Creek schist, limestone, phyllite, and chlorite schist (179 is micaceous). Downstream (at about the same elevation), samples 180 and 181 are derived from Totatlanika schist, which includes altered rhyolite and metasediments. Samples 176 and 177, 250 through 271, and 321 are likely in undifferentiated Tertiary mapped by Capps (1940). This sequence (mainly Eocene and younger) comprises shales, sandstones with interbedded coals, some arkosic sandstones, and conglomerates. These latter samples are uniformly calcareous and argillaceous.
250		control sample for 251	Light gray (N 7/0), dry); strongly calcareous. Insufficient sample for texture determination. Coarse fraction: strong platy light greenish gray (5GY 7/1) talc; very minor white (N 9/0) calcite.	
251			Light gray (5Y 7/2, dry); weakly calcareous. Insufficient sample for texture determination. Coarse fraction: dominantly angular light greenish gray (5GY 7/1) claystone (montmorillonite?); accessory subrounded and rounded dark-gray (N 3/0) shale; minor subangular light-gray (N 7/0) chert (?).	
252			White (5Y 8.2, dry). Insufficient sample for texture determination. Coarse fraction: dominantly angular light-gray (5Y 7/1) claystone (montmorillonite?); minor subangular light-gray (N 7/0) siltstone (?).	
253			Light gray (5Y 7/2, dry); moderately calcareous. Insufficient sample for texture determination. Coarse fraction: dominantly angular medium light-gray (N 6/0) siltstone (?); accessory yellowish gray (5Y 8/1) claystone (montmorillonite?).	
254			White (5Y 8/2, dry). Insufficient sample for texture determination. Coarse fraction: mixed medium-gray (N 5/0) and medium light-gray (N 6/0) chert; angular white (N 9/0) claystone; thin continuous white (N 9/0) and very light-gray (N 8/0) coatings on chert, botryoidal in part.	

St./prov.; sample no.	Location			Species using lick	Lick-earth description	Geologic interpretation
	twp.	rng.	sec.			
Alaska *(continued)*						
					DRY CREEK LICK COMPLEX *(continued)*	
255					Light gray (5Y 7/2, dry); very weakly calcareous. Insufficient sample for texture determination. Coarse fraction: dominantly angular and subangular medium light-gray (N 6/0) chert (?); accessory angular light-gray (2.5Y 7/2) clay aggregates, hard, porous.	
256					Light olive-gray (5Y 6/2, dry) with minor light gray (5Y 7/2); weakly calcareous. Insufficient sample for texture determination. Coarse fraction: mixed angular light olive-gray (5Y 6/2) clay (montmorillonite?) aggregates, porous; subangular medium light-gray (N 6/0) chert.	
257					Mixed light gray (5Y 7/2, dry) and white (5Y 8/2); moderately calcareous. Insufficient sample for texture determination. Coarse fraction: subangular medium dark-gray (N 4/0) chert. From preferred location in complex.	
258					Light brownish gray (2.5Y 6/2). Insufficient sample for texture determination. Coarse fraction: mixed pale yellow (2.5Y 7/4) and light-gray (N 7/0) friable saprolite.	
259					Light olive-gray (5Y 6/2, dry); weakly calcareous. Insufficient sample for texture determination. Coarse fraction: angular white (N 9/0) chert (?).	
260					Medium light gray (N 6/0, dry); weakly calcareous. Insufficient sample for texture determination. Coarse fraction: angular white (N 9/0) aphanitic mineral or rock, softer than Mohs' 5, microcrystalline in part under microscope.	

261		Light gray (5Y 7/2, dry); weakly calcareous. Insufficient sample for texture determination. Coarse fraction: angular medium light-gray (N 6/0) mineral or rock like 260. From preferred location in complex.
321		White (5Y 8/2); weakly calcareous. Insufficient sample for texture determination. Coarse fraction: angular light olive-gray (5Y 6/2) claystone (montmorillonite?); minor subrounded very dark-gray (N 3/0) shale; minor subangular light-gray (N 7/0) chert. From preferred location in complex.
262	control for 261 and 321	Olive-gray (5Y 5/2, dry); weakly calcareous. Insufficient sample for texture determination. Coarse fraction: moderate platy subangular light-gray (N 7/0) phyllite.
263		Dominantly white (N 9/0, dry) with accessory light gray (N 7/0, dry); weakly calcareous. Insufficient sample for texture determination. Coarse fraction: angular light-gray (N 7/0) microcrystalline rock, very thin (< 1 mm) lamellae of slightly lighter or darker gray or white (N 9/0), occasional mottle or streak of red (7.5YR 5/8).
264		White (N 9/0, dry). Insufficient sample for texture determination. Coarse fraction: angular white (N 9/0) claystone.
265		Mixed very light-gray (N 8/0, dry), light-gray (N 7/0), and reddish yellow (7.5YR 8/6). Insufficient sample for texture determination. Coarse fraction: like 263, some reddish yellow (7.5YR 6/8) weathered faces.
266		Light gray (10YR 7/1, dry). Insufficient sample for texture determination. Coarse fraction: like 263, but massive.
267		Mixed light-gray (5Y 7/2) and dark olive-gray (5Y 8/2) silty clay loam; sour odor. Coarse fraction: like 266.

St./prov.; sample no.	Location twp.	Location rng.	Location sec.	Species using lick	Lick-earth description	Geologic interpretation
Alaska *(continued)*						
					DRY CREEK LICK COMPLEX *(continued)*	
268					Mixed light olive-gray (5Y 6/2) and dark-gray (N 3/0) with very minor brown to dark-brown (7.5YR 4/4) loam (macerated rock); strongly calcareous; sticky. Coarse fraction: strong platy angular light-gray (N 7/0) talc and medium-gray (N 5/0) phyllite; minor medium dark-gray (N 4/0) quartzite.	
269					Very pale brown (10YR 7/4, dry). Insufficient sample for texture determination. Coarse fraction: angular yellow (10YR 7/8) and pale brown (10YR 6/3) siltstone or very fine-grained sandstone; rare white (N 9/0) on face.	
	(63°44′ N, 145°30′ W)			Dall sheep	GRANITE CREEK LICK COMPLEX	Granite Creek and Sheep Creek licks occur in like geologic settings (Moffitt 1954). In each case, lick is at or near contact of Mesozoic-age intrusive rocks and undifferentiated early Paleozoic or Precambrian rocks, which are the dominant rocks of the eastern Alaska Range and consist mainly of schist derived mostly from sedimentary rocks, quartzose in general, and argillite. Granite Mountain, one of six large intrusions in area, is composed of coarse-grained granodiorite, which Moffitt describes (p. 173) as "considerably weathered on the surface." This describes coarse material in samples 302-6.
276					Medium light-gray (N 6/0) with minor dark-gray (N 3/0) loam; strongly calcareous. Coarse fraction: subangular light-gray (N 7/0) and very light-gray (N 8/0) limestone.	
277					Medium light-gray (N 6/0) with minor dark-gray (N 3/0) sandy loam; moderately calcareous. Coarse fraction: subrounded very light-gray (N 8/0) marble bearing biotite crystals.	
278					Very light gray (N 8/0, dry); strongly calcareous. Insufficient sample for determination. Coarse fraction: like 277. From preferred location in complex.	Marble and limestone in samples 276–79 probably is derived from undifferentiated Paleozoic sequence. Moffitt suspects that Granite Creek is located along a fault. Samples from Sheep Creek were uniformly weakly calcareous. We speculate that one of rare limestones in the Paleozoic sequence outcrops in Creek's basin or one of schists (note micaceous nature of samples) is

279			Light-gray to gray (5Y 6/1) sandy loam; strongly calcareous; faint sour odor. Coarse fraction: like 277.	
303			Medium-gray (N 5/0) sandy loam; strongly calcareous. Coarse fraction: like 302.	
302			Dominantly medium light-gray (N 6/0) with minor very light-gray (N 8/0) and dark-gray (N 3/0) loamy, very coarse sand; moderately calcareous. Coarse fraction: angular very light-gray (N 8/0), medium light-gray (N 6/0), and black (N 1/0) saprolite; somewhat friable; biotite bearing. From preferred location in complex.	
306			Dominantly medium light-gray (N 6/0) with minor white (10YR 8/2) and dark-gray (N 2/0) loamy sand; strongly calcareous. Coarse fraction: like 303. From preferred location in complex.	
304			Olive-gray (5Y 5/2) with rare dark-gray (N 3/0) sandy loam, moderately calcareous; argillaceous odor. Coarse fraction: dominantly subangular and subrounded light-gray (N 7/0) and very light-gray (N 8/0) granite, saprolitic in part (white, 10YR 8/2).	
305			Olive-gray (5Y 5/2) with minor white (5Y 8/2) and pale yellow (5Y 7/4) coarse sandy loam; strongly calcareous. Coarse fraction: subangular light-gray (N 7/0), black (N 1/0), and yellow (10YR 7/8) saprolite; friable in part, like 302 and 303.	
	(63°55′ N, 147°45′ W)	Dall sheep	SHEEP CREEK LICK COMPLEX	See 276 (Granite Creek lick complex).
244			Dark olive-gray (5Y 3/2) loamy sand; weakly calcareous; micaceous.	
245			Dark olive-gray (5Y 3/2) coarse sand; weakly calcareous.	
246			Dark olive-gray (5Y 3/2) very coarse sand; weakly calcareous; micaceous (rare).	
247			Dark olive-gray (5Y 3/2) loamy sand; very weakly calcareous.	

St./prov.; sample no.	Location twp.	rng.	sec.	Species using lick	Lick-earth description	Geologic interpretation
Alaska *(continued)*						
					SHEEP CREEK LICK COMPLEX *(continued)*	
248					Olive-gray (5Y 4/2) loamy coarse sand; micaceous.	
249					Dark olive-gray (5Y 3/2) loamy coarse sand; very weakly calcareous.	
	(62°30′ N, 143°00′ W)			Dall sheep	LOST CREEK LICK	The following description of rocks in area of lick is given by Moffitt (1954): "The upper Triassic rocks are conspicuous on Lost and Trail Creeks, northern tributaries of Jack Creek, where they include massive limestone and calcareous thin-bedded shale and argillite units. Not less than 400 feet of limestone is exposed on Lost Creek, where it appears to be overlain by a much greater thickness of thin-bedded argillite." Moffitt provides a good view of Lost Creek (1954: Fig. 30).
374					Dark-gray (N 3/0) sandy clay loam; strongly calcareous. Coarse fraction: angular and subangular grayish black (N 2/0) dolomite (?); accessory massive very light-gray (N 8/0) calcium carbonate; occasional thin calcium carbonate coatings. Most preferred site.	
375					Medium dark-gray (N 4/0) silty clay loam; strongly calcareous. Coarse fraction: mixed subangular dark-gray (N 3/0) and medium light-gray (N 6/0) limestones; massive very light-gray (N 8/0) calcium carbonate.	
376					Medium dark-gray (N 4/0) sandy loam; strongly calcareous. Coarse fraction: angular medium-gray (N 5/0) very fine-grained rock, weakly calcareous; thin (< 1 mm) very light-gray (N 8/0) calcium carbonate veins cut rock.	
175	5N	2W	20	moose	Very dark grayish brown (10YR 3/2) peat. Coarse fraction: very fissile thin-bedded dark-gray (N 4/0) shale, may be slightly metamorphosed.	Peat probably overlies sediments derived from two north-south mountain ranges. Mountains on east are made up of undifferentiated Mesozoic-age rocks including Jurassic and Cretaceous shales, limestones, thick argillites, graywacke, slate, and Triassic limestone. Mountains on west side of valley unmapped but just to north are composed of basic lavas, tuffs, and greenstone, all younger than Cretaceous (Capps 1940).

274 (?)[a]	(68°55′ N, 143°40′ W)			mountain sheep	Medium dark-gray (N 4/0) sandy loam; very strongly calcareous. Coarse fraction: subangular medium light-gray (N 6/0) fine-grained limestone; rare subangular black (N 1/0) chert.	Lisburne formation (Mertie 1929) consisting of dark-gray to black fine-grained limestone, thin seams of pyrite, and abundant black chert in lower part.
275	...[b]	...	...	mountain sheep	Very pale brown (10YR 7/3) silt; earthy odor.	
307	(65°50′ N, 143°55′ W)			Dall sheep	Very dark grayish brown (10YR 3/2) loam; strongly calcareous. Coarse fraction: subangular medium dark-gray (N 4/0) asphanitic rock, very thin bedded in part, occasional micro vugs (basalt?), minor white (N 9/0) massive calcite; minor twigs and bark fragments. From wet portion of lick (cf. 308).	Alluvium of terrace deposits (Brabb and Churkin 1969). Upper Paleozoic argillite may outcrop in Jefferson Creek as in nearby drainages.
308					Very dark grayish brown (10YR 3/2) loam; pineapple aroma on rubbing; hydrophobic. Coarse fraction: weak platy angular medium-gray (N 5/0) siltstone (?).	See 307.
Arkansas						
132	1N	4E	16	white-tailed deer	Mixed light yellowish brown (10YR 6/4) and brownish yellow (10YR 6/6) silt loam.	Situated in Crowleys Ridge, which consists of thick loess (> 10 m) (Thorp and Smith 1952) overlying Tertiary gravels (Fenneman 1938). Sample description suggests that lick is located in loess.
134	2N	5W	5	white-tailed deer	Yellow red (5YR 5/6) silty clay loam.	Terrace sediments (Quaternary) exposed in headwaters of Hurricane Creek (pers. commun. J. D. McFarland III 1977).
217	1N	25W	5	white-tailed deer	Light yellowish brown (2.5Y 6/4) loam or silt loam. Coarse fraction: dominantly moderate platy subrounded and rounded brown to dark-brown (10YR 4/3) shale; minor subangular variegated red (2.5YR 5/6) and reddish yellow (7.5YR 6/6) fine-grained sandstone, rare twigs.	On or near thrust fault that brings Jackforth sandstone (Pennsylvanian) into contact with lower Atoka formation (Pennsylvanian), which comprises sandstones, siltstones, and shales. Coal seams of minor extent, pyrite, and marcasite occur in lower Atoka. J. D. McFarland III (pers. commun. 1977) advised that clay mineral dickite and sulfide mineralization occur along fault.

[a]Suspected lick.
[b]Location in Brooks Range not recorded.

St./prov.; sample no.	Location twp.	rng.	sec.	Species using lick	Lick-earth description	Geologic interpretation
California						
398	2N	6W	28	mountain sheep (desert)	Greenish gray (5G 6/1) clay loam with rare reddish yellow (10YR 7/6) sand-size grains; very strongly calcareous; pineapple aroma. Coarse fraction: subangular mottled light greenish gray (5GY 7/1) and very light-gray (N 8/0) limestone; minor subangular white (N 9/0) limestone mottled with dark greenish gray (5G 6/1), silicious.	At or near concealed fault in gray-buff granite of Mesozoic age (Rogers 1967). Limestone occurrence does not conform with mapped lithology of area.
155	27N	8W	33	deer, raccoon, birds	Grayish black (N 2/0) loam or sandy loam; hydrophobic; abundant organic matter. Coarse fraction: dominantly subrounded greenish gray (5GY 6/1) aphanitic rock (quartzite?); minor angular light-gray (N 7/0) quartzite; rare twigs and wood fragments.	Spring apparently associated with fault along which mudstone, siltstone, conglomerate, graywacke, and limestone of Ono and Rector formations (lower Cretaceous, undivided) are in contact with older metasediments including metachert, metagraywacke, and mica-quartz schist (Strand 1962).
Colorado						
163	79W	16S	...	mountain goat	Dark-brown (10YR 3/3) sand.	Granite rocks and felsic or hornblende gneisses, metabasalt, and interbedded graywacke (Tweto 1969). High pH, trace of calcareous material, and proximity of site to 104 suggest spring influence. Rocks similar to 116.
164	79W	15S	...	mountain sheep	Medium light-gray (N 6/0) with minor very light-gray (N 8/0) and dark-gray (N 3/0) sand; strongly calcareous.	Inactive hot spring in area of intrusive rocks of middle Tertiary-age (Tweto 1969), ranging from quartz- and feldspar-rich to intermediate composition. Calcareous nature of lick related to earlier hot spring activity; active hot springs nearby are depositing travertine (Feth and Barnes 1979).
165	79W	15S	...	mountain sheep	Pale brown (10YR 6/3) with minor very light-gray (N 8/0) sandy loam; very strongly calcareous; micaceous.	See 164.
117	79W	18S	...	deer, elk, mountain sheep	Dark-brown (10YR 3/3) loamy sand; rare yellow coarse-sand grains.	Area of felsic or hornblende gneisses, metabasalt, and interbedded metagraywacke (Tweto 1969).
169	74W	6S	...	elk, mountain sheep	Very dark grayish brown (10YR 3/3) fine sand; hydrophobic; micaceous; earthy odor.	Probably located in area of high basin debris sheets and moraines (Spurr and Garrey 1908). Idaho Springs formation outcrops locally and may represent source of lick minerals as it bears both biotite-sillimanite and biotite schists (note micaceous nature of sample).

173	74W	6S	. . .	mountain sheep	Dark yellowish brown (10YR 3/4) sandy loam; moderately hydrophobic.	Lick occurs in area of moraines, whose material was derived from quartz monzonites and biotite granites (Spurr and Garrey 1908).
116	75W	8N	. . .	deer, elk, mountain sheep	See 117.	Lick is in area where several rock types are mapped. One consists of granitic rocks, quartz monzonites, and granodiorites; the other consists of felsic or hornblende gneisses, metabasalts, and interbedded graywackes (Tweto 1969). Rocks are similar to 165.
118	75W	6N	. . .	deer, elk, mountain sheep	Light brownish gray (2.5Y 6/2) silty clay; expanding clay mineral type, sticky.	May be outcrop of Pierre shale (Cretaceous), which occurs in several small areas locally (Tweto 1969).
119	75W	6N	. . .	deer, elk, mountain sheep	Brownish black (5YR 2/1) sandy loam; moderately calcareous; micaceous.	Probably one of Tertiary-age volcanic sequences (intermediate in composition) (Tweto 1969) that occur in this area.
123	75W	6N	. . .	mountain sheep	Olive-gray (5Y 4/2) loamy sand; strongly calcareous.	Washings from ready-mix concrete trucks.
167	75W	6N	. . .	deer, elk, mountain sheep	Light brownish gray (10YR 6/2) clay loam; argillaceous aroma.	See 118.
168	75W	6N	. . .	deer, elk, mountain sheep	Grayish black (N 2/0) coarse sand; strongly aggregated.	May be like 119.
174	75W	6N	. . .	elk, mountain sheep	Dominantly brown to dark-brown (10YR 4/3) coarse sand with various colored sand grains; hydrophobic; micaceous. Coarse fraction: dominantly subangular dark-gray (N 3/0) biotite schist; common angular and subangular very light-gray (N 8/0) feldspar.	From lithology of coarse fraction, lick appears to be in area of Precambrian metamorphic rocks (Tweto 1969).
166	103W	8N	. . .	deer, mountain sheep	Weak red (10R 4/4) sand.	No published account found. From study to south (Hansen 1976), lick probably located in Uinta Mountain group (Precambrian age). Composed of light- to dark-red coarse- to medium-grained pebbly quartzitic sandstone that is locally friable and several beds of red to gray silty shale. Lick may be associated with fault that occurs along or otherwise perhaps controls alignment of Pot Creek (where lick occurs).

St./prov.; sample no.	Location twp.	rng.	sec.	Species using lick	Lick-earth description	Geologic interpretation
Colorado *(continued)*						
120	73W	26S	...	mountain sheep	Dark reddish brown (5YR 2/2) loamy sand.	This sample and 121, 122, and 170–72 from Saguache County appear to have a similar origin; all are brownish colored sands or loamy sands from the vicinity of Great Sand Dunes National Park. Sites appear to be at or near contact of aeolian sands of Dunes area and the west slope of the Sangre de Cristo Mountains, which are composed of felsic and hornblende gneisses, metabasalt, and interbedded metagraywacke. A fault underlies these sands and possibly mineralized waters have affected area (Tweto 1969).
121	73W	26S	...	mountain sheep	Very dark grayish brown (10YR 3/2) loamy sand.	See 120.
122	73W	26S	...	mountain sheep	Same as 121.	See 120.
170	73W	26S	...	deer	Very dark grayish brown (10YR 3/2) fine sand; strongly calcareous.	See 120.
171	73W	26S	...	mountain sheep	Very dark grayish brown (10YR 3/2) fine sand; very weakly calcareous; somewhat hydrophobic.	See 120.
172	73W	26S	...	mountain sheep	Very dark grayish brown (10YR 3/2) fine sand.	See 120.
Georgia						
213	...[a]	...	...	white-tailed deer	Dark brown (10YR 3/3) with minor yellowish brown (10YR 5/6) loam; micaceous. Coarse fraction: dominantly subrounded pink (7.5YR 7/4) micaceous rock, weathered and somewhat friable; very minor very light-gray (N 8/0) feldspar granules; rare twigs.	This site could not be located precisely, but rocks in Jasper County, description of coarse fraction, and analytical data suggest that lithology at lick is similar to that at Jones County licks below.
209	(33°08′ N, 83°39′ W)			white-tailed deer	Mixed yellow (5Y 7/4) and red (2.5YR 5/6) silt loam; micaceous.	No detailed geologic report for area. Area underlain by granite gneiss, biotite gneiss, and schist of Precambrian age (LaMoreaux 1946) and local dolerite dikes. Nature of rocks accompanying samples is consistent with above general description of gneisses and schists. Deeply weathered landscape, common to this area, is suggested.

210	(33°07′30″ N, 83°44′00″ W)			white-tailed deer	Red (2.5YR 5/6) with minor brown (7.5YR 5/4) loam; very micaceous.	Either in hornblende gneiss/amphibolite or granite gneiss/amphibolite (Ga. Geol. Surv. 1979).
211	(30°00′30″ N, 83°39′30″ W)			white-tailed deer	Brown to dark-brown (10YR 4/3) with very minor strong brown (7.5YR 5/6) sandy loam; very micaceous.	Biotite-granite gneiss/feldspathic-biotite gneiss/hornblende gneiss undifferentiated (Ga. Geol. Surv. 1979).
212	(33°03′30″ N, 83°42′00″ W)			white-tailed deer	Mixed reddish brown (7.5YR 6/6) and strong brown (7.5YR 5/6) loamy sand. Coarse fraction: dominantly subangular and subrounded very light-gray (N 8/0) quartzite, thinly coated with yellowish brown (10YR 6/4), minor angular pink (5YR 8/4) granite and minor angular dark-gray (N 3/0) and very light-gray (N 8/0) biotite gneiss, weathered and friable in part.	Biotite gneiss (Ga. Geol. Surv. 1979).
214	(33°06′10″ N, 83°39′00″ W)			white-tailed deer	Strong brown (7.5YR 5/6) loam. Coarse fraction: dominantly subrounded somewhat friable yellow (10YR 7/6) saprolite; minor subrounded fine-grained medium dark-gray (N 4/0) biotite schist; minor subangular light-gray (N 7/0) and pale red (10YR 6/4) quartzite.	Similar to 211.
Illinois						
372	9S	8E	26	white-tailed deer	Very dark grayish brown (10YR 3/2) silt loam; strongly calcareous. Coarse fraction: mixed subangular light-gray (N 7/0) and pink (7.5YR 7/4) fine-grained sandstone; medium light-gray (N 6/0) fine-grained limestone; rare very light-gray (N 8/0) acid igneous rock, rare mollusk shell fragment; twigs, leaf fragments; one 2 × 2.5 cm brick or tile fragment. See also 373.	Salt spring appears to originate in Ste. Genevieve limestone (Mississippian) that outcrops at spring (Butts 1925). Limestone is thin bedded and cherty in lower 15 m. Above it is very pure, massive gray, blue-gray or nearly white, mostly oolitic limestone. Tradewater formation (Pennsylvania) overlies Ste. Genevieve and is composed of sandstone, shale, and several thin coal beds.
373	9S	8E	26	white-tailed deer	Brown to dark-brown (10YR 4/3) silty clay loam. At site, salt encrusts leaves and twigs; smell of hydrogen sulfide from decaying vegetation.	About 50 m below spring described in 372.
125	10S	2E	8	white-tailed deer	Dark yellowish brown (10YR 4/4) loam; very weakly calcareous. Coarse fraction: subrounded light-gray (10YR 7/2) chert.	Probably in Makanda sandstone, member of Pottsville formation. Massive in base, the Makanda becomes interbedded with sandstones and gray-black shales in top (Lamar 1925).

St./prov.; sample no.	Location twp. rng. sec.	Species using lick	Lick-earth description	Geologic interpretation
Indiana				
324	(38°46′18″ N, 86°50′00″ W)	white-tailed deer	Light yellowish brown (10YR 6/4) silt loam; weakly calcareous.	Probably located in alluvium derived from Mississippian-age rocks of Stephensport and West Baden groups that consist of limestone, shale, and sandstone (Gray et al. 1970).
325	(38°48′31″ N, 86°50′20″ W)	white-tailed deer	Light yellowish brown (10YR 6/4) silt loam; slippery consistence.	See 324.
326	(38°46′35″ N, 86°49′55″ W)	white-tailed deer	Pale brown (10YR 6/3) silt loam with reddish yellow (7.5YR 6/6) mottles; hard fine iron-manganese concretions.	See 324.
327	(38°48′31″ N, 86°50′11″ W)	white-tailed deer	Yellowish brown (10YR 5/4) silt loam with minor brownish yellow (10YR 6/6) mottles; slippery consistence.	See 324.
328	(38°52′48″ N, 86°47′25″ W)	white-tailed deer	Very pale brown (10YR 6/4) silt loam; very slippery consistence.	See 324.
Kentucky				
319	(36°30′ N, 83°30′ W)	deer	Very dark grayish brown (10YR 3/2) with accessory dark yellowish brown (10YR 3/4) silt loam. Coarse fraction: dominantly twigs, wood, and bark fragments; rare rounded very pale-brown (10YR 7/4) quartzite.	Probably in Lee formation (Pennsylvanian) at level of sandstone member C, which is overlain by shale and underlan usually by coal or shale (England 1964).
152	(38°14′26″ N, 82°57′57″ W)	white-tailed deer	Dark grayish brown (2.5Y 4/2) and minor grayish brown loam. Coarse fraction: dominantly subangular and platy light-gray (10YR 7/1) very fine-grained sandstone; minor dark reddish brown (2.5YR 2/4) ironstone.	No detailed report found. Probably either with Lee formation or with lower Breathitt formation (both lower and middle Pennsylvanian). Lee formation is light-gray quartzose sandstone that is locally cemented with calcite or siderite. Overlying Breathitt formation is, in lower part, largely siltstone and shale with two discontinuous coal beds; carbonaceous shales are associated with coal beds (Whittington and Ferm 1967).
157	(36°10′54″ N, 88°06′36″ W)	fallow and white-tailed deer	White (2.5Y 8/2) silty clay with very minor fine-sand-size reddish yellow grains.	Tuscaloosa formation (upper Cretaceous), which consists of light-gray and poorly stratified gravel, clay, and silt (Seeland and Wilshire 1965). Chert is common.

153	(36°41′00″ N, 88°00′31″ W)			fallow and white-tailed deer	Pinkish gray (5YR 8/1) silt loam. Coarse fraction: subangular very light-gray (N 8/0) chert, somewhat polished.	See 157.
147	(37°30′ N, 87°45′ W)			white-tailed deer	Yellowish brown (10YR 5/6) silt loam.	Loess up to 6.1 m thick in area (Kehn 1975). Below 3.0 m depth, loess is calcareous. Scattered small outcrops of limestone (top of Sturgis formation, upper Pennsylvanian) in ridges and creeks nearby; probably underlies loess at lick site.
Louisiana						
194	17N	8W	13	deer	Light yellowish brown (10YR 6/4) fine sandy loam; somewhat hydrophobic.	No specific data available. Lick is in same township as Vacherie Salt Dome (Spooner 1929: Fig. 1).
189	20N	11W	8	white-tailed deer	Red (2.5YR 4/6) silty clay loam.	Prairie formation (Pleistocene) composed of quartz sand with occasional gravel and clay (Jones 1960). Lick site is immediately west of Interior Salt Dome Basin; two domes are in Webster Parish to the east.
131	6N	8W	12	white-tailed deer	Brown to dark-brown (7.5YR 4/2) and rare red (2.5YR 4/8) silt loam.	Nearest detailed geologic report deals with Vernon Parish, 19.3 km to south (Welch 1942), where Carnahan Bayou member of Fleming formation (Miocene) occurs at surface. Carnahan Bayou member is composed of a succession of fluviatile silts and sands with bentonite bed that sometimes creates a spring zone. The Interior Salt Dome Basin lies north and east of lick site; Chestnut Dome occurs at T13N and R6W (La. Dep. Conserv. Geol. Surv. 1975).
197	22N	3W	32	white-tailed deer	Yellowish red (5YR 5/8) sandy loam.	No specific data for Union Parish available. Area is northeast of Interior Salt Dome Basin of Louisiana.
198	21N	3W	20	white-tailed deer	Light yellowish brown (10YR 6/4) silt.	No specific data for Union Parish available. Area is northwest of Interior Salt Dome Basin of Louisiana.
Maryland						
219	(39°41′30″ N, 78°27′30″ W)			white-tailed deer	Dominantly very dark grayish brown (2.5Y 3/2) loamy coarse sand with dark reddish brown (2.5YR 3/4) and light yellowish brown (10YR 6/4) sand grains; weakly calcareous. Coarse fraction: dominantly moderate platy subrounded medium dark-gray (N 4/0) shale; accessory subrounded reddish brown (5YR 5/3) and light reddish brown (5YR 6/3) very fine-grained sandstone.	Jennings formation (Devonian, Berryhill et al. 1956), consisting of dark-gray to black platy shale in base and olive-gray platy silicious shales and interbedded siltstones and conglomeratic sandstones (Chemung member) in top. Lick is probably in middle or lower member. Little Pine and Indian licks are nearby.

St./prov.; sample no.	Location twp.	rng.	sec.	Species using lick	Lick-earth description	Geologic interpretation
Maryland *(continued)*						
220	(39°38′ N, 78°24′ W)			white-tailed deer	Brown (7.5YR 5/4) silt loam. Coarse fraction; subrounded light-gray (5YR 6/1) and gray (5YR 5/1) fine-grained sandstone. From dry portion of lick (see also 221).	Jennings formation. See 219.
221	(39°38′ N, 78°24′ W)			white-tailed deer	Yellowish brown (10YR 5/4) silt loam. Coarse fraction: (10YR 6/3) fine-grained sandstone, weathered. From wet portion of lick (see also 220).	Jennings formation. See 219.
228	(39°38′45″ N, 78°29′30″ W)			white-tailed deer	Very dark grayish brown (10YR 3/2) silt loam; hydrophobic; abundant organic matter. Area of Weikert-Gilpin soil association. Hydrogen sulfide odor at lick.	At or near contact of Romney shale (Devonian) with overlying Jennings formation (see 219). Romney shale is silty mudstone with interbedded siltstone in upper part (Hamilton member).
Michigan						
232	33N	1W	3	deer, elk	Very dark grayish brown (10YR 3/2) medium sand; weakly calcareous. Coarse fraction: dominantly subrounded pinkish gray (7.5YR 6/2) silty sandstone, soft; accessory subangular very light-gray (N 8/0) chert; minor very light-gray (N 8/0) quartzite; rare charcoal.	Abandoned and filled brine pit from oil well.
108	63N	58W	29	moose	Very dark grayish brown (10YR 3/2) loam; hydrophobic; odor of soybean oil on rubbing. Coarse fraction: well rounded dusky red (10YR 3/4) jasper, polished.	Quaternary alluvium underlain by Coral Harbour conglomerate (Huber 1973).
109	65N	36W	27	moose	Dark-brown (7.5YR 3/2) loamy coarse sand. Coarse fraction: mixed subrounded to rounded dark red (10YR 3/6), siltstone (?), pale red (10R 6/5) and weak red (10R 5/3) quartzite, and biotite-rich arkose pebbles.	Quaternary alluvium is underlain by Portage Lake volcanics (Huber 1973) that consist of basaltic and andesitic flows with interbedded sandstones, conglomerates, and pyroclastic rocks.
110	64N	38W	21	moose	Very dark grayish brown (10YR 3/2) loamy sand; moderately calcareous. Coarse fraction: like 109, with granite.	Quaternary alluvium overlying Portage Lake (see 109); may be associated with north-trending fault. Granite found in sample is an erratic.
111	67N	33W	34	moose	Very dark grayish brown (10YR 3/2) sandy loam; hydrophobic; abundant organic matter. Coarse fraction: subangular to subrounded dark brown (7.5YR 3/2) fine-grained quartzose sandstone.	Probably a section of Portage Lake volcanics (Huber 1973); contains tuffs, breccias, and tuffaceous sedimentary rocks.

113	64N	38	34(?)	moose	Dark-brown (7.5YR 3/2) sandy loam; hydrophobic. Coarse fraction: mixed subangular to subrounded very light-gray (N 9/0) dolomite with manganese oxide dendrites and yellow (10YR 7/6) and dark yellowish brown (10YR 4/4) sandstone, fine grained.	Probably in Quaternary alluvium over Portage Lake volcanics (see 109). Faulting occurs in area.
114	66N	34W	13	moose	Very dark grayish brown (10YR 3/2) loamy sand; hydrophobic. Coarse fraction: dominantly subangular light-gray (N 7/0) basic igneous rock (gabbro?).	Portage Lake volcanics (see 109).
323	64N	37W	33	moose	Very dark grayish brown (10YR 3/2) sandy loam; abundant organic matter; hydrophobic. Coarse fraction: twigs, wood fragments.	See 108.
104	22N	4W	17	white-tailed deer	Brown to dark-brown medium-grained sand; moderately calcareous.	Site of abandoned oil well.
399	4S	5E	1	cattle, deer (?)	Very dark grayish brown (10YR 3/2) loam; strongly calcareous; hydrophobic; odor of soil with active actinomycete flora (newly plowed). Sloan loam soil series.	Saline spring adjacent to Saline River.
Minnesota						
158	47N	24W	29	white-tailed deer	Dark-brown (10YR 3/3) sandy loam. Coarse fraction: dominantly rounded medium-gray (N 5/0) quartzite (?) and light-gray (N 7/0) quartzite; minor subrounded pale red (10R 6/3) potassium feldspar.	Middle Precambrian argillite and graywacke (Sims and Morey 1972, Plate 1).
159	139N	39W	1	white-tailed deer	Dark grayish brown (2.5Y 4/2) loam. Coarse fraction: mixed subangular and subrounded gray (10YR 5/1) aphanitic rock; subrounded reddish yellow (5YR 7/6) and light gray (N 7/0) quartzite.	Undivided Precambrian metasediments including graywacke, slate, conglomerate, quartzite, and clastic volcanic rocks of felsic to intermediate composition (Sims and Morey 1972, Plate 1).
415	59N	8W	34	moose	Black (10YR 2/1) peat. Coarse fraction: twigs and rare angular medium-gray (N 5/0) aphanitic igneous rock fragments. The pH of water-peat mixture was 4.3.	North Shore Volcanic group consisting of lavas from olivine basalt to quartz latite composition, local rhyolite, and occasional interbedded conglomerate (Sims and Morey 1972, Plate 1).
Mississippi						
102	16N	15E	11	white-tailed deer	Dark grayish brown (10YR 4/2) sandy loam.	Alluvial plain of Noxubee River, which is underlain here by Prairie Bluff chalk (Cretaceous) (Stephenson and Monroe 1940). Chalk is rather pure (68.9–82.5% $CaCO_3$) except phosphatic fossils occur in base.
115	16N	15E	9	white-tailed deer	Brown to dark-brown (10YR 4/3) loam or silt loam.	See 102.

St./prov.; sample no.	Location twp.	rng.	sec.	Species using lick	Lick-earth description	Geologic interpretation
Missouri						
383	49N	17W	33	white-tailed deer	Dark grayish brown (10YR 4/2) with rare reddish brown (5YR 4/3) silt; faint algal odor. Coarse fraction: mixed angular medium light-gray (N 7/0) quartzite, accessory angular pink (7.5YR 7/4) and pinkish white (7.5YR 8/2) cherts; minor subangular brown to dark-brown (7.5YR 4/4) ferruginous fine-grained sandstone.	Boone's Lick State Historical Site. Cherokee group, which is composed of limestones, shales, and coal (Mo. Geol. Surv. 1979). Site is inside glacial boundary.
377	48N	20W	17	cattle, formerly bison, white-tailed deer, elk	Very dark grayish brown (10YR 3/2) silt loam; weakly calcareous; very slippery consistence when wet.	Sulphur Springs Group (Mo. Geol. Surv. 1979), which is composed of shale, limestone, and sandstone. Sulphur spring at lick.
378	48N	22W	16	white-tailed deer, formerly bison, elk	Like 377, moderate algal odor.	Osagean series (Mississippian), which is predominantly cherty limestone (Mo. Geol. Surv. 1979). Southern limit of glaciation.
379	49N	21W	28	white-tailed deer	Like 378, moderately calcareous.	Similar to 378, but inside glacial boundary.
380	49N	21W	21	white-tailed deer	Like 377, but noncalcareous.	In alluvium, see 379.
381	50N	20W	32	white-tailed deer	Like 377.	See 380.
382	50N	20W	28	white-tailed deer	Very dark brown with rare yellowish brown (10YR 5/6) sandy loam; weakly calcareous; moderate algal odor. Coarse fraction: mixed subangular and subrounded dark-gray (N 3/0), very light-gray (N 8/0), and white (N 9/0) cherts.	See 379.
Montana						
333	28N	17W	26	elk	Mixed light-gray (N 7/0) and very pale brown (10YR 7/4 and 7/3) silty clay loam. Coarse fraction: angular white (10YR 8/2) claystone, somewhat friable; minor rounded very light-gray (N 8/0), very fine-grained sandstone.	Probably undifferentiated alluvial deposits of early Pleistocene and Tertiary age (Ross 1959), ranging from silts to gravels and local lignite beds.
334	29N	15W	24	mountain goat	Mixed medium light-gray (N 6/0) and very light-gray (N 8/0) very coarse loamy sand; strongly calcareous. Coarse fraction: subrounded and rounded very light-gray (N 8/0) fine-grained limestone; minor subangular white (N 9/0) coarse-grained calcite.	Missoula group (upper Belt series, upper Precambrian), which is, in part, limestone (Ross 1959).

140	18N	21W	11	bison	Pale brown (10YR 6/3) loam; very strongly calcareous.	No detailed map found. On state map (Ross et al. 1955), outcrop of Ravalli group (Belt series, Precambrian) is mapped here. Group consists of argillites, dolomite, and magnesian limestone (Altyn limestone that weathers to a distinctive yellowish brown; see 142).
141	19N	21W	36	bison	Brown to dark-brown (7.5YR 4/2) silt to loam.	Alluvium derived from Belt series (see 140).
142	19N	20W	31	bison	Light yellowish brown (10YR 6/4) silty clay loam; strongly calcareous.	Outcrop of Ravalli group or in alluvial sediments derived from Belt series (similar lithology). See 140.
North Dakota						
135	139N	103N	4	mule deer, mountain sheep	Grayish black (N 2/0) and minor dark grayish brown (10YR 4/2) loam or silt loam; very strongly calcareous; hydrophobic; abundant organic matter.	At seep in Sentinel Butte formation (Paleocene), which consists of gray silts, clays, and sands, with interbedded sandstones and occasional limestone and lignite (Bluemle 1977).
136 (?)[a]	139N	103W	4	mule deer, mountain sheep	Light olive-brown (2.5YR 5/4) and minor reddish yellow (7/5YR 6/8) loam; very strongly calcareous.	See 135.
137 (?)[a]	139N	103W	3	mule deer, mountain sheep	Brown to dark-brown (10YR 4/3) silt loam; very strongly calcareous.	See 135.
138 (?)[a]	139N	103W	4	mule deer, mountain sheep	Light brownish gray (2.5Y 6/4) silt loam; very strongly calcareous.	See 135.
133	145N	100W	20	mule deer, mountain sheep	Grayish black (N 2/0) loam or silt loam; hydrophobic, abundant organic matter.	See 135.
134	145N	100W	14	mule deer, mountain sheep	Very dark grayish brown (10YR 3/2) and minor brown to dark-brown (10YR 4/3) coarse sandy loam.	See 135.
Ohio						
107	12N	15W	11	white-tailed deer	Dark grayish brown (10YR 4/2) silt loam.	Wisconsinan lacustrine silts—nearby rocks that might influence groundwater chemistry are the Allegheny series (Sturgeon and Associates 1958), which at this site is four cyclothems each characterized by coal, clay, and shale. One 20-cm thick limestone occurs near middle of sequence.
182	8N	13W	27	cattle, white-tailed deer	Reddish brown (5YR 4/4) silty clay loam; weakly calcareous.	Monongehela series (Pennsylvanian) is a sequence of sandstones, limestones, and sandy shales (Nordling 1958). Calcareous nature of samples suggests that lick is near level of one of limestones.

[a]Suspected lick.

St./prov.; sample no.	Location twp.	rng.	sec.	Species using lick	Lick-earth description	Geologic interpretation
Ohio *(continued)*						
183	8N	13W	27	cattle, white-tailed deer	Reddish brown (5YR 5/4) silty clay loam; weakly calcareous.	One hundred meters south of 182.
156	16W	11N	10	white-tailed deer	Grayish brown (2.5Y 5/2) loam with very minor, very small yellow grains.	No detailed account; probably lower and middle Conemaugh series (Pennsylvanian) and upper Allegheny series (Stout 1927; Sturgeon and Associates 1958). The Conemaugh consists of shales, sandstones, and thin limestones and the Allegheny of shales, sandstones, and thin, patchy limestones (to level of middle Kittanning coal).
231	16W	10N	19	white-tailed deer	Dark-brown (10YR 3/3) loam.	See 156.
105	(38°46′18″ N, 87°12′44″ W)			white-tailed deer	Dark yellowish brown (10YR 4/4) silt loam.	No detailed report found. State map (Bownocker 1947) shows Waverly group outcropping in area. To the south in Kentucky, Erickson (1966) mapped rocks in the same topographic position (also a Lick Creek heading therein) as lower shale and middle siltstone members of the Borden formation, the modern equivalent (in part) of the Waverly. Immediately across the Ohio River and 6 km west of Portsmouth, Lower and Upper Lick Creeks occur in topographic positions that also must place these drainages in the lower Borden. Erickson (1966) noted local glauconitic siltstone and limestone in the siltstone member, and the shale bears local siderite concretions up to 12 cm in diameter.
106	(38°47′30″ N, 83°07′30″ W)			white-tailed deer	Dark grayish brown (2.5Y 4/2) sandy loam; weakly calcareous.	Seep in alluvium. See 105.
Oregon						
222	39S	4E	22	black-tailed deer	Dark yellowish brown (10YR 3/4) silty clay; moderately calcareous, strongly aggregated. Overlies 206.	Spring in area described by Callaghan and Buddington (1938) as high Cascade lavas. The rocks (Pliocene and Pleistocene ages) of the region are characterized by common light-gray olivine basalts, gray basalts, and andesites. Glassy lavas also occur. Pumice from eruption of Mount Mazama covers the area and almost certainly is represented in 206 (see description). Deer eating

206	39S	4E	22	black-tailed deer	Mixed dark yellowish brown (10YR 3/4) and light yellowish brown (10YR 6/4) loamy coarse sand (pumice); strongly calcareous. Underlies 222.	
Pennsylvania						
185	(41°32′30″ N, 78°12′30″ W)			white-tailed deer	Dark-brown (10YR 3/3) loam; hydrophobic; weak musty odor.	Alluvium in Catskill formation (Devonian) (Gray and Shepps 1960), which is red and brown shales and sandstones; adjacent slopes are underlain by Oswayo formation (Devonian, see 146).
186	(40°46′30″ N, 77°57′30″ W)			white-tailed deer, squirrels, woodchucks	Brown to dark-brown (10YR 4/3) and minor very dark grayish brown (10YR 3/2) sandy loam. Coarse fraction: mixed subrounded very dark grayish brown (10YR 4/3) coarse-grained sandstone, manganese stained; platy and subrounded dark reddish brown (2.5YR 3/4) ironstone.	May be artificial. Ordovician and Cambrian limestones and dolomites outcrop in this area (Gray and Shepps 1960). Galesburg formation (Cambrian) contains many sandstone beds, and sample texture and coarse fragment identification suggest it may occur at site.
187	(41°13′59″ N, 77°37′30″ W)			white-tailed deer, other animals	Dark grayish brown (10YR 4/2) silt loam.	May be artificial. Allegheny group (Pennsylvanian) underlies most of hills, although some summits consist of rocks of Pottsville Group (Pennsylvanian). Both groups are composed of cyclical sequences of sandstones, shales, limestones, and coal (Gray and Shepps 1960).
103	(41°35′00″ N, 78°47′30″ W)			white-tailed deer	Dark brown (10YR 3/3) loam; hydrophobic.	Below abandoned oil-well sump pit.
184	. . .[a]	. . .	. . .	white-tailed deer	Dark yellowish (10YR 4/4) loam.	See 185.
146 (?)[b]	(41°40′ N, 79°08′ W)			white-tailed deer	Dark grayish brown (10YR 4/2) loamy sand; hydrophobic. Coarse fraction: common subrounded light-gray (10YR 7/2) limestone, friable and weathered; very weakly calcareous; common rounded very light-gray (N 8/0) quartzite; minor wood fragments.	Adjacent to stream located in Cattaraugus formation (Devonian), which is composed of red, gray, and brown shales and sandstones (Gray and Shepps 1960). Adjacent slopes are evidently Oswayo formation (Devonian), which is composed of brown and greenish gray sandstones, some shale, and scattered calcareous lenses. Physical nature of sample suggests that lick is in Oswago formation.

[a]Jones Township, Elk County.
[b]Suspected lick.

St./prov.; sample no.	Location twp. rng. sec.	Species using lick	Lick-earth description	Geologic interpretation
South Carolina				
188	(35°42′22″ N, 81°26′08″ W)	white-tailed deer	Yellow (10YR 7/6) loam; micaceous.	Probably underlain by Carboniferous-age metamorphic rocks (Am. Assoc. Pet. Geol. no date).
226	(34°45′00″ N, 81°28′49″ W)	white-tailed deer	Brown to dark-brown (7.5YR 4/4) silt; micaceous.	Occurs in alluvium in area of bedrock mapped as Carboniferous-age metamorphic rocks (Am. Assoc. Pet. Geol. no date). Broad River drains a terrane of varied metamorphic rocks and acid igneous rocks.
Tennessee				
154	(36°17′30″ N, 88°02′30″ W)	white-tailed deer	Mixed pinkish gray (7.5YR 7/2) and reddish yellow (7.5YR 7/6) and minor red (2.5YR 4/8) silty clay. Coarse fragments: subrounded very light-gray (N 8/0), very fine-grained silicious rock.	Probably in Coffee sand (upper Cretaceous) (Parks and Russell 1975), which consists of quartz sand with lenses of brown clay. Sand is glauconitic in places.
195	(36°21′13″ N, 88°04′30″ W)	white-tailed deer	Light brownish gray (10YR 6/2) silty clay loam.	Either in Coffee sand (see 154) or Coon Creek formation (upper Cretaceous) (Parks and Russell 1975), which consists of quartz sand and glauconitic clay. There are also rare limestones and sandy and calcareous basal clays.
233	(36°04′20″ N, 84°54′13″ W)	white-tailed deer	Dark yellowish brown (10YR 4/4) medium sand; very weakly calcareous.	Seep at gas well established about 50 yr ago.
196	(36°23′24″ N, 88°03′16″ W)	white-tailed deer	Dark-brown (10YR 3/3) and minor dark-red (2.5YR 3/6) silty clay; very sticky.	See 195.
Vermont				
216	(42°14′32″ N, 73°04′19″ W)	white-tailed deer	Very dark grayish brown (2.5Y 3/2) loam or silt loam; moderate to strongly calcareous; hydrophobic.	At or very near contact of Shelburne marble (lower Ordovician) with Danby-Clarendon Springs formation (upper Cambrian) in area of complex folded rocks (Hewitt 1961). Clarendon Springs formation is a gray dolomite, sometimes calcitic, with quartzite veins and also bearing dark chert.
Virginia				
160	(38°20′20″ N, 79°12′03″ W)	deer, honeybees	Dark-brown (7.5YR 3/2) loamy sand.	Hampshire formation (Devonian), consisting of red sandstones—some arkosic—and shales (Calver 1963).

162	(38°28′01″ N, 77°44′51″ W)	deer	Yellowish brown (10YR 5/4) silt loam. Coarse fraction: subangular white (10YR 8/1 and 8/2) quartzite.	Probably in Wissahickon schist, which is dominantly a chlorite-muscovite schist but contains a quartzite in its lower part (Calver 1963).
208	(38°40′57″ N, 78°50′41″ W)	deer	Dark-brown (10YR 3/3) silt loam.	In Martinsburg shale along Paddy Run. Butts and Edmundson (1966) noted calcareous beds near top of Martinsburg in outcrop on Paddy Run. See also sample 207.
161	(38°30′00″ N, 79°07′15″ W)	deer	Dark reddish brown (2.5YR 3/4) sandy loam.	Pocono sandstone (Mississippian) (Brent 1960).
207	(39°04′10″ N, 78°31′28″ W)	deer	Brown to dark-brown (10YR 4/3) silt loam.	Lick in Martinsburg shale (Ordovician) (Calver 1963). Also note sample 208.
West Virginia				
230	(38°02′20″ N, 80°05′52″ W)	deer	Very dark grayish brown (10YR 3/2) loam; hydrophobic; abundant organic matter. Coarse fraction: moderate platy subrounded grayish brown (2.5YR 5/2) and brownish black (5YR 2/1) shale; very minor twigs and bark.	Braillier formation (Devonian), which consists of olive-gray to dark-gray shale, siltstones, and thin sandstone lenses (Cardwell et al. 1968).
223	(39°11′ N, 78°50′ W)	deer	Brown to dark-brown (10YR 4/3) loam; hydrophobic.	At or near contact of Chemung group with underlying Catskill group (both Devonian) (Tilton 1926). Chemung consists of gray to brown siltstones, sandstones, shales, and interbedded conglomerate. Catskill consists of red shales, sandstones, and siltstones.
229	(38°25′32″ N, 80°04′49″ W)	white-tailed deer	Dark reddish brown (5YR 3/2) silt loam; hydrophobic; abundant organic matter; weak tobacco aroma on rubbing. Coarse fraction: dominantly moderate platy subrounded by reddish brown (5YR 4/3) shale; accessory subangular and subrounded brown to dark-brown (7.5YR 4/3) fine-grained sandstone, micaceous; very minor twigs.	Mauch Chunk series (Mississippian), consisting predominantly of red shales, perhaps near contact with overlying sandstones of Pottsville series (Pennsylvanian) (Price 1929:99).
215	(38°46′ N, 79°47′ W)	deer	Grayish black (N 2/0) loam; very weakly calcareous; abundant organic matter. Coarse fraction: dominantly strong platy subrounded grayish black (N 2/0) shale; minor rounded white (N 9/0) and medium-gray (N 5/0) fine-grained limestones; rare angular coal; very minor twigs.	New River formation at level of Sewell coal (Pottsville group, Pennsylvanian). Physical description of sample conforms to outcrop of shales, limestone, and coal typical of rocks of vicinity (Reger 1931; Cardwell et al. 1968).

St./prov.; sample no.	Location twp.	rng.	sec.	Species using lick	Lick-earth description	Geologic interpretation
West Virginia (*continued*)						
224	(38°45′ N, 79°45′ W)			deer	Dark yellowish brown (10YR 3/4) silt or silt loam; abundant organic matter; aroma of freshly rubbed tobacco.	Either in Hinton formation or at contact with overlying Princeton formation (both Pennsylvanian). Princeton formation is predominantly sandstone; Hinton consists of shales, sandstone, and several thin limestones (Reger 1931; Cardwell et al. 1968).
225	(37°32′30″ N, 80°54′08″ W)			deer	Reddish brown (5YR 4/4) silt loam. Coarse fraction: weak platy subangular and subrounded reddish brown (5YR 5/2) shale.	At or near the contact of the Bluefield and Hinton formations (Mauch Chunk group, Mississippian). Both formations consist of red and green shales and sandstones with a few thin limestones; the Hinton also bears gray shales (Reger 1925; Cardwell et al. 1968).
Wyoming						
346	...[a]	...	...	pronghorn antelope	Mixed brown (10YR 5/3) and brown to dark-brown (10YR 4/3) medium-grained sand. Coarse fraction: subrounded light-gray (N 8/0) fine-grained limestone (dolomitic?), weathered to reddish yellow (7.5YR 6/6); rare rounded red (2.5YR 4/6) ironstone.	Earth saturated by sodium sulfate and bicarbonate water from artesian well at site (see analytical data).
128	41N	115W	9	elk, mountain sheep, deer (?)	Brown to dark brown (10YR 4/3) silt loam; very strongly calcareous. Coarse fraction: subrounded light-gray (5Y 7/1) fine-grained limestone; rare rounded weak red (10R 5/2) quartzite.	Valley floor in area influenced by Madison and/or Gallatin limestones (Love et al. 1952).
129	42N	115W	30	elk	Pale brown (10YR 6/3) loamy sand; very strongly calcareous.	Area of poorly consolidated sandstones, conglomerates, and tuffaceous marl, all of Pliocene or Miocene age (Love et al. 1952).
130	42N	116W	36	elk, deer (?)	Light brownish gray (10YR 6/2) loamy sand. Coarse fraction: subangular white (N 9/0) siltstone, very weakly calcareous.	Alluvium of Gros Ventre River, Jackson Hole. See 129.
139	42N	115W	33	elk	Brown to dark-brown (10YR 4/3) loam or silt loam. Coarse fraction: dominantly subangular light-gray (10YR 7/2) fine-grained limestone; common rounded gray (N 5/0) quartzite.	See 129.

[a]Clay Butte, Sweetwater County.

348	43N	111W	15	elk	Brown (10YR 5/3) silty clay loam; moderately calcareous. Coarse fraction: subangular and subrounded light-gray (10YR 7/1) fine-grained limestone, continuous coating of calcium carbonate on some faces, botryoidal in part.	Undivided Paleocene rocks (Love et al. 1952) consisting of rhyolitic tuff, freshwater limestone, quartzose conglomerate, brown sandstone, claystone, carbonaceous shale, and thin coals. Limestone and claystone members are probably present at lick.
349	43N	111W	14	elk, horse	Brown to dark-brown (10YR 4/3, dry) loam (?); very hydrophobic, aroma of molasses on rubbing; strongly aggregated; very low bulk density.	Wind River formation (Eocene) composed of variegated claystone, shale, sandstone, and lenticular conglomerate (Love et al. 1952). Origin of 349 cannot be deduced from this description.
350	43N	110W	18	elk	Reddish brown (2.5YR 4/4) with accessory medium light-gray (N 6/0) silt loam; strongly calcareous. Coarse fraction: dominantly subangular and subrounded pale red (5R 6/2) and light-brown (7.5YR 6/4) limestone; minor light-gray (N 7/0) limestone; minor medium-gray (N 5/0) and white (10YR 8/2) cherts.	See 349.
351 (?)[a]	43N	110W	7	elk	Brown to dark-brown (10YR 4/3) loam. Coarse fraction: subrounded white (10YR 8/2 and N 9/0) siltstone, soft.	See 349.
352 (?)[a]	43N	110W	6	elk	Olive-gray (5Y 5/2) with minor dark gray (5Y 4/1) loam. Coarse fraction: subrounded greenish gray (5GY 6/1) fine-grained sandstone containing biotite; accessory subangular yellowish gray (5Y 8/1) fine-grained claystone (?) containing biotite.	See 349.
353	43N	110W	6	elk	Light brownish gray (10YR 6/2) with minor light yellowish brown (10YR 6/4) silt loam; strongly calcareous. Coarse fraction: mixed angular light-gray (N 7/0) and very light-gray (N 8/0) fine-grained rocks; accessory light-gray (2.5Y 7/2) tuff.	Both Wind River formation and area of undivided upper and middle Eocene rocks occur (Love et al. 1952). Sample suggests that lick is in latter, which consists of green and gray tuff, claystone, sandstone, and conglomerate, and variegated tuffaceous claystone and sandstone in lower part.

[a]Suspected lick.

2. Composition of lick samples

Analytical results (oven-dried, 110°C) for saturation paste; normal, neutral, ammonium acetate extracts; and total sulfur and calcium-carbonate equivalent determinations for earth materials from licks. Water (under saturation paste extract) refers to moisture content at saturation. Data are arranged in order presented in descriptions of licks in App. 1

St./prov.; sample no.	Saturation extract									Ammonium acetate extract				Total	
	Ca	Mg	Na	K	Cl	SO_4	cond.	pH	H_2O	Ca	Mg	Na	K	S	$CaCO_3$
	me/100 g						mmho/cm		%	me/100 g				ppm	%
Alberta															
309	0.09	0.07	0.03	.008	0.010	0.14	0.61	7.1	28.4	13.35	5.80	0.28	0.59	970	1.5
329	1.00	4.70	0.29	.020	0.150	6.10	7.96	7.4	41.2	71.36	12.22	0.76	0.26	51000	21.4
330	...	...	...	...	...	...	...	...	...	...	...	...	...	360	9.6
332	...	...	...	...	...	...	...	...	...	...	...	...	...	58670	...
314	...	...	...	...	...	...	...	...	...	...	...	...	...	2390	...
412	0.14	0.03	0.63	.010	0.001	0.61	1.15	8.2	63.4	31.81	3.62	4.19	0.34	650	1.0
413	0.65	0.30	0.41	.030	0.001	1.12	1.51	8.0	73.6	49.40	8.14	1.57	0.64	1310	5.9
237	0.04	0.01	0.50	.007	0.020	0.20	1.30	8.4	35.7	10.82	2.32	3.13	0.33	362	...
310	7.08	3.30	14.52	.290	1.090	14.27	4.97	6.8	351.4	85.58	21.39	21.20	0.86	6720	...
316	...	...	...	...	...	...	...	...	...	...	...	...	...	3901	...
344	1.15	0.53	0.16	.060	0.050	0.27	1.48	6.3	110.2	44.41	6.56	0.74	0.65	1880	...
414	0.57	0.34	2.17	.040	0.040	3.54	8.63	7.9	30.7	15.34	2.94	5.09	0.72	2187	1.5
236	0.18	0.06	0.01	.010	0.020	0.01	0.69	8.1	34.2	41.17	3.17	0.33	0.46	288	44.0
240	...	...	...	...	...	...	...	...	...	...	...	...	...	811	4.0
243	...	...	...	...	...	...	...	...	...	...	...	...	...	439	...
342	...	...	...	...	...	...	...	...	...	...	...	...	...	505	62.8
343	0.63	5.31	1.61	.020	0.410	7.32	12.76	8.5	32.5	40.67	12.90	2.54	0.12	1440	63.5
312	3.69	2.08	1.28	.060	7.480	0.04	15.29	7.6	43.7	46.16	3.62	1.63	0.28	264	40.8
235	0.17	0.07	0.06	.004	0.080	0.02	0.35	7.0	63.9	23.08	4.98	0.33	0.22	537	...
238	0.95	14.62	3.98	.030	0.150	19.89	22.69	8.2	40.3	54.89	27.56	6.94	0.40	7260	33.0
241	0.84	0.44	0.59	.040	0.160	1.31	1.96	7.7	85.5	36.18	7.24	2.02	0.45	1360	1.1
311	2.25	1.10	0.46	.080	0.130	0.72	0.56	7.0	500.0	35.18	7.01	0.51	0.36	1040	6.1
394	1.80	1.13	0.45	.010	0.090	2.78	3.06	7.7	72.3	53.77	7.47	1.17	0.17	2848	55.2
395	0.04	1.22	0.12	.010	0.060	0.72	2.33	8.6	45.9	41.67	22.42	0.53	0.31	435	46.1
239	...	...	...	...	...	...	...	...	...	...	...	...	...	838	8.8
242	0.41	0.21	0.79	.020	0.050	0.59	1.30	6.7	86.4	30.69	9.50	3.98	0.37	2140	...
315	0.70	0.30	2.71	.052	0.410	0.33	1.35	8.0	217.7	85.58	20.14	10.53	0.50	1900	...

335	...	...	...	...	...	...	...	...	...	...	...	...	...	3147	...
322	...	...	...	...	...	...	...	...	...	...	...	...	...	7660	...
313	0.82	0.62	0.16	.009	0.060	1.20	1.31	8.0	100.0	...	...	...	...	574	...
218	0.16	0.07	0.69	.050	0.050	0.20	1.50	8.3	55.2	18.71	3.39	3.85	0.84	66	1.2
British Columbia															
397	0.98	22.32	2.11	.290	0.110	26.95	23.61	8.2	42.7	44.41	41.14	3.91	1.00	27450	31.1
396	0.59	35.20	4.01	.400	0.230	43.12	31.50	8.3	46.2	39.55	48.95	5.48	0.77	11200	16.2
234	0.28	0.16	0.12	.010	0.030	0.26	0.96	7.8	44.8	25.20	2.49	0.39	0.24	895	4.3
404	0.24	0.15	0.09	.003	0.270	0.10	1.31	8.1	35.8	44.41	1.73	0.39	0.10	151	49.7
405	0.40	2.19	0.77	.080	0.370	2.78	4.82	8.6	48.9	39.55	8.60	1.39	0.31	676	33.6
406	0.02	0.09	0.12	.006	0.001	0.08	0.81	8.4	27.9	26.32	3.17	0.42	0.16	49	41.3
400	0.15	0.65	2.23	.040	1.370	1.96	11.01	8.2	24.9	39.55	7.24	3.72	0.23	488	29.8
Manitoba															
126	...	...	...	...	...	...	...	...	...	...	...	...	...	5060	75.8
127	0.95	0.39	8.87	.21	11.18	1.14	30.01	7.6	34.0	21.04	1.83	14.25	0.77	548	...
148	0.03	0.01	0.56	.03	0.05	0.39	2.24	7.7	29.3	25.35	1.89	2.85	0.77	234	12.3
150	2.10	1.48	22.01	.56	9.73	12.94	7.48	7.9	299.4	65.87	19.00	45.56	1.62	7960	32.3
143	4.76	1.84	3.65	.08	0.57	8.39	5.91	7.5	113.7	104.31	13.80	6.39	0.64	12700	18.7
145	0.15	0.04	1.13	.05	0.21	0.03	0.22	7.4	500.0	15.46	4.49	3.64	0.62	231	0.3
149	...	...	...	...	...	...	...	...	...	...	...	...	...	3070	17.5
144	0.31	0.02	0.88	.08	0.31	0.43	1.79	8.1	62.2	51.51	3.87	2.72	0.80	1260	30.5
Northwest Territories															
100	3.40	2.80	11.04	.22	10.61	7.05	17.15	7.6	79.3	47.16	6.79	15.33	0.52	7630	37.7
101	1.22	0.78	1.26	.03	1.12	1.70	6.88	7.8	32.0	30.69	2.26	1.51	0.20	824	39.5
112	1.14	8.96	2.10	.07	0.09	12.63	9.21	4.5	75.9	3.09	15.67	2.59	0.39	3230	...
339	...	...	...	...	...	...	...	...	...	79.09	7.53	11.53	2.44	3443	2.2
340	...	...	...	...	...	...	...	...	...	85.58	13.80	33.71	3.19	22130	0.8
341	...	...	...	...	...	...	...	...	...	...	...	...	...	7730	13.8
347	...	...	...	...	...	...	...	...	...	49.90	29.82	1.20	0.85	...	23.5
354	...	...	...	...	...	...	...	...	...	48.90	7.28	1.56	0.25	...	87.5
272	...	...	...	...	...	...	...	...	...	...	...	...	...	201300	3.4
Nova Scotia															
151	3.44	0.1	0.19	.16	1.14	1.98	2.34	7.7	130.1	35.93	0.68	0.60	1.41	3030	...
Ontario															
199	0.24	0.12	0.18	.020	0.001	0.42	1.58	6.9	28.2	17.59	3.39	1.00	0.51	200	...
200	...	...	...	...	...	...	...	...	...	...	...	...	...	316	...
201	0.90	0.46	0.48	.100	0.001	0.88	0.74	5.4	205.1	39.55	10.41	1.41	1.09	1690	...
202	...	...	...	...	...	...	...	...	...	...	...	...	...	193	...
203	...	...	...	...	...	...	...	...	...	...	...	...	...	369	7.7
204	0.31	0.08	0.18	.006	0.380	0.14	2.64	7.6	21.5	26.32	1.28	0.70	0.22	166	2.8

St./prov.; sample no.	Saturation extract									Ammonium acetate extract				Total	
	Ca	Mg	Na	K	Cl	SO_4	cond.	pH	H_2O	Ca	Mg	Na	K	S	$CaCO_3$
	me/100 g						mmho/cm		%	me/100 g				ppm	%
Ontario *(continued)*															
205	0.04	0.02	0.09	.005	0.010	0.06	0.54	7.2	25.7	1.75	0.39	0.17	0.06	80	...
384	0.74	0.23	0.62	.060	0.130	0.12	0.19	7.3	500.0	46.16	9.28	1.14	0.40	1040	...
385	0.21	0.06	0.28	.020	0.100	0.07	0.84	8.0	49.4	43.91	2.94	1.28	0.55	371	23.4
386	0.55	0.14	0.88	.030	0.001	0.16	0.20	7.8	500.0	25.82	1.65	1.08	0.13	270	8.3
387	0.71	0.44	0.10	.020	0.050	0.74	1.21	7.0	83.1	25.82	7.92	0.39	0.52	675	...
388	0.08	0.04	2.22	.040	0.630	0.09	0.43	8.9	500.0	41.16	1.67	5.22	0.44	140	17.3
389	0.60	0.30	0.07	.050	0.130	0.06	0.15	6.6	500.0	30.19	8.14	0.20	0.38	500	...
390	0.28	0.08	0.11	.020	0.020	0.08	0.72	8.0	48.8	49.40	3.17	0.69	0.50	314	23.3
391	0.41	0.16	0.37	.030	0.001	0.06	0.99	6.9	500.0	17.34	4.28	0.22	0.18	322	0.6
392	0.29	0.10	0.05	.009	0.020	0.22	0.65	6.6	54.3	17.96	3.48	0.26	0.34	250	...
393	0.84	0.26	0.17	.040	0.130	0.03	0.21	7.4	500.0	26.32	3.39	0.39	0.21	306	...
Quebec															
190	0.14	0.08	0.55	.02	0.02	0.51	1.93	8.1	30.3	11.68	2.22	1.23	0.77	567	0.6
191	7.41	2.41	7.68	.04	0.31	13.97	2.24	3.4	500.0	9.13	2.82	0.99	0.16	5480	...
192	3.27	1.35	0.66	.08	0.20	4.46	6.03	6.6	53.3	43.78	3.13	1.20	0.45	4790	2.4
193	...	...	...	...	...	...	...	...	...	...	...	...	...	249	...
Alabama															
227	1.29	0.01	1.68	.05	3.37	0.08	4.84	3.7	74.8	12.60	2.00	3.85	0.65	272	...
Alaska: Canning River															
287	...	...	...	...	...	...	...	...	...	...	...	...	...	21550	18.9
317	...	...	...	...	...	...	...	...	...	...	...	...	...	363	...
288	...	...	...	...	...	...	...	...	...	...	...	...	...	698	...
289	...	...	...	...	...	...	...	...	...	...	...	...	...	10270	...
301	...	...	...	...	...	...	...	...	...	...	...	...	...	380	...
293	...	...	...	...	...	...	...	...	...	...	...	...	...	18680	...
294	...	...	...	...	...	...	...	...	...	...	...	...	...	2660	57.0
295	...	...	...	...	...	...	...	...	...	...	...	...	...	1614	...
296	...	...	...	...	...	...	...	...	...	...	...	...	...	6380	...
297	...	...	...	...	...	...	...	...	...	...	...	...	...	3750	...
299	...	...	...	...	...	...	...	...	...	...	...	...	...	101600	5.5
300	...	...	...	...	...	...	...	...	...	...	...	...	...	2868	11.5
280	...	...	...	...	...	...	...	...	...	...	...	...	...	27180	...
281	...	...	...	...	...	...	...	...	...	...	...	...	...	50940	0.1

282	...	...	...	...	...	...	...	...	...	...	...	...	...	16020	...
283	...	...	...	...	...	...	...	...	...	...	...	...	...	64470	...
284	...	...	...	...	...	...	...	...	...	...	...	...	...	73810	0.1
285	...	...	...	...	...	...	...	...	...	...	...	...	...	2306	...
298	...	...	...	...	...	...	...	...	...	...	...	...	...	17270	4.9
286	...	...	...	...	...	...	...	...	...	...	...	...	...	183200	...
290	...	...	...	...	...	...	...	...	...	...	...	...	...	29930	28.7
291	...	...	...	...	...	...	...	...	...	...	...	...	...	2405	...
292	...	...	...	...	...	...	...	...	...	...	...	...	...	2590	...
Alaska: LaSalle and Bluecloud															
355	...	...	...	...	...	...	...	...	...	...	...	...	...	7090	7.9
357	...	...	...	...	...	...	...	...	...	...	...	...	...	417	13.7
358	...	...	...	...	...	...	...	...	...	...	...	...	...	29740	12.3
361	...	...	...	...	...	...	...	...	...	...	...	...	...	554	11.6
367	...	...	...	...	...	...	...	...	...	...	...	...	...	335	56.7
369	...	...	...	...	...	...	...	...	...	...	...	...	...	338	15.8
356	...	...	...	...	...	...	...	...	...	...	...	...	...	296	13.8
359	...	...	...	...	...	...	...	...	...	...	...	...	...	610	15.6
362	...	...	...	...	...	...	...	...	...	...	...	...	...	568	33.2
363	...	...	...	...	...	...	...	...	...	...	...	...	...	5380	...
368	...	...	...	...	...	...	...	...	...	...	...	...	...	7610	...
370	...	...	...	...	...	...	...	...	...	...	...	...	...	952	0.2
360	...	...	...	...	...	...	...	...	...	...	...	...	...	1120	11.3
364	...	...	...	...	...	...	...	...	...	...	...	...	...	228	20.6
365	...	...	...	...	...	...	...	...	...	...	...	...	...	12800	10.3
366	...	...	...	...	...	...	...	...	...	...	...	...	...	119	40.3
371	...	...	...	...	...	...	...	...	...	...	...	...	...	104	7.8
Alaska: Dry Creek															
270	...	...	...	...	...	...	...	...	...	...	...	...	...	30736	...
271	...	...	...	...	...	...	...	...	...	...	...	...	...	148700	...
273	...	...	...	...	...	...	...	...	...	...	...	...	...	220	3.2
176	0.14	0.04	7.07	.15	0.001	0.62	...	7.0	500.0	34.06	4.30	24.36	4.08	314	0.8
177	0.78	0.16	0.11	.29	0.130	0.05	0.21	7.9	500.0	21.96	2.02	0.27	0.89	174	2.2
178	3.84	0.94	0.72	.54	0.250	1.72	0.94	7.8	500.0	13.18	1.87	0.90	0.77	1495	2.9
179	0.90	0.12	0.04	.19	0.130	0.08	0.19	7.9	500.0	8.87	0.70	0.12	0.14	286	...
180	12.48	15.68	0.11	.04	0.400	35.05	3.84	3.4	500.0	43.78	15.92	0.22	0.06	11800	...
181	2.48	0.46	0.10	.17	0.010	0.72	0.49	7.2	500.0	4.31	1.25	0.07	0.20	1505	...
250	...	...	...	...	...	...	...	...	...	...	...	...	...	422	8.3
251	...	...	...	...	...	...	...	...	...	...	...	...	...	299	...

St./prov.; sample no.	Saturation extract Ca	Mg	Na	K	Cl	SO_4	cond.	pH	H_2O	Ammonium acetate extract Ca	Mg	Na	K	Total S	$CaCO_3$
	me/100 g						mmho/cm		%	me/100 g				ppm	%
Alaska: Dry Creek *(continued)*															
252	...	...	...	...	...	...	...	...	...	...	...	...	...	2390	...
253	...	...	...	...	...	...	...	...	...	...	...	...	...	838	...
254	...	...	...	...	...	...	...	...	...	...	...	...	...	265	...
255	...	...	...	...	...	...	...	...	...	...	...	...	...	1130	...
256	...	...	...	...	...	...	...	...	...	...	...	...	...	430	...
257	...	...	...	...	...	...	...	...	...	56.01	14.71	2.03	2.07	361	7.5
258	...	...	...	...	...	...	...	...	...	...	...	...	...	3570	...
259	...	...	...	...	...	...	...	...	...	...	...	...	...	669	...
260	...	...	...	...	...	...	...	...	...	...	...	...	...	362	...
261	...	...	...	...	...	...	...	...	...	...	...	...	...	718	...
321	...	...	...	...	...	...	...	...	...	...	...	...	...	1920	...
262	...	...	...	...	...	...	...	...	...	...	...	...	...	1320	1.3
263	...	...	...	...	...	...	...	...	...	...	...	...	...	684	...
264	...	...	...	...	...	...	...	...	...	...	...	...	...	1180	...
265	...	...	...	...	...	...	...	...	...	...	...	...	...	286	...
266	...	...	...	...	...	...	...	...	...	...	...	...	...	6390	...
267	...	...	...	...	...	...	...	...	...	...	...	...	...	596	...
268	...	...	...	...	...	...	...	...	...	...	...	...	...	942	...
269	...	...	...	...	...	...	...	...	...	...	...	...	...	1090	...
Alaska: Granite Creek															
276	...	...	...	...	...	...	...	...	...	...	...	...	...	296	...
277	...	...	...	...	...	...	...	...	...	...	...	...	...	164	...
278	...	...	...	...	...	...	...	...	...	...	...	...	...	46	...
279	...	...	...	...	...	...	...	...	...	...	...	...	...	43	...
302	...	...	...	...	...	...	...	...	...	17.34	2.47	1.79	0.60	76	1.9
303	...	...	...	...	...	...	...	...	...	...	...	...	...	200	2.9
304	...	...	...	...	...	...	...	...	...	...	...	...	...	181	4.2
305	...	...	...	...	...	...	...	...	...	...	...	...	...	357	4.9
306	...	...	...	...	...	...	...	...	...	...	...	...	...	226	3.3
Alaska: Sheep Creek															
244	...	...	...	...	...	...	...	...	...	...	...	...	...	76	2.5
245	...	...	...	...	...	...	...	...	...	...	...	...	...	38	...
246	...	...	...	...	...	...	...	...	...	...	...	...	...	43	0.8
247	...	...	...	...	...	...	...	...	...	...	...	...	...	29	...

248	...	...	...	...	...	...	...	...	...	...	...	...	...	37	...
249	...	...	...	...	...	...	...	...	...	...	...	...	...	22	0.5
Alaska: Lost Creek															
374	0.50	0.12	3.07	.23	0.13	0.53	0.57	8.3	500.0	35.05	1.95	3.07	0.77	1260	7.1
375	0.38	0.04	1.67	.15	0.13	0.22	0.46	8.5	500.0	31.81	2.71	3.00	0.96	762	6.0
376	0.90	0.08	0.02	.11	0.13	0.09	0.21	8.4	500.0	32.93	1.25	0.46	0.42	555	6.2
Alaska: Other															
175	2.56	0.46	3.89	.17	3.86	1.22	1.39	5.8	514.4	79.09	8.37	7.14	0.64	4480	...
274	...	...	...	...	...	...	...	...	...	...	...	...	...	317	57.1
275	...	...	...	...	...	...	...	...	...	...	...	...	...	673	...
307	...	...	...	...	...	...	...	...	...	...	...	...	...	541	...
308	...	...	...	...	...	...	...	...	...	...	...	...	...	392	...
Arkansas															
132	0.03	0.010	0.39	.020	0.05	0.270	1.10	7.6	40.8	8.11	6.06	1.20	0.58	142	0.1
124	0.08	0.060	0.37	.003	0.32	0.080	1.18	5.0	42.0	7.35	6.16	1.79	0.38	153	...
217	0.01	0.001	0.06	.001	0.10	0.002	0.16	7.5	500.0	1.45	7.47	3.52	0.22	78	...
California															
398	0.08	0.02	0.001	.02	0.02	0.02	0.4	7.9	27.9	35.68	17.65	0.48	0.73	73	6.8
155	...	...	...	...	...	...	...	...	...	16.97	0.19	27.29	0.75	680	...
Colorado															
163	0.08	0.060	0.42	.010	0.09	0.150	1.56	7.9	24.4	7.86	3.66	1.59	0.22	110	0.1
164	...	...	...	...	...	...	...	...	...	...	...	...	...	55	2.5
165	0.94	1.380	2.16	.070	0.99	3.490	7.41	7.9	46.6	45.07	7.63	4.77	0.61	1130	30.6
117	0.03	0.008	0.58	.007	0.26	0.240	2.38	6.5	27.7	4.29	0.76	5.03	0.24	278	...
169	0.27	0.090	0.07	.090	0.13	0.003	0.12	6.0	500.0	7.10	1.09	0.15	0.54	171	...
173	...	...	...	...	...	...	...	...	...	...	...	...	...	342	...
116	...	...	...	...	...	...	...	...	...	...	...	...	...	115	...
118	0.01	0.010	2.97	.070	0.13	0.160	0.30	9.2	500.0	18.50	1.62	17.94	2.60	100	...
119	...	...	...	...	...	...	...	...	...	...	...	...	...	255	0.9
123	...	...	...	...	...	...	...	...	...	...	...	...	...	401	15.6
167	...	...	...	...	...	...	...	...	...	...	...	...	...	36	...
168	...	...	...	...	...	...	...	...	...	...	...	...	...	122	...
174	...	...	...	...	...	...	...	...	...	...	...	...	...	158	...
166	...	...	...	...	...	...	...	...	...	...	...	...	...	997	...
120	...	...	...	...	...	...	...	...	...	...	...	...	...	309	...
121	...	...	...	...	...	...	...	...	...	...	...	...	...	123	...
122	0.08	0.040	0.10	.010	0.02	0.020	0.48	7.4	28.7	5.32	1.48	0.37	0.53	91	...
170	0.55	0.580	1.94	.130	0.84	2.310	8.10	8.4	33.5	27.04	5.54	4.18	1.38	654	1.5
171	1.32	1.140	2.22	.100	0.92	3.050	8.08	8.2	37.6	18.00	5.96	3.38	0.93	1050	0.5
172	...	...	...	...	...	...	...	...	...	...	...	...	...	246	...

St./prov.; sample no.	Saturation extract									Ammonium acetate extract				Total	
	Ca	Mg	Na	K	Cl	SO_4	cond.	pH	H_2O	Ca	Mg	Na	K	S	$CaCO_3$
	me/100 g						mmho/cm		%	me/100 g				ppm	%
Georgia															
213	0.003	0.006	0.27	.003	0.18	0.020	0.69	5.1	48.9	5.84	6.95	8.97	0.25	24	...
209	0.001	0.001	0.48	.040	0.26	0.030	0.53	7.9	82.4	4.98	8.56	7.78	0.22	60	...
210	0.050	0.120	0.70	.005	0.88	0.010	1.76	6.5	56.4	6.49	10.88	3.13	0.23	65	...
211	...	...	...	...	...	...	...	...	...	...	...	...	...	68	...
212	0.001	0.001	0.12	.003	0.05	0.005	0.28	6.0	39.7	4.34	6.11	3.26	0.20	41	...
214	0.010	0.010	0.21	.004	0.08	0.060	0.52	6.7	47.3	5.40	5.53	2.22	0.22	56	...
Illinois															
372	5.32	1.84	19.30	.110	16.48	3.56	30.76	7.8	53.6	63.62	4.07	28.71	0.69	7140	6.3
373	...	...	...	...	...	...	...	...	...	...	...	...	...	3520	...
125	0.04	0.02	2.21	.009	0.31	1.30	5.96	8.4	35.9	3.24	1.09	6.04	0.15	335	0.2
Indiana															
324	0.010	0.005	0.39	.004	0.02	0.18	1.23	9.0	26.1	4.14	0.66	3.00	0.10	358	...
325	0.050	0.010	0.88	.020	0.13	0.12	0.16	7.7	500.0	3.39	2.10	3.78	0.16	175	...
326	0.005	0.002	0.16	.003	0.02	0.07	0.62	8.5	30.6	2.45	0.90	2.48	0.10	66	0.1
327	0.030	0.030	0.51	.003	0.07	0.51	1.56	6.0	36.0	4.34	2.14	2.54	0.23	255	...
328	0.340	0.230	0.60	.010	0.24	0.94	2.24	7.4	48.3	7.78	3.74	1.59	0.44	343	0.1
Kentucky															
319	...	...	...	...	...	...	...	...	...	...	...	...	...	357	...
152	0.36	0.34	0.30	.030	0.001	0.960	1.45	4.6	58.5	6.27	3.13	0.73	0.36	840	...
157	0.03	0.02	0.33	.003	0.370	0.001	0.61	4.9	74.5	1.90	1.23	0.72	0.12	22	...
153	0.03	0.09	0.59	.010	0.660	0.001	1.47	4.4	48.3	0.55	1.54	0.79	0.08	91	...
147	0.02	0.02	0.48	.004	0.030	0.090	1.25	8.3	42.1	5.07	3.24	2.32	0.17	82	0.2
Louisiana															
194	0.530	1.500	11.20	.050	12.69	2.45	35.49	3.4	29.0	0.70	2.09	14.57	0.10	675	...
189	0.060	0.110	1.46	.010	1.23	0.18	3.27	8.4	56.1	8.62	10.97	6.36	0.67	253	0.1
131	0.380	0.300	0.41	.060	0.21	0.90	1.78	5.0	56.2	5.83	2.19	2.72	1.18	442	...
197	0.002	0.003	0.10	.008	0.04	0.11	0.42	5.7	33.8	1.19	0.64	1.46	0.46	49	...
198	...	...	...	...	...	...	...	...	...	...	...	...	...	84	...
Maryland															
219	0.32	0.08	0.98	.010	0.24	0.57	2.24	7.1	29.9	15.34	1.15	1.20	0.37	471	1.1
220	...	...	...	...	...	...	...	...	...	...	...	...	...	164	...
221	0.01	0.02	0.20	.006	0.05	0.11	0.80	7.6	28.5	2.35	2.06	0.79	0.12	179	...
228	...	...	...	...	...	...	...	...	...	...	...	...	...	3014	...

Michigan															
232	0.16	0.01	0.33	.009	0.13	0.08	1.49	7.9	23.2	12.60	0.51	0.53	0.15	129	1.0
108	...	...	...	...	...	...	...	...	...	...	...	...	...	2220	...
109	0.06	0.02	0.37	.030	0.02	0.20	2.55	6.9	18.1	3.04	0.66	0.99	0.49	168	...
110	0.49	0.09	0.19	.006	0.63	0.08	3.41	7.5	22.8	20.03	0.88	0.53	0.13	173	0.9
111	7.50	3.07	2.18	.130	0.14	14.27	6.98	4.6	113.6	20.33	5.80	3.13	0.32	3220	...
113	...	...	...	...	...	...	...	...	...	...	...	...	...	289	0.2
114	0.48	0.19	0.20	.010	0.07	0.39	1.75	5.7	39.8	10.39	1.95	0.54	0.13	238	...
323	2.18	1.04	1.01	.080	0.73	3.39	4.85	5.4	64.6	29.44	5.08	1.98	0.40	4123	...
104	1.17	0.37	4.22	.050	4.69	0.49	1.30	6.9	500.0	20.53	1.77	5.03	0.10	215	6.0
399	...	...	...	...	...	...	...	...	...	...	...	...	...	611	11.6
Minnesota															
158	0.06	0.140	0.34	.006	0.02	0.11	2.04	8.3	27.0	7.35	4.60	1.00	0.19	111	0.4
159	0.02	0.005	0.42	.020	0.12	0.06	1.36	5.6	31.5	4.82	1.44	2.59	0.77	176	...
415	...	...	...	...	...	...	...	...	...	13.47	5.80	6.34	0.22	593	0.5
Mississippi															
102	0.02	0.005	0.14	.002	0.05	0.10	0.78	5.1	21.9	1.10	0.76	0.51	0.06	154	...
115	1.29	0.120	2.72	.020	0.21	3.77	7.88	7.7	41.9	16.48	1.17	5.37	0.20	1220	0.2
Missouri															
383	3.59	3.25	22.07	.27	33.16	2.75	55.64	4.6	37.9	10.98	4.52	25.88	0.58	4050	...
377	1.76	0.93	4.96	.12	8.13	1.67	15.15	7.6	52.0	21.46	3.85	7.76	0.86	898	0.8
378	2.39	1.18	6.84	.11	11.15	0.85	18.02	7.7	58.3	18.71	4.98	12.07	0.75	509	0.3
379	...	...	...	...	...	...	...	...	...	...	...	...	...	1290	0.9
380	...	...	...	...	...	...	...	...	...	...	...	...	...	821	...
381	...	...	...	...	...	...	...	...	...	...	...	...	...	527	...
382	...	...	...	...	...	...	...	...	...	...	...	...	...	602	...
Montana															
333	0.002	0.010	0.37	.020	0.001	0.001	0.02	7.0	500.0	10.43	8.37	5.84	3.27	44	...
334	0.030	0.020	0.01	.004	0.005	0.010	0.34	8.2	13.6	30.69	2.06	0.22	0.19	516	26.8
140	0.005	0.005	0.52	.004	0.500	0.250	1.42	7.5	36.7	2.79	4.91	3.45	0.40	174	2.2
141	0.030	0.060	2.72	.040	1.380	1.060	0.65	8.3	500.0	2.17	2.04	21.20	0.80	436	...
142	0.890	1.160	2.18	.020	0.720	3.810	5.70	7.8	57.8	44.91	5.96	21.53	0.50	3940	3.6
North Dakota															
135	1.73	1.97	2.98	.04	0.07	5.84	8.03	7.3	58.0	70.83	20.17	9.09	0.34	7720	8.0
136	...	...	...	...	...	...	...	...	...	...	...	...	...	12800	14.0
137	0.04	0.44	0.87	.02	1.02	0.24	3.78	7.0	37.0	36.06	16.98	2.19	0.55	259	12.9
138	...	...	...	...	...	...	...	...	...	...	...	...	...	1460	12.3
133	0.24	7.90	27.71	.19	1.00	36.94	36.08	4.0	57.4	15.46	14.63	46.43	0.83	10200	..
134	4.77	2.30	13.85	.23	0.31	19.27	3.57	6.9	500.0	39.92	12.74	25.55	0.83	10100	..

St./prov.; sample no.	Saturation extract									Ammonium acetate extract				Total	
	Ca	Mg	Na	K	Cl	SO_4	cond.	pH	H_2O	Ca	Mg	Na	K	S	$CaCO_3$
	me/100 g						mmho/cm		%	me/100 g				ppm	%
Ohio															
107	0.450	0.880	0.52	.030	0.310	1.33	1.95	7.3	82.3	11.68	9.28	1.30	0.40	1770	...
182	1.120	0.750	0.79	.030	0.010	2.63	3.10	7.8	63.5	30.91	9.02	2.25	1.06	465	0.8
183	...	...	...	...	...	...	...	...	...	...	...	...	...	1920	0.7
156	0.140	0.060	0.34	.030	0.001	0.44	1.15	4.7	44.6	5.07	2.40	0.91	0.45	564	...
231	...	...	...	...	...	...	...	...	...	...	...	...	...	598	...
105	0.005	0.003	0.14	.007	0.020	0.05	0.30	5.2	46.7	2.00	1.36	0.66	0.35	345	...
106	0.080	0.060	0.07	.020	0.020	0.10	0.48	5.0	47.6	2.40	1.13	0.09	0.22	574	1.0
Oregon															
206	0.36	0.22	1.19	.04	0.79	0.32	1.88	8.0	63.3	66.99	12.90	3.13	1.27	776	50.7
222	1.42	2.22	5.33	.14	10.32	1.66	7.70	7.7	94.9	46.66	33.94	14.35	3.02	696	2.4
Pennsylvania															
185	0.86	0.250	0.16	.080	0.06	0.71	2.76	4.6	46.3	4.56	0.78	0.27	0.31	359	...
186	0.04	0.020	0.87	.070	0.67	0.03	2.98	4.7	26.9	0.75	0.16	0.79	0.34	108	...
187	0.05	0.002	0.45	.008	0.48	0.04	1.27	3.9	47.6	1.00	0.27	0.61	0.05	209	...
103	0.20	0.020	0.25	.005	0.14	0.16	0.73	5.0	54.9	12.08	1.44	1.24	0.56	778	...
184	0.08	0.020	2.09	.050	2.13	0.04	3.91	4.4	59.7	0.40	0.14	2.85	0.49	487	...
146	...	...	...	...	...	...	...	...	...	...	...	...	...	174	0.2
South Carolina															
188	0.02	0.005	0.02	.004	0.03	0.01	0.06	4.4	54.6	0.05	0.16	0.09	0.28	100	...
226	0.18	0.030	0.68	.020	0.12	0.70	2.22	5.9	47.8	3.44	3.56	1.46	0.15	383	...
Tennessee															
154	0.29	0.54	1.62	.010	2.26	0.33	4.47	6.1	58.4	1.60	2.40	2.25	0.06	90	...
195	0.04	0.01	0.11	.005	0.21	0.04	0.73	3.9	33.4	2.20	3.74	0.57	0.13	60	...
233	0.01	0.02	0.21	.004	0.10	0.01	0.98	7.7	25.8	2.64	0.60	0.57	0.07	32	0.1
196	0.56	1.81	1.56	.008	0.24	3.76	4.67	4.5	64.3	3.44	8.14	3.39	0.10	720	...
Vermont															
216	1.24	0.22	0.18	.02	0.14	0.91	2.9	7.0	45.9	30.69	1.21	0.47	0.09	576	10.2
Virginia															
160	0.02	0.020	0.005	.010	0.001	0.010	0.16	4.5	32.3	0.85	0.31	0.13	0.15	87	...
162	0.10	0.040	20.10	.040	2.570	0.002	5.13	4.6	41.9	0.60	0.16	2.59	0.20	104	...
208	0.08	0.030	0.120	.030	0.090	0.130	0.50	5.2	56.9	6.05	1.11	0.37	0.31	400	...
161	...	...	...	...	...	...	...	...	...	...	...	...	...	129	...
207	0.02	0.005	0.160	.005	0.020	0.050	0.41	5.9	41.7	3.84	0.39	0.76	0.17	142	...

West Virginia															
230	1.93	0.04	0.20	.130	0.11	3.49	3.06	4.6	87.5	12.60	2.94	0.40	0.70	1640	...
223	1.03	0.04	0.09	.050	0.06	2.34	3.44	4.2	46.8	5.84	2.94	0.26	0.38	1640	...
229	0.51	0.04	0.11	.030	0.06	0.29	0.69	5.3	101.7	13.72	2.49	0.26	0.34	873	...
215	2.30	0.24	5.07	.060	5.46	0.82	8.07	6.4	85.6	32.93	1.40	6.52	0.40	1660	1.5
224	...	...	...	...	...	...	...	...	...	...	...	...	...	501	...
225	0.03	0.02	0.16	.002	0.02	0.05	0.49	8.0	36.4	8.00	3.83	1.14	0.30	64	...
Wyoming															
346	0.009	0.003	2.05	.02	0.050	1.38	8.62	9.5	21.1	4.74	0.60	10.53	1.20	364	0.3
128	0.220	0.440	0.44	.06	0.650	0.15	2.27	8.0	43.2	40.56	12.21	1.26	1.22	303	24.4
129	...	...	...	...	...	...	...	...	...	...	...	...	...	163	7.1
130	...	...	...	...	...	...	...	...	...	...	...	...	...	96	0.1
139	0.150	0.090	0.21	.04	0.180	0.11	1.05	8.1	43.3	47.65	6.27	0.86	1.18	262	22.7
348	0.060	0.020	0.26	.01	0.005	0.02	0.78	8.5	28.9	38.42	4.52	2.48	1.50	204	4.7
349	0.720	0.380	0.08	.06	0.005	4.30	4.16	3.3	84.1	2.79	0.84	0.12	0.49	3140	5.2
350	0.060	0.020	0.07	.01	0.010	0.02	0.38	8.2	33.0	41.67	4.52	0.69	0.92	175	8.6
351	...	...	...	...	...	...	...	...	...	...	...	...	...	150	...
352	...	...	...	...	...	...	...	...	...	...	...	...	...	82	...
353	0.150	0.020	0.88	.06	0.001	0.02	0.19	8.7	500.0	35.18	1.13	3.85	1.15	126	3.0

3. Nomenclature for plant names

Nomenclature for plant names cited in text follows the system in one or more of the following references: Bailey and Bailey (1976), Correll and Johnston (1970), Fernald (1950), Hultén (1968), Little (1979), Taylor and MacBryde (1977), and Terrell (1977).

Common name	Scientific name
Acacia, catclaw	*Acacia greggii*
Apple	*Malus*
Aspen, quaking	*Populus tremuloides*
Aster	*Aster*
Barley	*Hordeum vulgare*
Bearberry	*Arctostaphylos uva-ursi*
Birch, paper	*Betula papyrifera*
Bluegrass	*Poa*
Bromegrass	*Bromus*
smooth	*B. inermis*
Buckwheat	*Eriogonum*
Cedar	*Juniperus*
red	*J. virginiana*
Cherry, fire	*Prunus pensylvanica*
Chokecherry	*Prunus virginiana*
Clover	*Trifolium*
red	*T. pratense*
Cocksfoot	*Dactylis glomerata*
Corn	*Zea mays*
Cranberry, low-bush	*Vaccinium*
Crab, wild	*Malus coronaria*
Crowberry, black	*Empetrum nigrum*
Currant	*Ribes*
Dogwood	*Cornus*
flowering	*C. florida*
red-osier	*C. stolonifera*
Dryas	*Dryas*
False box	*Paxistima*
Fescue, Idaho	*Festuca idahoensis*
Fir, balsam	*Abies balsamea*
Geranium, sticky	*Geranium viscosissimum*
Gooseberry	*Ribes*
Grama	*Bouteloua*
blue	*B. gracilis*
needle	*B. aristidoides*
Grass	
beach rye	*Elymus arenarius*
buffalo	*Buchloë dactyloides*
grama	*Bouteloua*
orchard	*Dactylis glomerata*
rescue	*Bromus unioloides*
tundra	*Dupontia fischeri*
Greenbrier	*Smilax*
Guajillo	*Acacia berlandieri*
Hazel, beaked	*Corylus cornuta*
Honeysuckle	*Lonicera*
Japanese	*L. japonica*
Horsetail	*Equisetum*

Common name	Scientific name
Juniper	*Juniperus*
creeping	*J. horizontalis*
Lespedeza, Korean	*Lespedeza stipulacea*
Lichen (fructicose)	*Alectoria*
Lily, yellow pond	*Nuphar advena*
Locust, black	*Robinia pseudoacacia*
Maple, mountain	*Acer spicatum*
Mesquite	*Prosopis*
Mistletoe	*Phoradendron serotinum*
Mountain-ash	*Sorbus*
Oak	*Quercus*
live	*Q. virginiana*
turbinella	*Q. turbinella*
wavyleaf	*Q. undulata*
Pear, prickly	*Opuntia*
Persimmon	*Diospyros*
common	*D. virginiana*
Pine, longleaf	*Pinus palustris*
Pineberry	*Amelanchier*
Pondweed	*Potamogeton*
Poplar	*Populus*
Raspberry, red	*Rubus idaeus*
Rose	*Rosa*
Ryegrass	*Lolium*
Sagebrush	*Artemisia*
black	*A. nova*
common	*A. tridentata*
Nuttall's	*A. nuttallii*
silver	*A. cana*
Sagewort	*Artemisia frigida*
Saltbush	*Atriplex*
fourwing	*A. canescens*
Nuttall's	*A. nuttallii*
Serviceberry	*Amelanchier*
Saskatoon	*A. alnifolia*
Silk tassel	*Garrya flavescens*
Spruce, white	*Picea glauca*
Strawberry, Virginia	*Fragaria virginiana*
Sumac	*Rhus*
smooth	*R. glabra*
Sunflower, swamp	*Helianthus angustifolius*
Tea, Labrador	*Ledum palustre*
Timothy	*Phleum pratense*
Walnut	*Juglans*
Wheat	*Triticum aestivum*
Wheatgrass	*Agropyron*
crested	*A. desertorum*
western	*A. smithii*
Willow, arctic	*Salix arctica*
Winter fat	*Eurotia lanata*
common	*E. lanata*
Yew	*Taxus*
Yucca, banana	*Yucca baccata*

4. Nomenclature for animal names

Nomenclature for animals cited in text follows the system in one or more of the following; American Ornithologists' Union (1957), Hall and Kelson (1959), Walker (1975), and Sutherland (1978).

Common name	Scientific name
Antelope, pronghorn	*Antilocapra americana*
Bear	*Ursidae*
black	*Ursus americanus*
brown	*U. kenaiensis* or *alexandras*
Bison	*Bison bison*
Buffalo	*B. bison*
Caribou	*Rangifer*
barren-ground	*R. tarandus stonei* or *arcticus*
woodland	*R. t. montanus*
Cat (domestic)	*Felis*
Cattle	*Bos*
Chicken	*Gallus*
Chimpanzee	*Chimpansee troglodytes*
Deer	*Odocoileus*
black-tailed	*O. hemionus columbianus*
mule	*O. hemionus hemionus*
white-tailed	*O. virginianus*
Dove, mourning	*Zenaidura macroura*
Earthworm	*Oligochaeta*
Elk	*Cervus canadensis*
Elephant	*Loxodonta*
Fox	*Canidae*
Gazelle, Thompson's	*Gazella thomsoni*
Giraffe	*Giraffa camelopardalis*
Goat (domestic)	*Capra*
mountain	*Oreamnos americanus*
Grouse, ruffed	*Bonasa umbellus*
Hare	*Lepus americanus*
Horse	*Equus*
Honeybee	*Apis melifera*
Lemming	*Lemmus*
Macaw	*Ara*
red and green	*A. clouptera*
red and yellow	*A. macao*
Mammoth	*Mammuthus*
Mastodont	*Mammut americanum*
Moose	*Alces alces*
Musk-ox	*Ovibus moschatus*
Otter	*Lutra*
Parakeet, Carolina (parrot)	*Conuropsis carolinensis*
Pigeon	*Columba*
band-tailed	*C. fasciata*
white-crowned	*C. leucocephala*
Porcupine	*Erethizon dorsatum*
Rabbit	*Sylvilagus*
Rat, Norway	*Rattus norvegicus*
Raccoon	*Procyon lotor*

Common name	Scientific name
Sheep (domestic)	*Ovis*
bighorn	*O. canadensis canadensis*
Dall	*O. dalli dalli*
desert bighorn	*O. canadensis nelsoni* and *mexicana*
mountain	*O. canadensis*
Stone	*O. dalli stonei*
Sloth	Mylodontidae
ground	*Paramylodon harlani*
Squirrel	*Sciurus*
Swallowtail, tiger	*Papilo glaucus*
Swine	*Sus*
Tapir	*Tapirus*
Termite	Isoptera
Termite	*Macrotermes*
Topi	*Damiliscus korrigum jimela*
Weasel	*Mustela*
Wildebeest	*Connochaeles taurinus albojubatus*
Woodchuck	*Marmota monax*
Zebra	*Equus burchelli bohmi*

5. *Counties and regions of samples (Table 2.1)*

Counties and physiographic regions where surface soil samples were obtained for analyses in Table 2.1.

State	Counties and regions
New York	Allegany, Cayuga, Chautauqua, Chenango, Clinton, Columbia, Cortland, Dutchess, Erie, Genesee, Herkimer, Jefferson, Monroe, Oneida, Orange, Oswego, St. Lawrence, Tompkins, Washington, Wayne, and Yates counties.
Georgia A	Bacon, Bleckley, Burke, Dodge, Emanuel, Jeff Davis, Jefferson, Jenkins, and Warren counties
B	Cherokee, Fannin, Forsyth, Gilmer, Hall, Murray, Pickens, Towns, and Union counties
Kentucky A	Interior low plateaus (exclusive of Inner and Outer Bluegrass) and Appalachian Plateau
B	Extreme southern Pennyroyal
North Dakota	Cass and McHenry counties
Colorado	Front Range between Longmont and Fountain (Montane and plains situations included)
Wyoming, Montana	Powder River basin
Montana, Wyoming and Dakotas	Northern rolling high plains (Austin 1965)

6. *Samples from low extractable calcium areas (Table 7.10)*

Lick samples in discriminant analysis from licks in areas of topsoil with low Ca (< 10 me/100 g) from eastern United States. See Table 7.10.

State	Lick sample number
Alabama	227
Arkansas	217
Georgia	209, 210, 212, 213, 214
Illinois	125, 372
Indiana	324, 325, 326, 327, 328
Kentucky	147, 152, 153, 157
Louisiana	131, 189, 194, 197
Maryland	219, 221
Minnesota	158, 159
Ohio	105, 106, 107, 156, 182
Pennsylvania	184, 185
South Carolina	188, 226
Tennessee	154, 195, 196
Vermont	216
Virginia	160, 162, 207, 208
West Virginia	215, 223, 225, 229, 230

7. Samples from high extractable calcium areas (Table 7.13)

Lick samples used in discriminant analysis from licks in areas of topsoils with high Ca (> 10 me/100 g) from eastern United States. See Table 7.13.

State	Lick sample number
Arkansas	124, 132
Mississippi	102, 115
Missouri	377, 378, 383

8. *Weight data for fawn does (Fig. 8.1)*

Identification of and characteristics of weight data for fawn does used in constructing Fig. 8.1.

State and area	Period	*n*	Weight (kg)	Reference
New Hampshire	1951–54	568	26.5	Siegler 1968
Massachusetts	1948–50	246	27.3	Shaw and McLaughlin 1951
Rhode Island	1974–78	366	28.6	R.I. Dep. Environ. Manage. n.d.
New York				
Adirondacks	1954	173	23.3	Severinghaus 1955
southern tier	1939–43	76	26.2	
Lordville	1965–66	115	25.4	Hesselton and Sauer 1973
Millbrook	1967	62	25.4	
Ripley	1965	38	29.0	
Seneca Army Depot	1965, 1967	44	30.4	
Pennsylvania				
Allegheny Natl. Forest	1938–39	111	23.2	Park and Day 1942
New Jersey				
Deer Manage. Zones 3, 6	1974–76	n.d.	23.4	
Deer Manage. Zones 10, 11	1974–76	n.d.	29.6	
Deer Manage. Zone 12	1974–76	n.d.	29.2	
Deer Manage. Zone 21	1974–76	n.d.	17.0	
Deer Manage. Zone 14	1974–76	n.d.	29.3	
Ohio				
northeast	n.d.	71	27.5[a]	Nixon et al. 1970
northwest	n.d.	83	30.3[a]	
Hill	n.d.	62	27.3[a]	
West Virginia				
east	1951–54	282	19.5	Gill 1956
Allegheny	1951–54	707	24.0	
west	1951–54	568	27.7	
south	1951–54	136	20.9	
Indiana				
northern[b]	1977	47	26.4	H. P. Weeks, Jr. pers. commun. 1979
Atterbury	1974–77	312	27.7	
Naval Weapons Support Center–Crane	1974–77	351	28.8	
Mogan Ridge	1976–77	7	25.8	
Charleston Army Depot	1974, 1975 1977	15	19.8	

Kentucky				
Fort Knox	1958		30.5	Dechert 1967
Illinois				
north	1973	n.d.	30.0	Calhoun and Loomis 1974
central	1973	n.d.	31.1	
south	1973	n.d.	25.0	
Wisconsin				
northeast	1938–47	421	24.4	Dahlberg and Guettinger 1956
northwest	1938–47	328	25.2	
central	1938–47	271	23.9	
Minnesota				
Mud Lake	1951–54	278	32.3[c]	Krefting and Erikson 1956
Iowa	1954–62	288	32.5	Kline 1965
Missouri				
Deer Manage. Zone 2	1975–76	193	32.2[d]	Porath and Torgerson 1976, 1978
Deer Manage. Zone 3	1975–76	396	29.7[d]	
Deer Manage. Zone 5	1975–76	163	23.6[d]	
Deer Manage. Zone 6	1975–76	29	24.6[d]	
Deer Manage. Zone 7	1975–76	42	20.8[d]	
Deer Manage. Zone 8	1976–76	138	22.3[d]	
Deer Manage. Zone 9	1976–76	24	24.2[d]	
Deer Manage. Zone 11	1975–76	12	20.2[d]	
Deer Manage. Zone 12	1975–76	96	25.4[d]	

[a]Corrected for heart plus liver in original data by subtracting 1.6 kg.
[b]Data combined for Jasper-Pulaski, Kingsbury, LaSalle, Winimac, and Willow Slough, all shown in Kankakee River basin; Pigeon River Management Area in extreme northeastern Indiana.
[c]Corrected for heart plus liver; mean not weighted for annual sample.
[d]Mean not weighted for annual sample.

9. Identification of soil associations (Fig. 8.6)

Identification of soil associations in Fig. 8.6 (North Cent. Tech. Comm. 3 Soil Surv. 1965).

Map identification	Great soil group(s)	Soil associations
F12	Gray-Brown Podzolic	Wooster-Canfield-Massillon
F25		St. Clair-Blount-Pewamo
F28		Rittman-Wadsworth-Trumbull
F29		Mahoning-Trumbull-Ellsworth
F30		Venango-Mahoning
G1	Gray-Brown Podzolic and Humic Gley	Plainfield-Granby-Zimmerman
G2		Berrien-Waseon-Coloma
G4		Alexandria-Bennington-Marengo
G5		Miami-Crosby-Brookston
G6		Russell-Fincastle-Brookston
G7		Painesville-Canadea-Lorain
G8		Blount-Pewamo-Morley
H2	Gray-Brown Podzolic and Lithosol	Muskingum-Keene-Wellston
H3		Hanover-Muskingum
H5		Muskingum-Upshur-Brooke
I4	Gray-Brown Podzolic and Red-Yellow Podzolic	Cincinnati-Rossmoyne-Fairmount
I6		Bratton-Maddox
M6	Humic Gley and Low Humic Grey	Toledo-Colwood-Fulton
M7		Hoytville-Nappanee-Wauseon
M8		Paulding
06	Planosol	Bluford-Vigo-Clermont
P13	Regosol, Lithosol, and Rendzina	Fairmount-Switzerland
Q1	Alluvial	Huntington-Wheeling
Q2		Havre-Farland

10. Identification of soil associations (Fig. 8.9)

Identification of soil associations in Fig. 8.9 (North Cent. Tech. Comm. 3 Soil Surv. 1965).

Map identification	Great soil group(s)	Soil associations
A12	Brunizem and Humic Gley	Marshall-Knox
A16		Shelby-Sharpsburg
A20		Shelby-Grundy-Haig
A23		Shelby-Seymour-Edina
A24		Summit-Woodson-Labette
F2	Gray-Brown Podzolic	Dexter
F11		Boone-Bolivar
F19		Menfro-Alford-Hosmer
F21		Lindley-Weller-Gara
F26		Baxter-Eldon-Nixa
F27		Union-Weldon
I1	Gray-Brown Podzolic and Red-Yellow Podzolic	Clarksville-Taney
I2		Clarksville-Ozark
I5		Ashe-Tilsit-Hagerstown
I8		Lebanon-Hanceville
O1	Planosol	Oswego
O2		Parsons
O3		Cherokee-Parsons
O4		Putnam-Cowden
Q3	Alluvial	Genesee-Huntsville-Wabash
Q4		Waverly-Bonnie-Philo
Q6		Bottomlands, undifferentiated
Q7		Sharkey

11. Composition of earth from Boney Spring, Missouri

In Chap. 1 we discuss the occurrence of several late Pleistocene and Holocene springs that probably were licks. Among these was Boney Spring, which Saunders (1977) studied in detail and from which he provided a sample of earth from a mammoth femur. The sample was dark gray (5Y 4/1) and light gray (5Y 6/1), indistinctly laminated, noncalcareous compact clay. Bits of cancellous bone were incorporated on one side of the sample and this area was avoided in selecting material for analysis. An ammonium acetate extraction was carried out and results are reproduced in App. Table 11.

The composition of this sample is similar to geometric mean contents of alkaline earths from all the licks in this study. Boney Spring was enriched especially in magnesium, probably reflecting the dissolution of dolomites in nearby uplands. The levels of sodium and potassium are characteristic of soils in the region (e.g., Illinois data in Table 8.12).

It is regrettable that the chemistry and mineralogy of this prehistoric lick have not been scrutinized. Analysis of this single sample suggests that alkaline earths were sought; this conclusion is consistent with stratigraphic relations of the lick that include a central tufa deposit that may have represented the most desired material sought by animals visiting the spring.

App. Table 11. Contents (me/100 g) of extractable bases in earth materials from Boney Spring, Missouri, compared with geometric mean contents from all licks in this study (Table 7.7) and from deer licks in eastern United States (Table 7.10).

	Ca	Mg	Na	K
Boney Spring	15.3	7.3	0.11	0.63
All licks	13.3	3.6	2.0	0.41
Eastern licks	4.0	2.0	1.5	0.25

12. Mortality in caged Carolina parakeets

In Chap. 2 the common visits of parakeets to licks in Kentucky are described. McKinley (1980*a, b, c*) recently published a thorough review of literature of the Carolina parakeet. His account of diseases of the parakeet is particularly significant when considered in the perspective of mammal use of licks and the benefits that we speculate geophagy gives. McKinley (1980*b*:60–61) cites several records of both adult and young birds dying of "fits," "meningitis," and "strokes of apoplexy." In one case the birds' keeper "believed that something had frightened them and *scared them to death*" (McKinley 1980*b*:60, our italics).

These observations could be interpreted as nutritionally induced tetany. In one case (p. 60) the fancier attributed death to eating "too much hemp seed." Carolina parakeets evidently were strict granivores, being special epicures of cocklebur seed when available (Wilson and Bonaparte 1878) and very adept at extracting it (Plate 26 in Audubon 1946). This diet may have placed a particular stress for magnesium and calcium on these birds. This need for calcium may have caused parakeets to seek out the licks in Kentucky, which (see Chap. 1) contain substantial calcium sulfate. Other evidence of this association of parakeets with calcium sulfate is found in the notes of Nicolas Biddle in the journal of the Lewis and Clark ascent of the Missouri River in 1804 (Jackson 1978:509): "June 26. A few miles up the Blue river are quarries of Plaster of Paris since worked and brought to St. Louis. Parroquets a small kind of parrots." Evidently parakeets were common in the area of the quarries, for the observations are made as contiguous entries.

The American goldfinch, another strict granivore, often is observed pecking at mud and soil at the margins of puddles (R. R. Graber, pers. commun.) where it may be eating earth to satisfy a nutritional imbalance. Recently Miller (1978) studied nesting of goldfinches in relation to food exploitation in east central Illinois. She noted the birds' use of a creek bed and an eroded bank near the study area. Analyses of the creek bed and the eroded-bank earths revealed them (particularly the latter) to be rich in calcium and magnesium. The amount of sodium was not much different from levels in thistle seeds that made up a large part of their diet. H. G. Lumsden (pers. commun.) also has told us of occasional large numbers of crossbills colliding with autos along highways in Ontario during winter when the birds seem to be gleaning salt used to de-ice the pavement. It is not clear whether sodium or calcium chloride was used at these localities, although it is probably sodium chloride (R. Whittam 1981, pers. commun.). Dawson et al. (1965) provided red crossbills sodium chloride solutions of from 100 to 300 me/l concentrations. When offered concentrations of 100 or 200 me/l in the presence of distilled water, the birds expressed no significant preference among the three choices and particularly shunned the 200 me/l concentration. The authors considered that their evidence (p. 617) "provide[d] no support for the hypothesis that crossbills have a requirement for sodium chloride beyond that which they can obtain in their food." For months these birds had been fed piñon nuts, which the authors state were "relatively high in potassium and low in sodium" (p. 616).

These disparate observations suggest that strictly and nearly granivorous birds other than the parakeet also seek out and eat earth materials to countervail deficiencies. Future research may reveal the deficient elements to be among the alkaline earths.

13. Approximate soil taxonomic equivalents

Approximate equivalents in 1938 soil classification system (Baldwin et al. 1938), as revised in 1959 (Soil Survey Staff 1960), to current classification system (Soil Survey Staff 1975).

Great soil group of 1938 classification revised	Equivalent great groups of current classification
Bog soils	Fibrists, Hemists, and Saprists
Brown Forest soils	Eutrochrepts
Chernozems	Cryoborolls, in part Argiustolls, Haplustolls Argiborolls, and Haploborolls
Gray-Brown Podzolic soils	Hapludalfs
Gray-Wooded soils	Eutroboralfs
Humic Gley soils	Argiaquolls, Cryaquolls, Haplaquolls, Humaquents, and Umbraquults
Lithosols	Lithic subgroups of many taxa
Low-Humic Gley soils	Aquults, Haplaquents, Haplaquepts, and Ochraqualfs
Podzol soils	Cryorthods, Fragiorthods, and Haplorthods
Planosols	Albaqualfs, Argialbolls, Fragiaqualfs, and Glossaqualfs
Regosols	Vitrandepts, in part taxa within Psamments and Orthents
Solonetz soils	Natric great groups of Alfisols, Aridisols and Mollisols

LITERATURE CITED

Aandahl, Andrew R. 1972. Soils of the Great Plains. Lincoln, Nebr. Map.

Aikawa, Jerry K. 1981. *Magnesium: Its biologic significance.* Boca Raton: CRC Press.

Albrecht, William A. 1946. The soil as the basis of wildlife management. Trans. 8th Midwest Wildl. Conf., 1–8.

Aleksandrowicz, J., and J. Stachura. 1980. Trial of ecologic prophylaxis of human and cattle leukemias by supplementation of deficiencies of magnesium and other bioelements by means of mining salts. In Cantin, Marc, and Mildred S. Seelig, 225–31.

Allred, Warren J. 1942. A study of the Crystal Creek bighorn sheep range. In Honess, Ralph F., and Nedward M. Frost, 62–87.

American Association of Petroleum Geologists. No date. Geologic highway map of the mid-continent region. Map publ. by assoc.

American Ornithologists' Union. 1957. *Check-list of North American birds,* 5th ed. New York: American Ornithologists' Union.

Anderson, A. E., D. E. Medin, and D. C. Bowden. 1972. Blood serum electrolytes in a Colorado mule deer population. *J. Wildl. Dis.* 8(2):183–90.

Anderson, Bruce L., Rex D. Pieper, and Volney W. Howard, Jr. 1974. Growth response and deer utilization of fertilized browse. *J. Wildl. Manage.* 38(3):525–30.

Archer, E. E. 1956. The determination of small amounts of sulfate by reduction to hydrogen sulfide and titration with mercuric or cadmium salts with dithizone as an indicator. *Analyst* 81:181–82.

Arms, K., P. Feeny, and R. C. Lederhouse. 1973. Sodium: Stimulus for puddling behavior by tiger swallowtail butterflies, *Papilo glaucus. Science* 185:372–74.

Arnold, G. W. 1964. Some principles in the investigation of selective grazing. *Proc. Aust. Soc. Anim. Prod.* 5:258–71.

Arnold, G. W., and M. L. Dudzinski. 1978. *Ethology of free ranging domestic animals.* Vol. 2, *Developments in Animal and Veterinary Sciences.* New York: Elsevier Sci. Publ.

Audubon, John J. 1946. *The birds of America from original drawings.* (Publ. by author, 1826–30.) New York: Macmillan.

Audubon, J., and J. Bachman. 1851. *The viviparous quadrupeds of North America.* Vol. 2. New York: V. G. Audubon.

Aurbach, Gerald D., ed. 1976. *Parathyroid gland.* Vol. 7. *Handbook of Physiology.* Washington, D.C.: Am. Physiol. Soc.

Austin, Morris E. 1965. Land resource regions and major land resource areas. USDA Handb. 296.

Baestrup, F. W. 1940. The periodic die-off in certain herbivorous mammals and birds. *Science* 92(2390):354–55.

Bailey, C. B. 1977. Influence of aluminum hydroxide on the solubility of silicic acid in rumen fluid and the absorption of silicic acid from the digestive tract of ruminants. *Can. J. Anim. Sci.* 57(6):239–44.

Bailey, J. A. 1967. Mineral content of deer browse on the Huntington wildlife forest. *New York Fish Game J.* 14(1):76–78.

Bailey, Liberty H., and Ethel Z. Bailey. 1976. *Hortus third: A concise dictionary of plants cultivated in the United States and Canada.* Revised, expanded by staff of L. H. Bailey Hortorium, Cornell Univ. New York: Macmillan.

Baker, T. 1953. Food habit study of animals (antelope). *Wyo. Wildl.* 17(11):24–26.

Baldwin, M., Charles E. Kellogg, and J. Thorp. 1938. Soil classification. *In* USDA, 1938, 979–1001.

Barkley, Sylvia A., George O. Batzli, and Boyd D. Collier. 1980. Nutritional ecology of microtine rodents: A simulation model of mineral nutrition for brown lemmings. *Oikos* 34:103–14.

Barksdale, Jelks. 1929. Possible salt deposits in the vicinity of the Jackson fault, Alabama. Ala. Geol. Surv. Circ. 10.

Barnett, D. M., S. A. Edlund, and L. A. Dredge. 1977. Terrain characterization and evaluation: An example from eastern Melville Island. Geol. Surv. Can. Pap. 76–23.

Barrett, Morley W., and Gordon A. Chalmers. 1977. Clinicochemical values for adult free-ranging pronghorns. *Can. J. Zool.* 55(8):1252–60.

Beale, Donald M., and Arthur D. Smith. 1970. Forage use, water consumption, and productivity of pronghorn antelope in western Utah. *J. Wildl. Manage.* 34(3):570–82.

Bear, Firman E., ed. 1964. *Chemistry of the soil,* 2d. ed. Am. Chem. Soc. Monogr. Ser. New York: Reinhold Publ.

Bear, I. B., and R. G. Thomas. 1964. Nature of argillaceous odour. *Nature* 201:993–95.

Beath, O. A. 1942. Biological significance of mineral licks. *In* Honess, Ralph F., and Nedward M. Frost, 88–94.

Beaumont, A. B., and M. E. Snell. 1935. The effect of magnesium deficiency on crop plants. *J. Agric. Res.* 50:553–62.

Beckett, P. H. T., and R. Webster. 1971. Soil variability: A review. *Soils Fert.* 34(1):1–15.

Beeson, Kenneth C. 1941. The mineral composition of crops with particular reference to the soils in which they were grown. USDA Misc. Publ. 369.

Beetle, A. A. 1950. Buffalograss: Native of the shortgrass plains. Univ. Wyo. Agric. Exp. Stn. Bull. 293.

———. 1956. Range survey in Wyoming's Big Horn Mountains. Wyo. Agric. Exp. Stn. Bull. 341.

Beikman, H. M., and E. H. Lathram. 1976. Preliminary geologic map of northern Alaska. U.S. Geol. Surv. Misc. Field Stud. Map MF-789.

Bell, R. H. V. 1971. A grazing system in the Serengeti. *Sci. Am.* 225(1):86–93.

Berryhill, Henry L., Jr., George W. Colton, Wallace deWitt, Jr., and John E. Johnston. 1956. Geologic map of Allegany County. Md. Dep. Geol., Mines, Water Resour.

Best, D. A., G. M. Lynch, and O. J. Rongstad. 1977. Annual spring movements of moose to mineral licks in Swan Hills, Alberta. Proc. 13th North Am. Moose Conf. Workshop. Abstr.

Bever, W. 1951. What about South Dakota's sagebrush? *S.Dak. Conserv. Dig.* 18(1):12–13.

Bhattacharya, A. N., and R. G. Warner. 1968. Voluntary feed intake of pelleted diets for cattle, sheep, and rabbits as affected by different alkali supplements. *J. Anim. Sci.* 27(5):1418–25.

Bizzell, J. A. 1930. The chemical composition of New York soils. N.Y. Agric. Exp. Stn. Bull. 513.

Blair, R. M., and E. A. Epps, Jr. 1969. Seasonal distribution of nutrients in plants of seven browse species in Louisiana. USDA For. Serv. Res. Pap. SO-51.

Blair-West, J. R., Elspeth Bott, G. W. Boyd, J. P. Coghlan, D. A. Denton, J. R. Goding, S. Weller, M. Wintour, and R. D. Wright. 1965. General biological aspects of salivary secretion in ruminants. *In* Dougherty, R. W., 198–220.

Blair-West, J. R., J. P. Coghlan, D. A. Denton, J. F. Nelson, E. Orchard, B. A. Scoggins, R. D. Wright, K. Meyers, and C. L. Junquiera. 1968. Physiological, morphological, and behavioral adaptation to a sodium deficient environment by wild native Australian and introduced species of animals. *Nature* 217(5132):922–28.

Blean, Kathleen M., ed. 1977. The United States Geological Survey in Alaska: Accomplishments during 1976. U.S. Geol. Surv. Circ. 751-B.

Blood, D. C., J. A. Henderson, and O. M. Radostits. 1979. *Veterinary medicine.* Philadelphia: Lea and Febiger.

Bluemle, John P. 1977. Geologic highway map of North Dakota. N.Da. Geol. Surv. Misc. Map 19.

Bois, P. 1964. Tumor of the thymus in magnesium-deficient rats. *Nature* 204(4965):1316.

Botkin, D. B., P. A. Jordan, A. S. Dominiski, H. S. Lowendorf, and D. E. Hutchinson. 1973. Sodium dynamics in a northern ecosystem. *Proc. Natl. Acad. Sci.* 70(10):2745–48.

Bouchard, Rodrique. 1970. Etude chimique préliminaire des vasières de la resérve Matane, 1965. Ministère du Tourisme, de la Chasse et de la Pêche du Québec, Service de la Faune. *Rapport* 5:343–52.

Bowen, H. J. M. 1966. *Trace elements in biochemistry*. New York: Academic Press.

Bownocker, J. A. 1947. Geologic map of Ohio. Ohio Geol. Surv.

Boyd-Orr, J. 1929. *Minerals in pastures and their relation to animal nutrition*. London: H. K. Lewis.

Brabb, Earl E., and Michael Churkin, Jr. 1969. Geologic map of the Charley River Quadrangle, east-central Alaska. U.S. Geol. Surv. Misc. Geol. Invest. Map I-573.

Brent, William B. 1960. Geology and mineral resources of Rockingham County. Va. Div. Miner. Resour. Bull. 76.

Brock, Alec R., Forrest L. Cunningham, Donald F. Rapparlie, John N. Thatcher, E. Cecil Flesher, Kenneth L. Stone, and Lawrence A. Tornes. 1974. Soil survey of Putnam County, Ohio. USDA Soil Conserv. Surv., Ohio Dep. Nat. Resour. Div. Lands Soil, and Ohio Agric. Res. Dev. Cent.

Brown, D. A., V. E. Nash, A. G. Caldwell, Lindo J. Bartelli, R. C. Carter, and O. R. Carter. 1973. A monograph of the soils of the southern Mississippi River Valley alluvium. South. Coop. Ser. Bull. 178.

Brown, Jerry, ed. 1975. Ecological investigations of the tundra biome in the Prudhoe Bay region, Alaska. Biol. Pap. Univ. Alaska Spec. Rep. 2.

Bryant, H. C. 1918. Deer licks of the Trinity National Forest Game Refuge. *Calif. Fish Game* 4(1):21–25.

Buck, P. D. 1947. The biology of the antelope (*Antilocapra americana*) in Montana. Master's thesis, Mont. State Coll., Bozeman.

Buechner, Helmut K. 1950. Range ecology of the pronghorn on the Wichita Mountains Wildlife Refuge. *Trans. 15th North Am. Wildl. Conf.* 15:627–44.

Bunge, G. von, 1873. Ueber die Bedentung des Kochsalzes und das Verhalten der Kalisalze im menschlichen Organismus. *Z. Biol.* 9:104–43.

Buol, S. W., ed. 1973. Soils of the southern states and Puerto Rico. South. Coop. Ser. Bull. 174.

Burchett, Raymond R. 1969. Geologic bedrock map of Nebraska. Nebr. Geol. Surv.

Burke, David, George P. Howard, Robert C. Lund, Patricia McConnell, and Robert McDowell. 1975. New Jersey's white-tailed deer: A report on New Jersey's deer management program for fiscal year 1974–1975. N.J. Div. Fish, Game, Shellfish. Deer Rep. 2.

Burke, David, Robert Byrne, Robert Eriksen, George P. Howard, Robert C. Lund, Patricia McConnell, Robert C. Winkel, and Leonard Wolgast. 1976. New Jersey's white-tailed deer: A report on New Jersey's deer management program for fiscal year 1975–1976. N.J. Div. Fish, Game, Shellfish. Deer Rep. 3.

Burke, David, Robert Eriksen, George P. Howard, Robert C. Lund, Patricia McConnell, and Robert P. Winkel. 1977. New Jersey's white-tailed deer: A report on New Jersey's deer management program for fiscal year 1976–1977. N.J. Div. Fish, Game, Shellfish. Deer Rep. 4.

Burke, David, Robert Byrne, Robert Eriksen, George P. Howard, Robert C. Lund, Patricia McConnell, and Douglas Roscoe. 1978. New Jersey's white-tailed deer: A report on New Jersey's deer management program for fiscal year 1977–1978. N.J. Div. Fish, Game, Shellfish. Deer Rep. 5.

Burke, E., and H. E. Morris. 1933. Nutrient elements used by leaves and growth of apple trees. *Plant Physiol.* 8:537–44.

Butts, Charles. 1910. Birmingham, Alabama. U.S. Geol. Surv. Folio Geol. Atlas of the United States. Folio 175.

———. 1925. Geology and mineral resources of the Equality-Shawneetown area. Ill. State Geol. Surv. Bull. 47.

Butts, Charles, and Raymond S. Edmundson. 1966. Geology and mineral resources of Frederick County. Va. Div. Miner. Resour. Bull. 80.

Calef, George W., and Grant M. Lortie. 1975. A mineral lick of the barren-ground caribou. *J. Mamm.* 56(1):240–42.

Calhoun, John, and Forrest Loomis. 1974. Prairie whitetails. Ill. Dep. Conserv.
Callaghan, Eugene, and A. F. Buddington. 1938. Metalliferous mineral deposits of the Cascade Range in Oregon. U.S. Geol. Surv. Bull. 893.
Calver, James L. 1963. Geologic map of Virginia. Va. Div. Miner. Resour.
Campbell, R. B., E. W. Mountjoy, and F. G. Young. 1973. Geology of McBride map-area, British Columbia. Geol. Surv. Can. Pap. 72-35.
Cantin, Marc, and Mildred S. Seelig, eds. 1980. *Magnesium in health and disease*. New York: SP Medical and Scientific Books.
Capps, Stephen R. 1940. Geology of the Alaska railroad region. U.S. Geol. Surv. Bull. 907.
Cardwell, Dudley H., Robert B. Erwin, and Herbert P. Woodward. 1968. Geologic map of West Virginia. W.Va. Geol. Econ. Surv.
Causey, Lawson V., and John G. Newton. 1972. Geology of Clarke County, Alabama. Geol. Surv. Ala. Map 95.
Chamberlin, L. C., H. R. Timmermann, B. Snider, F. Dieken, B. Loescher, and D. Fraser. 1977. Physical and chemical characteristics of some natural licks used by big game animals in northern Ontario. Proc. 13th North Am. Moose Conf. Workshop., 200–14.
Chatelain, Edward F. 1950. Bear-moose relationships on the Kenai peninsula. Trans. 15th North Am. Wildl. Conf., 224–34.
Cheatum, E. L., and C. W. Severinghaus. 1950. Variations in fertility of white-tailed deer related to range conditions. Trans. 15th North Am. Wildl. Conf., 170–89.
Christian, John J. 1950. The adreno-pituitary system and population cycles in mammals. *J. Mamm.* 31(3):247–59.
———. 1963. Endocrine adaptive mechanisms and the physiologic regulation of population growth. *In* Mayer, W. V., and R. G. Van Gelder, 189–353.
Clarke, S. E., and E. W. Tisdale. 1945. Chemical composition of native forages of southern Alberta and Saskatchewan in relation to grazing practices. Can. Dep. Agric. Publ. 769.
Cline, Marlin G. 1961. Soils and soil associations of New York. N.Y. State Coll. Agric. Cornell Ext. Bull. 930.
Cline, Marlin G., and D. J. Lathwell. 1963. Physical and chemical properties of soils of northern New York. N.Y. Coll. Agric. Cornell Ext. Bull. 981.
Cohen, A. C., Jr. 1959. Simplified estimators for the normal distribution when samples are singly censored or truncated. *Technometrics* 1(3):217–37.
Cole, G. F. 1956. The pronghorn antelope, its range use and food habits in central Montana with special reference to alfalfa. Mont. State Coll. Agric. Exp. Stn. Tech. Bull. 516.
Cole, G. F., and B. T. Wilkins. 1958. The pronghorn antelope, its range use and food habits in central Montana with special reference to wheat. Mont. Fish Game Dep. Tech. Bull. 2.
Collett, R. 1911–12. *Norges Pattedyr*. Oslo.
Condit, D. Dale. 1912. Conemaugh formation in Ohio. Ohio Geol. Surv. 4th Ser. Bull. 17.
Connor, Jon J., and Hansford T. Shacklette. 1975. Background geochemistry of some rocks, soils, plants, and vegetables in the conterminous United States. U.S. Geol. Surv. Prof. Pap. 574-F.
Cooley, J. R., and W. Burroughs. 1962. Sand additions to high-concentrate beef cattle rations. *J. Anim. Sci.* 21:991.
Correll, Donovan S., and Marshall C. Johnston. 1970. *Manual of vascular plants of Texas*. Renner, Texas: Texas Research Foundation.
Cowan, Ian McT. 1956. Life and times of the coast black-tailed deer. *In* Taylor, Walter P., 523–617.
Cowan, Ian McT., and V. C. Brink. 1949. Natural game licks in the Rocky Mountain National Parks of Canada. *J. Mamm.* 30(4):379–87.
Crawford, James C., and D. C. Church. 1971. Response of black-tailed deer to various chemical taste stimuli. *J. Wildl. Manage.* 35(2):210–15.
Crisp, D. J., ed. 1964. *Grazing in terrestrial and marine environments*. 1962. Oxford: Blackwell.
Cushwa, Charles T., Robert L. Downing, Richard F. Harlow, and David F. Urbston. 1970. The importance of woody twig ends to deer in the southeast. USDA For. Serv. Res. Pap. SE-67.

Dagg, Anne I., and J. Bristol Foster. 1976. *The giraffe, its biology, behavior, and ecology.* New York: Van Nostrand Reinhold.

Dahlberg, Burton L., and Ralph C. Guettinger. 1956. The white-tailed deer in Wisconsin. Wis. Cons. Dep. Tech. Wildl. Bull. 14.

Dalke, P. D., R. D. Beeman, F. J. Kindel, R. J. Robel, and T. R. Williams. 1965. Use of salt by elk in Idaho. *J. Wildl. Manage.* 29(2):319–32.

Darling, F. Fraser. 1937. *A herd of red deer.* London: Oxford Univ. Press.

Dawson, William R., Vaughan H. Shoemaker, Harrison B. Tordoff, and Arieh Borut. 1965. Observations on the metabolism of sodium chloride in the red crossbill. *Auk* 82(4):606–23.

Dechert, James A. 1967. The effects of over population and hunting on the Fort Knox deer herd. *Proc. 21st Annu. Conf. Southeast Assoc. Game Fish Comm.* 21:15–23.

Denney, Arthur H. 1944. Wildlife relationships to soil types. Trans. 9th North Am. Wildl. Conf., 316–23.

Denton, D. A. 1965. Evolutionary aspects of the emergence of aldosterone secretion and salt appetite. *Physiol. Rev.* 45(2):245–95.

Dixon, Joseph S. 1939. Some biochemical aspects of deer licks. *J. Mamm.* 20(1):109.

Dougherty, R. W., ed. 1965. *Physiology of digestion in the ruminant.* Washington: Butterworths.

Douglas, G. A. 1953. Antelope restoration. Colo. Dep. Fish Game Q. Prog. Rep., Apr., Fed. Aid Div., 35–37.

Douglas, R. J. W., and D. K. Norris. 1960. Virginia Falls and Sibbeston Lake map-areas, Northwest Territories. Geol. Surv. Can. Pap. 60-19.

Duarte, C. G. 1980. Magnesium metabolism in potassium-adapted rats. *In* Cantin, Marc, and Mildred S. Seelig, 93–103.

Duncan, Don A., and E. A. Epps, Jr. 1958. Minor mineral elements and other nutrients on forest ranges in central Louisiana. La. Agric. Exp. Stn. Bull. 516.

Eckhart, Richard A. 1951. Gysipferous deposits on Sheep Mountain, Alaska. U.S. Geol. Surv. Bull. 989-C.

Eckles, C. H., and T. W. Gullickson. 1932. Phosphorus deficiency in the rations of cattle. Minn. Agric. Exp. Stn. Tech. Bull. 91.

Edmonds, C. J. 1976. Effects of aldosterone on the colon. *In* Robinson, J. W. L., 269–79.

Edmundson, Raymond S. 1958. Industrial limestones and dolomites in Virginia: James River District west of the Blue Ridge. Va. Div. Miner. Resour. Bull. 73.

Edwards, C. A., and J. R. Lofty. 1977. *Biology of earthworms.* London: Chapman and Hall.

Edwards, R. Y., and R. W. Ritcey. 1960. Foods of caribou in Wells Gray Park, British Columbia. *Can. Field Nat.* 74:3–7.

Elton, Charles. 1942. *Voles, mice, and lemmings: Problems in population dynamics.* London: Oxford.

Emery, R. S., L. D. Brown, and J. W. Thomas. 1964. Effect of sodium and calcium carbonates on milk production and composition of milk, blood, and rumen contents of cows fed grain *ad libitum* with restricted roughage. *J. Dairy Sci.* 47:1325–29.

England, Kenneth J. 1964. Geology of the Middlesboro South quadrangle, Tennessee-Kentucky-Virginia. U.S. Geol. Surv. GQ 301.

Erdman, O. A. 1950. Alexo and Saunders map-areas, Alberta. Geol. Surv. Can. Mem. 254.

Erdman, James A., and Larry P. Gough. 1975. Trace elements in soil, lichen, and grama grass of the Powder River basin. U.S. Geol. Surv. Open-File Rep. 75-436.

Erickson, Ralph L. 1966. Geologic map of part of the Friendship quadrangle, Lewis and Greenup counties, Kentucky. U.S. Geol. Surv. GQ 526.

Faux, W. 1823. *A journal of a tour to the United States.* London: W. Simpkin and R. Marshall.

Fenneman, Nevin M. 1938. *Physiography of eastern United States.* New York: McGraw-Hill.

Fernald, Merritt L. 1950. *Gray's manual of botany,* 8th ed. New York: American Book Co.

Ferrel, C. M., and H. R. Leech. 1950. Food habits of the pronghorn antelope of California. *Calif. Fish Game* 36(1):21–26.

Feth, J. H., and Ivan Barnes. 1979. Spring-deposited travertine in eleven western states. U.S. Geol. Surv. Water Resour. Invest. 79-35. Open File Rep. Map with text (unnum.).

Fichter, E., and A. E. Nelson. 1962. Study of pronghorn population. Idaho Dep. Fish Game. Pittman Robertson Proj. W85-R-13.

Finerty, James P. 1980. *The population ecology of cycles in small mammals: Mathematical theory and biological fact.* New Haven: Yale Univ. Press.

Fiskell, John G. A., and H. F. Perkins, eds. 1970. Selected Coastal Plain soil properties. South. Coop. Bull. 148.

Flock, Mark A. 1979. Genesis of selected soils in the lower Illinois River Valley. Master's thesis, Univ. Ill., Urbana.

Flynn, A., A. W. Franzmann, and D. P. Arneson. 1975. Sequential hair shaft analysis as an indicator of prior mineralization in the Alaskan moose. *J. Anim. Sci.* 41(3):906–10.

Fontenot, J. P. 1979. Animal nutrition aspects of grass tetany. *In* Rendig, Victor V., and David L. Grunes, 51–62.

Franzmann, Albert W., Arthur Flynn, and Paul D. Arneson. 1975. Levels of some mineral elements in Alaskan moose hair. *J. Wildl. Manage.* 39(2):374–78.

Fraps, G. S., and V. L. Cory. 1940. Composition and utilization of range vegetation of Sutton and Edwards counties. Texas Agric. Exp. Stn. Bull. 586.

Fraps, G. S., and J. F. Fudge. 1937. Chemical composition of soils of Texas. Tex. Agric. Exp. Stn. Bull. 549.

Fraser, D., and E. Reardon. 1980. Attraction of wild ungulates to mineral-rich springs in central Canada. *Holarct. Ecol.* 3(1):36–40.

Freddy, David J., and Albert W. Erickson. 1975. Status of the Selkirk Mountain caribou. Proc. 1st Int. Reindeer Caribou Symp., 221–27.

French, M. H. 1945. Geophagia in animals. *East Afr. Med. J.* 22(4):103–10; 22(5):152–61.

French, C. E., C. B. Smith, H. R. Fortmann, R. P. Pennington, G. A. Taylor, W. W. Hinish, and R. W. Swift. 1957. Survey of 10 nutrient elements in Pennsylvania forage crops. I. Red clover. Penn. Agric. Exp. Stn. Bull. 624.

Frison, George C. 1978. *Prehistoric hunters of the High Plains.* New York: Academic Press.

Fulton, Linda P. 1979. Structure and isopach map of the New Albany-Chattanooga-Ohio shale (Devonian and Mississippian) in Kentucky: Eastern sheet. Ky. Geol. Surv. Ser. 11. Map.

Gabrielse, H. 1962*a*. Geology Kechika British Columbia. Geol. Surv. Can. Prelim. Map 42.

———. 1962*b*. Geology Rabbit River British Columbia. Geol. Surv. Can. Prelim. Map 46.

Garrison, George A. 1972. Carbohydrate reserves and response to use. *In* McKell, Cyrus M., et al., 271–78.

Gastler, George, Alvin L. Moxon, and William T. McKean. 1951. Composition of some plants eaten by deer in the Black Hills of South Dakota. *J. Wildl. Manage.* 15(4):352–57.

Geist, V. 1971. *Mountain sheep: A study in behavior and evolution.* Chicago: Univ. Chicago Press.

———. 1975. On the management of mountain sheep: Theoretical considerations. *In* Rikhoff, J. C., and E. Zern, 77–98.

Georgia Geological Survey. 1979. Geologic map of Georgia.

Gerloff, G. C., D. G. Moore, and J. T. Curtis. 1964. Mineral content of native plants of Wisconsin. Wis. Agr. Exp. Stn. Resour. Rep. 14.

Giguere, J. F. 1975. Geology of St. Ignace Island and adjacent islands: District of Thunder Bay. Ont. Div. Mines GR 118.

Gill, John. 1956. Regional differences in size and productivity of deer in West Virginia. *J. Wildl. Manage.* 20(3):286–92.

Gill, Hiram V., Fred C. Larance, Hardy Cloutier, and H. Wade Long. 1972. Soil Survey of Desha County, Arkansas. USDA Soil Conserv. Serv. Rep. 9.

Goatcher, W. D., and D. C. Church. 1970. Review of some nutritional aspects of the sense of taste. *J. Anim. Sci.* 31(5):973–81.

Goff, John H. 1975. *Place names of Georgia.* Athens: Univ. Ga. Press.

Grace, N. D. 1970. Magnesium absorption in the digestive tract of sheep. *Proc. N.Z. Soc. Anim. Prod.* 30:21.

Gray, Carlyle, and V. C. Shepps. 1960. Geologic map of Pennsylvania. Pa. Geol. Surv. Ser. 4.

Gray, Henry., William J. Wayne, and Charles E. Wier. 1970. Geologic map of the 1° x 2° Vincennes quadrangle and parts of adjoining quadrangles, Indiana, and Illinois, showing bedrock and unconsolidated deposits. Ind. Geol. Surv. Reg. Geol. Map.

Green, R. 1970. Geological map of Alberta. Res. Counc. Alberta Map 35.

Green, R. G., and C. L. Larson. 1937. Shock disease and wild snowshoe rabbits. *Am. J. Physiol.* 119:319–20.

———. 1938*a*. A description of shock disease in the showshoe hare. *Am. J. Hyg.* 28(2):190–212.

———. 1938*b*. Shock disease and the snowshoe hare cycle. *Science* 87(2257):298–99.

Green, R. G., C. L. Larson, and J. F. Bell. 1939. Shock disease as the cause of the periodic decimation of the snowshoe hare. *Am. J. Hyg.* 30(3):83–102.

Grigal, D. F., L. F. Ohmann, and R. B. Brander. 1976. Seasonal dynamics of tall shrubs in northeastern Minnesota: Biomass and nutrient element changes. *For. Sci.* 22:195–208.

Groff, Joseph. 1978. A study of the New Jersey deer herd: IV. Location, evaluation and classification of deer range with emphasis on winter range. N.J. Div. Fish, Game, Shellfish. Final Rep. Proj. W-45 R-14. Mimeogr.

Grunes, D. L., P. R. Stout, and J. R. Brownell. 1970. Grass tetany of ruminants. *Adv. Agron.* 22:331–74.

Günther, T., R. Averdunk, and H. Ising. 1980. Biochemical mechanisms in magnesium deficiency. *In* Cantin, Marc, and Mildred S. Seelig, 57–65.

Gupta, A. K. 1952. Estimation of the mean and standard deviation of a normal population from a censored sample. *Biometrika* 39:260–73.

Haines, Francis. 1970. *The buffalo.* New York: Thomas Y. Crowell.

Hall, E. Raymond, and Keith R. Kelson. 1959. *The mammals of North America.* Vol. 1: 1–546; vol. 2: 547–1078. New York: Ronald Press.

Halls, Lowell, K. 1970. Nutrient requirements of livestock and game. Range and Wildlife Habitat Evaluation: A Research Symposium. USDA Misc. Pub. 1147:10–17.

Hamada, T., S. Maeda, and K. Kameoka. 1970. Effects of minerals on formation of colour in rumen epithelium of kids. *J. Dairy Sci.* 53:588–91.

Hamilton, J. W. 1958. Chemical composition of certain native forage plants. Wyo. Agric. Exp. Stn. Bull. 356.

Hamilton, J. W., and Carl S. Gilbert. 1972. Composition of Wyoming range plants and soils. *Wyo. Agric. Exp. Stn. Res. J.* 55.

Hammarsten, James F., and William O. Smith. 1957. Symptomatic magnesium deficiency in man. *N. Engl. J. Med.* 256:897–99.

Handfield, R. C. 1968. Sekwi formation, a new lower Cambrian formation in the southern Mackenzie Mountains, District of Mackenzie. Geol. Surv. Can. Pap. 68-47.

Hansen, Wallace R. 1976. Geologic map of the Canyon of Lodore south quadrangle, Moffat County, Colorado. U.S. Geol. Surv. GQ 1403.

Hanson, Harold C., and Robert L. Jones. 1976. *The biogeochemistry of blue, snow, and Ross's geese.* Carbondale: South. Ill. Univ. Press.

Harborne, J. B., and C. F. Van Sumere, eds. 1975. *The chemistry and biochemistry of plant proteins.* New York: Academic Press.

Harlow, R. F. 1961. Fall and winter foods of Florida white-tailed deer. *Q. J. Fla. Acad. Sci.* 24(1):19–38.

Harper, James A., Joseph H. Harn, Wallace W. Bentley, and Charles F. Yocum. 1967. The status and ecology of the Roosevelt elk in California. *Wildl. Monogr.* 16.

Healy, William B. 1970. Ingested soil as a possible source of elements for grazing animals. *Proc. N.Z. Soc. Anim. Prod.* 30:11–19.
———. 1973. Nutritional aspects of soil ingestion by grazing animals. *Chem. Biochem. Herb.* 1:567–88.
Healy, William B., and G. F. Wilson. 1971. Deposits on rumen epithelium associated with ingestion of soil. *N.Z. J. Agric. Res.* 14(1):122–31.
Healy, William B., G. Crouchley, R. L. Gillett, P. C. Rankin, and H. M. Watts. 1972. Ingested soil and iodine deficiency in lambs. *N.Z. J. Agric. Res.* 15(4):778–82.
Healy, William M. 1971. Forage preferences of tame deer in a northwest Pennsylvania clear-cutting. *J Wildl. Manage.* 35(4):717–23.
Heaton, F. W. 1980. Magnesium in intermediary metabolism. *In* Cantin, Marc, and Mildred S. Seelig, 43–55.
Hebert, Daryl, and I. McTaggart Cowan. 1971. Natural salt licks as a part of the ecology of the mountain goat. *Can. J. Zool.* 49(5):605–10.
Heimer, Wayne E. 1973. Dall sheep movements and mineral lick use. Alaska Dep. Fish Game, Fed. Aid Wildl. Restoration, Final Rep. Proj. W-17-2, W-17-3, W-17-4, W-17-5, Job 6.1R. Mimeogr.
Hesse, P. R. 1955. A chemical and physical study of the soils of termite mounds in East Africa. *J. Ecol.* 43(2):449–61.
Hesse, Richard. 1921. Über den Einfluss des Untergrundes auf das Gedeihen des Rehes. *Zool. Jahrb. Physiol.* 38:203–42.
Hesselton, William T., and Peggy R. Sauer. 1973. Comparative physical condition of four deer herds in New York according to several indices. *N.Y. Fish Game J.* 20(2):77–107.
Hewitt, Philip C. 1961. The geology of the Equinox quadrangle and vicinity, Vermont. Vt. Geol. Surv. Bull. 18.
Hibbs, L. Dale. 1966. A literature review on mountain goat ecology. Colo. Game, Fish, Parks Dep. Spec. Rep. 8.
Hilgard, E. W. 1906. *Soils, their formation, properties, composition, and relations to climate and plant growth.* New York: Macmillan.
Hill, Ralph R., and Dave Harris. 1943. Food preferences of Black Hills deer. *J. Wildl. Manage.* 7(2):233–35.
Hinders, R. G., Gy. Vidacs, and G. M. Ward. 1961. Effects of feeding dehydrated alfalfa pellets as the only roughage to dairy cows. *J. Dairy Sci.* 44(6):1178.
Hofmann, Reinhold R. 1973. *The ruminant stomach: Stomach structure and feeding habits of East African game ruminants.* Nairobi: East Afr. Lit. Bur.
Hole, Francis D. 1976. *Soils of Wisconsin.* Madison: Univ. Wis. Press.
Honess, Ralph F., and Nedward M. Frost, eds. 1942. A Wyoming bighorn sheep study. Wyo. Game Fish Dep. Bull. 1.
Hoover, W. H. 1978. Digestion and absorption in the hindgut of ruminants. *J. Anim. Sci.* 46(6):1789–99.
Hopkins, Cyril G. 1910. *Soil fertility and permanent agriculture.* Boston: Ginn.
Hopper, T. H., and L. L. Nesbitt. 1930. The chemical composition of some North Dakota pasture and hay grasses. N.Dak. Agric. Exp. Stn. Bull. 236.
Hopper, T. H., L. L. Nesbitt, and A. J. Pinckney. 1931. The chemical composition of some Chernozem-like soils of North Dakota. N.Dak. Agric. Exp. Stn. Tech. Bull. 246.
Hosley, Neil W. 1956. Management of the white-tailed deer in its environment. *In* Taylor, Walter P., 187–261.
Huber, N. King. 1973. Geologic map of Isle Royale National Park, Keweenaw County, Michigan. U.S. Geol. Surv. Misc. Geol. Invest. Map I-796.
Hultén, Eric. 1968. *Flora of Alaska and neighboring territories.* Stanford: Stanford Univ. Press.
Hundley, Louis R. 1959. Available nutrients in selected deer browse species growing on different soils. *J. Wildl. Manage.* 23(1):81–90.
Hungate, Robert E. 1966. *The rumen and its microbes.* New York: Academic Press.
Hungerford, C. R. 1970. Response of Kaibab mule deer to management of summer range. *J. Wildl. Manage.* 34(4):852–62.
Hyams, Edward. 1952. *Soil and civilization.* New York: Thames and Hudson.

Iggo, Ainsley. 1977. Activity of peripheral nerves and junctional regions. *In* Swenson, Melvin J., 549–67.

Imlay, G. 1792. *A topographical description of the western territory of North America.* London: J. Debrett.

Irish, E. J. W. 1965. Geology of the Rocky Mountain Foothills, Alberta. Geol. Surv. Can. Mem. 334.

Jackson, Donald. 1978. *Letters of the Lewis and Clark expedition, with related documents.* Urbana: Univ. Ill. Press.

Jackson, M. L. 1973. Soil chemical analysis: Advanced course. Madison, Wis.

Jacobsen, Harry A., David C. Guynn, Jr., Larry F. Castle, and Edward J. Hackett. 1977. Relationships between soil characteristics and body weights, antler measurements, and reproduction of white-tailed deer in Mississippi. Proc. J. Northest-Southeast Deer Study Group Meet., 47–55.

Jakle, John A. 1968. Salt-derived place names in the Ohio Valley. *Names* 16(1):1–5.

———. 1969. Salt on the Ohio Valley frontier, 1770–1820. *Ann. Assoc. Am. Geogr.* 59(4):687–709.

Jeffery, Paul G. 1975. *Chemical methods of rock analysis.* New York: Pergamon Press.

Jeffery, Robert G. 1980. Band-tailed pigeon (*Columba fasciata*). *In* Sanderson, Glen C., 211–45.

Jencks, Everett M. 1969. Some chemical characteristics of the major soil series of West Virginia. W.Va. Univ. Agric. Exp. Stn. Bull. 582T.

Jensen, Rue, and Donald R. Mackey. 1979. *Diseases of feedlot cattle.* Philadelphia: Lea and Febiger.

Jillson, W. P. 1936. *Big Bone Lick.* Louisville, Ky.: Standard Printing.

Johnson, C. M., and H. Nishita. 1952. Microestimation of sulfur in plant materials, soils, and irrigation waters. *Anal. Chem.* 24(4):736–42.

Jones, Douglas E. 1960. Geologic map of Bossier Parish, Louisiana. La. Dep. Conserv. Geol. Bull. 37.

Jones, Ian C., and I. W. Henderson, eds. 1976. *General, comparative, and clinical endocrinology of the adrenal cortex.* Vol 1. New York: Academic Press.

Jones, J. B., and Orlo L. Musgrave. 1963. Fertility status of Ohio soils as shown by soil tests in 1961. Ohio Agric. Exp. Stn. Res. Circ. 118.

Jones, J. B., Jr., M. C. Blount, and S. R. Wilkinson, eds. 1972. Magnesium in the environment: Soils, crops, animals, and man. Proc. Symp. Fort Valley State Coll., Fort Valley, Ga.

Jung, G. A., R. S. Adams, S. B. Guss, D. C. Kradel, D. E. Baker, and W. W. Hinish. 1975. Animal health problems considered as related to agronomic practices. *Sci. Agric.*, Penn. State Agric. Exp. Stn. 22(4):12–13.

Kare, Morley R., and G. K. Beauchamp. 1977. Taste, smell, and hearing. *In* Swenson, Melvin J., 713–30.

Karns, Patrick D., and Charles Kinsey. 1972. Establishment of physiological baselines for white-tailed deer. II. Calcium, phosphorus, and magnesium in serum of white-tailed deer. Minn. Wildl. Res. Job Completion Rep. Big Game Job 29, 53–75. Mimeogr.

Kehn, Thomas M. 1975. Geologic map of the Bordley quadrangle, Union and Webster counties, Kentucky. U.S. Geol. Surv. GQ 1275.

Keith, Lloyd B. 1963. *Wildlife's ten-year cycle.* Madison: Univ. Wis. Press.

———. 1974. Population dynamics in mammals. *In* Kjerner, Ingbritt, and Per Bjurholm, 2–58.

Keith, Lloyd B., and Lamar A. Windberg. 1978. A demographic analysis of the snowshoe hare cycle. *Wildl. Monogr. 58.*

Kilmer, V. J., S. E. Younts, and N. C. Brady, eds. 1968. *The role of potassium in agriculture.* Am. Soc. Agron., Crop Sci. Soc. Am., and Soil Sci. Soc. Am. Madison, Wis.

Kitchen, David W. 1974. Social behavior and ecology of the pronghorn. *Wildl. Monogr.* 3.

Kjerner, Ingbritt, and Per Bjurholm, eds. 1974. 11th Int. Congr. Game Biol. Stockholm. Nat. Swed. Environ. Prot. Board.

Kline, Paul D. 1965. Status and management of the white-tailed deer in Iowa, 1954–1962. *Proc. Iowa Acad. Sci.* 72:207–17.
Knight, Richard R., and M. R. Mudge. 1967. Characteristics of some natural licks in the Sun River area, Montana. *J. Wildl. Manage.* 31(2):293–98.
Korschgen, L. J. 1954. A study of food habits of Missouri deer. Mo. Conserv. Comm.
———. 1962. Foods of Missouri deer with some management implications. *J. Wildl. Manage.* 26(2):164–72.
Krausman, Paul R., and John A. Bissonette. 1977. Bone chewing behavior of desert mule deer. *Southwest Nat.* 22(1):149–50.
Krefting, Laurits W., and Arnold B. Erikson. 1956. Results of special deer hunts on the Mud Lake National Wildlife Refuge, Minnesota. *J. Wildl. Manage.* 20(3):297-302.
Küchler, A. W. 1970. Potential natural vegetation. *In* National Atlas of the United States of America. U.S. Geol. Surv., 90–91.
Kufeld, R. C. 1973. Foods eaten by the Rocky Mountain elk. *J. Range Manage.* 26(2):106–13.
Kufeld, R. C., O. C. Wallmo, and Charles Feddema. 1973. Foods of Rocky Mountain mule deer. USDA For. Serv. Res. Pap. RM-111.
Lamar, J. E. 1925. Geology and mineral resources of the Carbondale quadrangle. Ill. State Geol. Surv. Bull. 48.
LaMoreaux, Philip E. 1946. Geology and ground-water resources of the Coastal Plain of east central Georgia. Ga. Div. Mines, Min., Geol. Bull. 52.
Lane, William B., and James D. Sartor. 1966. Exchangeable calcium content of United States soils. *Soil Sci.* 101(5):390–91.
Langman, V. A. 1978. Giraffe pica behavior and pathology as indicators of nutritional stress. *J. Wildl. Manage.* 42(1):141–47.
Laragh, John H., and Jean E. Sealey. 1973. The renin-angiotensin-aldosterone hormonal system and regulation of sodium, potassium, and blood pressure homeostasis. *In* Orloff, Jack, and Robert W. Berliner, 831–908.
Lay, Daniel W. 1957. Some nutritional problems of deer in the southern pine type. Proc. 10th Annu. Conf. Southeast Assoc. Game Fish Comm., 53–58.
Leffingwell, Ernest deK. 1919. The Canning River region, northern Alaska. U.S. Geol. Surv. Prof. Pap. 109.
Leite, Edward A. No date. Fall fescue toxicities and their effects on domestic and wild herbivores. Ohio Dep. Nat. Resour. Div. Wildl. Inserv. Doc. 68(1076). Mimeogr.
Leopold, A. Starker. 1950. Deer in relation to plant succession. Trans. 15th North Am. Wildl. Conf., 571–80.
Lessig, Heber D., Thomas N. Rubel, Dennis L. Brown, and Thornton J. F. Hole. 1977. Soil Survey of Washington County, Ohio. USDA Soil Conserv. Serv., USDA For. Serv., Ohio Dep. Nat. Resour. Div. Lands Soil, and Ohio Agric. Res. Dev. Cent.
Levin R. J. 1976. The adrenal cortex and the alimentary tract. *In* Jones, Ian C., and I. W. Henderson, 207–91.
Liebig, Justus. 1885. Principles of agricultural chemistry with special reference to the late researches made in England. *In* Pomeroy, Lawrence R., 11–28.
Little, Elbert L., Jr. 1979. Checklist of United States trees. USDA Handb. 541.
Livingston, David M., and Warren E. Wacker. 1976. Magnesium metabolism. *In* Aurbach, Gerald D., 215–23.
Livingstone, D. A. 1963. Chemical composition of rivers and lakes. U.S. Geol. Surv. Prof. Pap. 440G.
Lommasson, Thomas. 1930. The value of salt on alkali ranges in southeastern Montana. *Northwest Sci.* 4:74–76.
Louisiana Department of Conservation, Louisiana Geological Survey. 1975. Louisiana Salt Domes. Unnum. map.
Lovass, Allan, L. 1958. Mule deer food habits and range use. Little Belt Mountains, Montana. *J. Wildl. Manage.* 22(3):275–83.
Love, J. D., J. L. Weitz, and R. K. Hose. 1952. Geologic map of Wyoming. U.S. Geol. Surv.
Loveless, Charles M. 1959. The Everglades deer herd life history and management. Fla. Game Fresh Water Fish Comm. Tech. Bull. 6.

McBride, James K. 1946. Deer lick improvements. Ohio Conserv. Bull. 10(10):20–21.
McCreary, O. C. 1939. Phosphorus in Wyoming pasture, hay, and other feeds. Wyo. Agric. Exp. Stn. Bull. 233.
McCulloch, Clay Y. 1973. Seasonal diets of mule and white-tailed deer. Ariz. Fish Game Dep. and U.S. For. Serv. Rocky Mt. For. Range Exp. Stn. Spec. Rep. 3, Pt. 1.
McCrory, Wayne. 1967. Absorption and excretion by mountain goats of minerals found in a natural lick. Can. Wildl. Serv. Mimeogr. Rep. CWSC 1138.
McFarlan, Arthur C. 1943. *Geology of Kentucky.* Lexington: Univ. Ky. Press.
McGerrigle, H. W., and W. B. Skidmore. 1967. Geological map, Gaspé Peninsula. Geol. Explor. Serv., Que. Dep. Nat. Resour. Geol. Map 1642.
McKell, Cyrus M., James P. Blaisdell, and Joe R. Goodin, eds. 1972. Wildland shrubs: Their biology and utilization. USDA For. Serv. Gen. Tech. Rep. INT-1.
McKinley, Daniel. 1980*a*. The balance of decimating factors and recruitment in extinction of the Carolina parakeet. I. *Ind. Audubon Q.* 58(1):8–18.
_______. 1980*b*. The balance of decimating factors and recruitment in extinction of the Carolina parakeet. II. *Ind. Audubon Q.* 58(2):50–61.
_______. 1980*c*. The balance of decimating factors and recruitment in extinction of the Carolina parakeet. III. *Ind. Audubon Q.* 58(3):103–14.
McLaughlin, F. Loy, Lawrence E. Garland, and Benjamin W. Day. 1971. Delineation of deer management zones in Vermont based on physical characteristics. Trans. Northeast Sect. Wildl. Soc. 28:47–66.
Magruder, N. D., C. E. French, L. C. McEwen, and R. W. Swift. 1957. Nutritional requirements of white-tailed deer for growth and antler development. II. Pa. Agric. Exp. Stn. Bull. 628.
Maianu, Alexandru. 1981. The regularities of salt accumulation in surface waters, groundwaters, and soils in North Dakota. I. Rivers. Agron. Abstr. 1981 Annu. Meet. Am. Soc. Agron., Crop Sci. Soc. Am., and Soil Sci. Soc. Am.
March, G. L., and R. M. F. S. Sadleir. 1972. Studies on the band-tailed pigeon (*Columba fasciata*) in British Columbia. II. Food resource and mineral-gravelling activity. *Syesis* 5:279–84.
Marshall, C. Edmund. 1964. *The physical chemistry and mineralogy of soils.* Vol. 1, *Soil materials.* New York: Wiley.
Martinka, C. J. 1967. Mortality of northern Montana pronghorns in a severe winter. *J. Wildl. Manage.* 31(1):159–64.
Mason, Ellis. 1952. Food habits and measurements of Hart Mountain antelope. *J. Wildl. Manage.* 16(3):387–89.
Mattson, William J., Jr. 1980. Herbivory in relation to plant nitrogen content. *Annu. Rev. Ecol. Syst.* 11:119–61.
Mausel, P. W., E. C. A. Runge, and S. G. Carmer. 1975. Soil productivity indexes for Illinois counties and soil associations. Ill. Agric. Exp. Stn. Bull. 752.
Mayer, H., H. Scholz, and Fr. W. Busse. 1980. Investigations on the magnesium metabolism of the brain in hypomagnesemic sheep. *In* Cantin, Marc, and Mildred S. Seelig, 801–6.
Mayer, W. V., and R. G. Van Gelder, eds. 1963. *Physiological mammalogy.* Vol. 1. New York: Academic Press.
Mayland, H. F., and D. L. Grunes. 1979. Soil-climate-plant relationships in the etiology of grass tetany. *In* Rendig, Victor V., and David L. Grunes, 123–75.
Mayland, H. F., G. E. Shewmaker, and R. C. Bull. 1977. Soil ingestion by cattle grazing crested wheatgrass. *J. Range Manage.* 30(4):264–65.
Meagher, Margaret M. 1973. The bison of Yellowstone National Park. U.S. Natl. Park Serv. Sci. Monogr. 1.
Mertie, J. B., Jr. 1929. The Chandalar-Sheenjek district of Alaska. U.S. Geol. Surv. Bull. 810-B.
Metson, A. J. 1974. Magnesium in New Zealand soils. I. Some factors governing the availability of soil magnesium. A review. *N.Z. J. Exp. Agric.* 2:277–319.
Miesch, A. T. 1967. Methods of computation for estimating geochemical abundance. U.S. Geol. Serv. Prof. Pap. 574-B.
Miller, J. K., F. C. Madsen, and E. W. Swanson. 1977. Effects of ingested soil on ration utilization by dairy cows. *J. Dairy Sci.* 60(4):618–22.

Miller, J. K., E. W. Swanson, and F. C. Madsen. 1978. Moderate soil eating by livestock shown to be nutritionally beneficial. *Feedstuffs* 50(43):30–32.

Miller, Linda J. 1978. The spatial and temporal dispersion of nests as an adaptation to food exploitation and nest predation in the American goldfinch. Master's thesis, Univ. Ill., Urbana.

Missouri Geological Survey. 1979. Geologic map of Missouri.

Moffitt, Fred H. 1954. Geology of the eastern part of the Alaska Range and adjacent area. U.S. Geol. Surv. Bull. 989-D.

Mosby, Henry S., and Charles T. Cushwa. 1969. Deer "madstones" or bezoars. *J. Wildl. Manage.* 33(2):434–37.

Mountjoy, E. W. 1963. Mount Robson Geology (southeast quarter). Alta.-B.C. Geol. Surv. Can. Map 47-1963.

Muckenhirn, R. J., L. T. Alexander, R. S. Smith, W. D. Shrader, F. F. Riecken, P. R. McMiller, and H. H. Krusekopf. 1955. Field descriptions and analytical data of certain loess-derived Gray-Brown Podzolic soils in the Upper Mississippi Valley. Ill. Agric. Exp. Stn. Bull. 587.

Murie, Olaus J. 1951. *The elk of North America.* Harrisburg, Pa.: Stackpole.

Murphy, Dean A. 1970. Deer range appraisal in the Midwest. *In* USDA For. Serv., 2–10.

National Cooperative Soil Survey. 1970. Lickdale series. USDA Soil Conserv. Serv.

National Cooperative Soil Survey. 1971. Mahoning series. USDA Soil Conserv. Serv.

National Cooperative Soil Survey. 1979*a*. Licking series. USDA Soil Conserv. Serv.

National Cooperative Soil Survey. 1979*b*. Stonelick series. USDA Soil Conserv. Serv.

National Geographic Society. 1976. Portrait U.S.A. Photomosaic map.

National Research Council, Subcommittee on Beef Cattle Nutrition. 1963. Nutrient requirements of beef cattle. Rev. ed. Natl. Acad. Sci.

National Research Council, Subcommittee on Dairy Cattle Nutrition. 1978. Nutrient requirements of dairy cattle. 5th ed. Natl. Acad. Sci.

National Research Council, Subcommittee on Sheep Nutrition. 1975. Nutrient requirements of sheep. 5th ed. Natl. Acad. Sci.

Nesbitt, William H., and Jack S. Parker. 1977. North American big game. Boone Crockett Club and Natl. Rifle Assoc. Am., Washington, D.C.

New Jersey Division of Fish, Game, and Shellfisheries. 1978. 1978–1979 New Jersey Deer Guide. Bur. Wildl. Manage.

Newson, Robin, and Antoon de Vos. Population structure and body weights of snowshoe hares on Manitoulin Island, Ontario. *Can. J. Zool.* 42(6):975–86.

Nicolai, Jürgen. 1975. *Bird life.* New York: G. P. Putnam's Sons.

Nie, Norman H., C. Hadlai Hull, Jean G. Jenkins, Karin Steinbrenner, and Dale H. Bent. 1975. *SPSS: Statistical package for the social sciences.* New York: McGraw-Hill.

Nixon, Charles M., Milford W. McClain, and Kenneth R. Russell. 1970. Deer food habits and range characteristics in Ohio. *J. Wildl. Manage.* 34(4):870–86.

Nockels, Cheryl F., L. D. Kintner, and W. H. Pfander. 1966. Influence of ration on morphology, histology, and trace mineral content of sheep rumen papillae. *J. Dairy Sci.* 49(9):1068–74.

Nordling, Donald M. 1958. Geology and mineral resources of Morgan County, Ohio. Ohio Geol. Surv. Bull. 56.

Norris. D. K. 1958. Geology, Carbondale River, Alta.-B.C. Geol. Surv. Can. Map 5.

North Central Regional Technical Committee 3 on Soil Surveys. 1965. Productivity of soils in the north central region of the United States. Univ. Ill. Agric. Exp. Stn. Bull. 710.

Northeastern Soil Research Committee. 1954. The changing fertility of New England soils. USDA Inf. Bull. 133.

Nriagu, Jerome O., ed. 1980. *Zinc in the environment. II. Health effects.* New York: (Wiley Interscience) Wiley.

Ollerenshaw, N. C. 1966. Geology, Burnt Timber Creek, Alberta. Geol. Surv. Can. Map 11-1965.

Olmstead, Donald L. 1970. Effect of certain environmental factors on white-tailed deer productivity in the eastern United States and Texas. Conn. Agric. Exp. Stn. Proj. 340. Mimeogr.

Ontario Department of Lands and Forests. 1965. Thunder Bay surficial geology. Geol. Compil. Ser. Map S265.
Ontario Department of Mines. 1965. Tashota-Geraldton Sheet. Geol. Compil. Ser. Map 2102.
———. 1972. Nipigon-Schreiber Sheet. Geol. Compil. Ser. Map 2232.
Orloff, Jack, and Robert W. Berliner, eds. 1973. *Renal physiology*. Sec. 8. *Handbook of physiology*. Washington, D.C.: Am. Physiol. Soc.
Oschwald, W. R., F. F. Riecken, R. I. Dideriksen, W. H. Scholtes, and F. W. Schaller. 1965. Principal soils of Iowa. Iowa State Univ. Coop. Ext. Serv. Spec. Rep. 42.
Packard, Fred M. 1946. An ecological study of the bighorn sheep in Rocky Mountain National Park, Colorado. *J. Mamm.* 27(1):3–28.
Paquay, R., R. de Baere, and A. Lousse. 1970*a*. Statistical research on the fate of water in the adult cow. I. Dry cows. *J. Agric. Sci.* 74(3):423–32.
———. 1970*b*. Statistical research on the fate of water in the adult cow. II. The lactating cow. *J. Agric. Sci.* 75(2):251–55.
Park, Barry C., and Besse B. Day. 1942. A simplified method of determining the condition of white-tailed deer herds in relation to available forage. USDA Tech. Bull. 840.
Parker, Gerald R. 1978. The diets of musk-oxen and Peary caribou on some islands in the Canadian High Arctic. Can. Wildl. Serv. Occas. Pap. 35.
Parks, William S., and Ernest E. Russell. 1975. Geologic map showing upper Cretaceous, Paleocene, and lower and middle Eocene units and distribution of younger fluvial deposits in western Tennessee. U.S. Geol. Surv. Misc. Invest. Ser. Map I-916.
Pease, James L., Richard H. Vowles, and Lloyd B. Keith. 1979. Interaction of snowshoe hares and woody vegetation. *J. Wildl. Manage.* 43(1):43–60.
Peden, D. G., G. M. VanDyne, R. W. Rice, and R. M. Hansen. 1974. The trophic ecology of *Bison bison* L. on shortgrass plains. *J. Appl. Ecol.* 11(2):489–97.
Peek, J. M. 1974. A review of moose food habits studies in North America. *Nat. Can.* 101(1, 2):195–215.
Penzhorn, B. L. 1982. Soil-eating by Cape Mountain zebras *(Equus zebra zebra)* in the Mountain Zebra National Park. *Koedoe* 25:83–88.
Peterson, Randolph L. 1953. Studies of the food habits and habitat of moose in Ontario. R. Ont. Mus. Zool. Paleontol. Contrib. 36.
Pitelka, F. A. 1964. The nutrient-recovery hypothesis for arctic microtines. I. Introduction. *In* Crisp, D. J., 55–56.
Pohlman, G. G. 1937. Land-class maps of West Virginia. W.Va. Agric. Exp. Stn. Bull. 285.
Pomeroy, Lawrence R., ed. 1974. *Cycles of essential elements*. Vol. 1, *Benchmark Papers in Ecology Series*. Stroudsburg, Pa.: Dowden, Hutchinson, and Ross.
Porath, Wayne R., and Oliver Torgerson. 1976. White-tailed deer population measurements and harvest recommendations. Harvest analysis and recommendations. Mo. Dep. Conserv. Study Completion Rep., Fed. Aid Proj. W-13-R-30.
———. 1978. White-tailed deer population measurement, harvest analysis and season recommendations. Deer harvest and hunter pressure analysis. Mo. Dep. Conserv. Performance Rep., Fed. Aid Proj. W-13-R-31.
Potter, Paul E. 1978. Structure and isopach map of the New Albany–Chattanooga-Ohio shale (Devonian and Mississippian) in Kentucky: Central sheet. Ky. Geol. Surv. Ser. 10.
Prasad, Anada. 1980. Manifestations of zinc abnormalities in human beings. *In* Nriagu, Jerome O., 29–59.
Pratt, D. J., and M. D. Gwynne. 1977. Rangeland management and ecology in East Africa. Huntington, N.Y.: R. E. Krieger Publ.
Prenzlow, E. J., D. L. Gilbert, and F. A. Glover. 1968. Some behavior patterns of the pronghorn. Colo. Dep. Game, Fish, Parks Spec. Rep. 17.
Price, Paul H. 1929. Pocahontas County. W.Va. Geol. Surv. Cy. Rep.
Price, R. A. 1971*a*. Geology, Barrier Mountain (east half), Alberta. Geol. Surv. Can. Map 1273A.
———. 1971*b*. Geology, Barrier Mountain (west half), Alberta. Geol. Surv. Can. Map 1274A.

Price, R. A., and E. W. Mountjoy. 1971. Geology, Scalp Creek (east half), Alberta. Geol. Surv. Can. Map 1275A.

Price, R. A., and N. C. Ollerenshaw. 1971. Geology, Scalp Creek (west half), Alberta. Geol. Surv. Can. Map 1276A.

Purdue University Agriculture Experiment Station/USDA Soil Conservation Service. 1977. Soil characterization in Indiana. II. 1967–1973 data. Purdue Agric. Exp. Stn. Bull. 174.

Quackenbush, Granville A. 1955. Our New Jersey land. N.J. Agric. Exp. Stn. Bull. 775.

Quinton, Dee A., and Ronald G. Horejsi. 1977. Diets of white-tailed deer on the rolling plains of Texas. *Southwest Nat.* 22(4):505–9.

Rassmussen, H., D. P. B. Goodman, N. Friedmann, J. E. Allen, and K. Kurokawa. 1976. Ionic control of metabolism. *In* Aurbach, G. D., 225–64.

Rausch, Robert. 1950. Observations on a cyclic decline of lemmings (*Lemmus*) on the arctic coast of Alaska during the spring of 1949. *Arctic* 3(3):166–77.

Rausch, R., and J. D. Tiner. 1949. Studies on the parasitic helminths of the north central states. II. Helminths of voles (*Microtus* spp.). Preliminary report. *Am. Midl. Nat.* 41(3):665–94.

Ray, C. E., B. N. Cooper, and W. S. Benninghoff. 1967. Fossil mammals and pollen in a Late Pleistocene deposit at Saltville, Virginia. *J. Paleontol.* 41(3):608–22.

Redmond, Charles E., William H. Brug, John R. Van, and E. L. Milliron. 1975. Soil survey of Richland County, Ohio. USDA Soil Conserv. Serv., Ohio Dep. Nat. Resour. Div. Lands Soil, and Ohio Agric. Res. Dev. Cen.

Reeder, Neil E., Victor L. Riemenschneider, and Paul W. Reese. 1973. Soil survey of Ashtabula County, Ohio. USDA Soil Conserv. Serv., Ohio Dep. Nat. Resour. Div. Lands Soil, and Ohio Agric. Res. Dev. Cen.

Reesor, J. E. 1973. Geology of the Lardeau map-area (east half), British Columbia. Geol. Surv. Can. Mem. 369.

Reger, David B. 1925. Summers County general and economic geology. W.Va. Geol. Surv. Map 6.

———. 1931. Randolph County general and economic geology. W.Va. Geol. Surv. Map 2.

Rendig, Victor V., and David L. Grunes, eds. 1979. Grass tetany. Proc. Symp. Am. Soc. Agron. Spec. Pub. 35.

Renfro, H. B. 1973. Geological highway map of Texas. Am. Assoc. Pet. Geol. U.S. Geol. Highw. Map Ser. Map 7.

Rhode Island Department of Environment Management. No date. Rhode Island deer information sheet. Div. Fish Wildl. Mimeogr.

Rice, P. R., and D. C. Church. 1974. Taste responses of deer to browse extracts, organic acids, and odors. *J. Wildl. Manage.* 38(4):830–36.

Richards, L. A. 1954. Diagnosis and improvement of saline and alkali soils. USDA Handb. 60.

Richie, William F. 1970. Regional differences in weight and antler measurements of Illinois deer. *Trans. Ill. Acad. Sci.* 63(2):189–97.

Richter, Curt P. 1937. Increased salt appetite in adrenalectomized rats. *Am. J. Physiol.* 115:155–61.

Rigg, T., and H. O. Askew. 1934. Soil and mineral supplements in the treatment of bush-sickness. *Emp. J. Exp. Agric.* 2:1–8.

Rikhoff, J. C., and E. Zern, eds. 1975. The wild sheep of North America. Proc. Workshop Manage. Biol. North Am. Wild Sheep.

Rindsig, R. B., and L. H. Schultz. 1970. Effect of bentonite on nitrogen and mineral balances and ration digestibility of high-grain rations fed to lactating dairy cows. *J. Dairy Sci.* 53(7):888–92.

Ritchie, A., J. R. Bauder, R. L. Christman, and P. W. Reese. 1978. Soil Survey of Portage County, Ohio. USDA Soil Conserv. Serv., Ohio Dep. Nat. Resour. Div. Lands Soil, and Ohio Agric. Res. Dev. Cen.

Robinson, J. W. L., ed. 1976. *Intestinal ion transport.* Baltimore: Univ. Park Press.

Roe, Frank G. 1970. *The North American buffalo: A critical study of the species in its wild state.* 2d ed. Toronto: Univ. Toronto Press.

Rogers, Lynn L., Jack J. Mooty, and Deanna Dawson. 1981. Foods of white-tailed deer in the upper Great Lakes region: A review. USDA For. Serv. Gen. Tech. Rep. NC-65.
Rogers, Thomas H. 1967. Geologic map of California, San Bernardino sheet. Calif. Div. Mines Geol. Unnum. map.
Rook, J. A. F., and C. C. Balch. 1959. The physiological significance of the fluid consistency of faeces from cattle grazing spring pasture. *Proc. Nutr. Soc.* 18:35.
Ross, Clyde P. 1959. Geology of Glacier National Park and the Flathead Region, northwestern Montana. U.S. Geol. Surv. Prof. Pap. 296.
Ross, Clyde P., David A. Andrews, and Irving J. Witkind. 1955. Geologic map of Montana. U.S. Geol. Surv. Map.
Russell, Dan M. 1974. *The dove shooter's handbook*. New York: Winchester Press.
Russo, John P. 1956. The desert bighorn sheep in Arizona. Ariz. Game Fish Dep.
Sabbe, Wayne E., Les Hileman, David Scott, and Richard Maples. 1972. Status of magnesium in certain Arkansas field and forage crops. *In* Jones, J. B., Jr., et al., 291–308.
Sand, L. B., and F. A. Mumpton, eds. 1978. *Natural zeolites: Occurrence, properties, use*. New York: Pergamon Press.
Sanderson, Glen C., ed. 1980. *Management of migratory shore and upland game birds in North America*. Lincoln: Univ. Nebr. Press.
Sather, J. H., and G. Schildman. 1955. Nebraska pronghorn. Nebr. Game, Forestation, Parks Comm. Pittman-Robertson Proj. 15R Publ.
Saunders, Jeffrey J. 1977. Late Pleistocene vertebrates of the western Ozark highland. Ill. State Mus. Rep. Invest. 33.
Scarvie, O., and J. A. Arney. 1957. Food habits of pronghorn antelope, *Antilocapra americana*, in October in northern Colorado. Fort Collins: Colo. State Univ.
Schollenberger, C. J., and R. H. Simon. 1945. Determination of exchange capacity and exchangeable bases in soil-ammonium acetate method. *Soil Sci.* 59(1):13–24.
Schorger, A. W. 1937. The range of the bison in Wisconsin. *Trans. Wis. Acad. Sci., Arts, Lett.* 30:117–30.
———. 1954. The elk in early Wisconsin. *Trans. Wis. Acad. Sci., Arts, Lett.* 43(1):5–23.
Schultz, Arnold M. 1964. The nutrient-recovery hypothesis for Arctic microtine cycles. II. Ecosystem variables in relation to Arctic microtine cycles. *In* Crisp, D. J., 57–68.
Schultz, A. M. 1969. A study of an ecosystem: The Arctic tundra. *In* Van Dyne, G. M., 77–93.
Scott, E. M., Ethel L. Verney, and Patricia D. Morissey. 1950. Self selection of diet. XI. Appetites for Ca, Mg, and K. *J. Nutr.* 41(2):187–202.
Seal, U. S., L. J. Verme, J. J. Ozoga, and A. W. Erickson. 1972. Nutritional effects on thyroid activity and blood of white-tailed deer. *J. Wildl. Manage.* 36(4):1041–52.
Seeland, David A., and Howard G. Wilshire. 1965. Geologic map of part of the Rushing Creek quadrangle in southwestern Kentucky. U.S. Geol. Surv. GQ 445.
Seelig, Mildred S. 1980. *Magnesium deficiency in the pathogenesis of disease*. New York: Plenum Medical Book Co.
Seelig, Mildred S., and George E. Bunce. 1972. Contribution of magnesium deficit to human disease. *In* Jones, J. B., et al., 61–107.
Selye, H. 1946. The general adaptation syndrome and the diseases of adaptation. *J. Clin. Endocrin. Metab.* 6:17-230.
———. 1947. *Textbook of endocrinology*. Montreal: Acta Endocrin.
———. 1950. *The physiology and pathology of exposure to stress: A treatise based on the concepts of the general-adaptation-syndrome and the diseases of adaptation*. Montreal: Acta Inc.
Severinghaus, C. W. 1955. Deer weights as an index of range conditions on two wilderness areas in the Adirondack Region. *N.Y. Fish Game J.* 2(2):154–60.
Severinghaus, C. W., H. F. Maguire, R. A. Cookingham, and J. E. Tanck. 1950. Variation by age class in the antler beam diameters of white-tailed deer related to range conditions. Trans. 15th North Am. Wildl. Conf., 551–68.

Severson, R. C., L. P. Gough, and J. M. McNeal. 1977. Availability of elements in soils to native plants, Northern Great Plains. U.S. Geol. Surv. Open File Rep. 77-847.

Severson, R. C., J. M. McNeal, and L. P. Gough. 1978. Total and extractable element composition of some Northern Great Plains soils. U.S. Geol. Surv. Open File Rep. 78-1105, 87–116.

Sharp, Geoffrey W. G., and Alexander Leaf. 1973. Effects of aldosterone and its mechanism of action on sodium transport. *In* Orloff, Jack, and Robert W. Berliner, 815–30.

Shaw, S. P., and C. L. McLaughlin. 1951. The management of white-tailed deer in Massachusetts. Mass. Div. Fish Game Resour. Bull. 13.

Shay, C. Thomas. 1978. Late prehistoric bison and deer use in the eastern prairie-forest border. Plains Anthropol., Mem. 14, 194–212.

Shiras, George, III. 1912. The white sheep, giant moose, and smaller game of the Kenai Peninsula, Alaska. *Natl. Geogr. Mag.* 23:423–94.

Short, Henry L. 1971. Forage digestibility and diet of deer on southern upland range. *J. Wildl. Manage.* 35(4):698–706.

Short, H. L., R. M. Blair, and E. A. Epps, Jr. 1975. Composition and digestibility of deer browse in southern forests. U.S. For. Serv. Resour. Pap. SO-111.

Siegler, Hilbert R., ed. 1968. The white-tailed deer of New Hampshire. N.H. Fish Game Dep. Surv. Rep. 10.

Sileo, Louis, Jr. 1973. Fertility analyses of the ranges of the white-tailed deer in the eastern United States. Conn. Agric. Exp. Stn. Proj. 340. Mimeogr. Rep.

Simpson, George G. 1945. The principles of classification and a classification of mammals. Bull. Am. Mus. Nat. Hist., vol. 85.

Sims, P. K., and G. B. Morey, eds. 1972. *Geology of Minnesota: A centennial volume.* St. Paul: Minn. Geol. Surv.

Singer, Francis J. 1975. Behavior of mountain goats, elk, and other wildlife in relation to U.S. Highway 2, Glacier National Park. Report for Fed. Highw. Adm. and Glacier Natl. Park.

Skinner, M. P. 1922. The pronghorn. *J. Mamm.* 3(2):82–105.

Skoog, R. O. 1968. Ecology of the caribou (*Rangifer tarandus granti*) in Alaska. Ph.D. diss., Univ. Calif., Berkeley.

Skoryna, S. C., and D. Waldron-Edward, eds. 1971. *Intestinal absorption of metal ions, trace elements, and radionuclides.* Oxford: Pergamon Press.

Smith, H. M., and J. W. Huckabee, Jr. 1943. Soil survey of Dimmit County, Texas. USDA and Tex. Agric. Exp. Stn. Ser. 1938. No. 4.

Smith, H. M., M. H. Layton, J. T. Miller, T. W. Glassey, and R. M. Marshall. 1940. Soil survey of Zavala County, Texas. USDA and Tex. Agric. Exp. Stn., Ser. 1934. No. 21.

Smith, H. M., R. M. Marshall, and I. C. Mowery. 1942. Soil survey of Maverick County, Texas. USDA and Tex. Agric. Exp. Stn., Ser. 1936. No. 10.

Smith, J. B., and B. E. Gilbert. 1945. Rhode Island soil types: Texture and chemical composition, and a utility index. R.I. Agric. Exp. Stn. Bull. 296.

Smith, J. C., Jr., and James A. Halsted. 1970. Clay ingestion (geophagia) as a source of zinc for rats. *J. Nutr.* 100(8):973–80.

Soil Conservation Service and Rutgers University. 1974. Interpretations for New Jersey soils. Natl. Coop. Soil Surv.; Soil Conserv. Serv.; USDA; and Coll. Agric. Environ. Sci., Rutgers Univ. Unnum. mimeogr.

Soil Survey. 1968. Soil survey laboratory data and descriptions for some soils of Illinois. USDA Soil Conserv. Serv. Soil Surv. Invest. Rep. 19.

———. 1974. Soil Survey laboratory data and descriptions for some soils of New York. USDA Soil Conserv. Serv. Soil Surv. Invest. Rep. 25.

Soil Survey Staff. 1951. Soil Survey Manual. USDA Handb. 18.

———. 1960. Soil classification, a comprehensive system. 7th approx. USDA Soil Conserv Serv.

———. 1975. Soil taxonomy, a basic system of soil classification for making and interpreting soil surveys. USDA Handb. 436.

Sotala, D. J., and C. M. Kirkpatrick. 1973. Foods of white-tailed deer, *Odocoileus virginianus,* in Martin County, Indiana. *Am. Midl. Nat.* 89(2):281–86.

Spencer, David L., and Calvin J. Lensink. 1970. The muskox of Nunivak Island, Alaska. *J. Wildl. Manage.* 34(1):1–15.
Spooner, W. C. 1929. Interior salt domes of Louisiana. Am. Assoc. Pet. Geol. Bull. 10(3):217–92.
Spurr, Josiah E., and George S. Garrey. 1908. Geology of the Georgetown quadrangle, Colorado. U.S. Geol. Surv. Prof. Pap. 63.
Stephenson, Lloyd, and Watson Monroe. 1940. The upper Cretaceous deposits. Miss. State. Geol. Surv. Bull. 40.
Stevenson, I. M. 1958. Truro map-area, Colchester and Hants counties, Nova Scotia. Geol. Surv. Can. Mem. 297.
Stiteler, William M., Jr., and Samuel P. Shaw. 1966. Use of woody browse by white-tail deer in heavily forested areas of northeastern United States. Trans. 31st North Am. Wildl. Conf., 205–12.
Stockstad, D. S., Melvin S. Morris, and Earl C. Lory. 1953. Chemical characteristics of natural licks used by big game animals in western Montana. Trans. 18th North Am. Wildl. Nat. Resour. Conf., 247–58.
Stout, Wilbur. 1927. Geology of Vinton County. Ohio Geol. Surv. Bull. 31.
———. 1943. Geology of water in Ohio. Ohio Geol. Surv. 4th ser. Bull. 44.
Stout, W. L., D. C. Kradel, G. A. Jung, and C. G. Smiley. 1976. Blood composition of well-managed high-producing Holstein cows in Pennsylvania. Pa. State Agric. Exp. Stn. Prog. Rep. 358.
Stout, W. L., D. P. Belesky, G. A. Jung, R. S. Adams, and B. L. Moser. 1977. A survey of Pennsylvania forage mineral levels with respect to dairy and beef cow nutrition. Pa. State Agric. Exp. Stn. Prog. Rep. 364.
Strand, Rudolph G. 1962. Redding Sheet, Geologic Map of California. Calif. Div. Mines Geol. Unnum. map.
Stransky, J. J., and L. K. Halls. 1967. Woodland management trends that affect game in coastal plain forest types. Proc. 21st Annu. Conf. Southeast Assoc. Game Fish Comm., 104–8.
Struever, Stuart, and Felicia A. Holton. 1979. *Koster: Americans in search of their prehistoric past.* Garden City, New York: Anchor Press/Doubleday.
Sturgeon, Myron T., and Associates. 1958. The geology and mineral resources of Athens County, Ohio. Ohio Geol. Surv. Bull. 57.
Sutherland, Douglas W. S. 1978. Common names of insects and related organisms. Entomol. Soc. Am. Spec. Pub. 78-1.
Suttle, N. F. 1975. An effect of soil ingestion on the utilization of dietary copper by sheep. *J. Agric. Sci. Camb.* 84:249–54.
Suttle, N. F., and A. C. Field. 1967. Studies on magnesium in ruminant nutrition. 8. Effect of increased intakes of potassium and water on the metabolism of magnesium, phosphorus, sodium, potassium, and calcium in sheep. *Br. J. Nutr.* 21(3):819–32.
Swartz, F. M. 1939. Keyser Limestone and Helderberg Group. *In* Willard, Bradford, 29–91.
Swenson, Melvin J., ed. 1977. *Duke's physiology of domestic animals.* 9th ed. Ithaca: Cornell Univ. Press.
Swift, R. W. 1948. Deer select most nutritious forages. *J. Wildl. Manage.* 12(1):109–10.
Tailleur, I. L., I. F. Ellersieck, and C. F. Mayfield. 1977. Southwestern Brooks Range–Ambler River quadrangle AMRAP. *In* Blean, Kathleen M., B22–B24.
Tansy, Martin F. 1971. Intestinal absorption of magnesium. *In* Skoryna, S. C., and D. Waldron-Edward, 193–210.
Tatsuoka, Maurice M. 1970. *Selected topics in advanced statistics: An elementary approach.* No. 6, *Discriminant analysis.* Champaign, Ill.: Inst. Pers. Ability Test.
Taylor, Roy L., and Bruce MacBryde. 1977. Vascular plants of British Columbia. Univ. B.C. Bot. Gard. Tech. Bull. 4.
Taylor, Walter P., ed. 1956. *The deer of North America.* Harrisburg, Pa.: Stockpole Co.; Washington, D.C.: Wildl. Manage. Inst.
Tedrow, J. C. F. 1977. *Soils of the polar landscapes.* New Brunswick: Rutgers Univ. Press. 628 pp.
———. 1978. New Jersey soils. N.J. Agric. Exp. Stn. Cir. 601.

Tener, J. S. 1954. A preliminary study of the musk-oxen of Fosheim Peninsula, Ellesmere Island, N.W.T. Can. Wildl. Serv. Wildl. Manage. Bull. Ser. 1, No. 9.

———. 1965. Musk-oxen in Canada. Can. Wildl. Ser. Monogr. 2.

Terrell, Edward E. 1977. A checklist of names for 3,000 vascular plants of economic importance. USDA Handb. 505.

Terwilliger, C. 1946. Food habits of antelope. Fort Collins: Colo. State Univ. Game Manage. Probl.

Thomas, Grant W., and Billy W. Hipp. 1968. Soil factors affecting potassium availability. *In* Kilmer, V. J., et al., 269–91.

Thomas, J. R., H. R. Cosper, and W. Bever. 1964. Effects of fertilizers on the growth of grass and its use by deer in the Black Hills of South Dakota. *Agron. J.* 56(2):223–26.

Thompson-Seton, Ernest. 1901. *Lives of the hunted.* New York: Charles Scribner's Sons.

Thorp, James, and H. T. U. Smith. 1952. Pleistocene eolian deposits of the United States, Alaska, and parts of Canada. Geol. Soc. Am. Map.

Thorsland, Oscar A. 1966. Nutritional analyses of selected deer foods in South Carolina. Proc. 20th Annu. Conf. Southeast. Assoc. Game Fish Comm., 84–104.

Tidball, Ronald R. 1974. Average composition of agricultural soils in Missouri counties. U.S. Geol. Surv. Open File Rep. 74-66.

———. 1976. Chemical variation of soils in Missouri associated with selected levels of the soil classification system. U.S. Geol. Surv. Prof. Pap. 954B.

Tileston, J. V., and L. E. Yeager. 1962. A resume of Colorado big game research projects 1939–1957. Colo. Dep. Game Fish Tech. Bull. 9.

Tilton, John L. 1926. Map II of Hampshire County showing general and economic geology. W.Va. Geol. Surv. Map.

Todd, Jeffrey. 1972. A literature review on bighorn sheep food habits. Colo. Div. Game Fish Parks Spec. Rep. 27.

Torgerson, Oliver, and William H. Pfander. 1971. Cellulose digestibility and chemical composition of Missouri deer foods. *J. Wildl. Manage.* 35(2):221–31.

Torii, Kazuo. 1978. Utilization of natural zeolites in Japan. *In* Sand, L. B., and F. A. Mumpton, 441–50.

Towry, R. K., Jr., E. D. Michael, R. L. Reid, and T. J. Allen. 1974. Quality of deer forages from West Virginia. Proc. 28th Annu. Conf. Southeast Assoc. Game Fish Comm., 574–80.

Turner, S. F., T. W. Robinson, and W. N. White. Rev. by D. E. Outlaw, W. O. George et al. 1960. Geology and ground-water resources of the Winter Garden district, Texas. 1948. U.S. Geol. Surv. Water Supply. Pap. 1481.

Twenhofel, W. H. 1936. The greensands of Wisconsin. *Econ. Geol.* 31(5):472–87.

Tweto, Ogden. 1969. Preliminary geologic map of Colorado. U.S. Geol. Surv. Misc. Field Study Map 788.

Ullrey, D. E., W. G. Youatt, H. E. Johnson, L. D. Fay, B. L. Schoepke, W. T. Magee, and K. K. Keahey. 1973. Calcium requirements of weaned white-tailed deer fawns. *J. Wildl. Manage.* 37(2):187–94.

Ullrey, D. E., W. G. Youatt, H. E. Johnson, A. B. Cowan, L. D. Fay, R. L. Covert, W. T. Magee, and K. K. Keahey. 1975. Phosphorus requirements of weaned white-tailed deer fawns. *J. Wildl. Manage.* 39(3):590–95.

USDA. 1938. Soils and men. USDA Yearb. Agric.

USDA Forest Service. 1948. Forest Regions of the United States. For. Serv. Unnum. map.

———. 1970. White-tailed deer in the Midwest. USDA For. Serv. Res. Pap. NC-39.

USDA Soil Conservation Service. 1972. General Soil Map of Pennsylvania. Unnum. map.

USDA Soil Conservation Service and Kentucky Agricultural Experiment Station. 1975. General soil map of Kentucky.

USDA Soil Conservation Service and Purdue University Agricultural Experiment Station. 1977. Soil Associations of Indiana. Purdue Univ. Coop. Ext. Serv. Publ. AY209. Map.

U.S. Geological Survey. 1977. Geochemical survey of the western energy regions. Open File Rep. 77-872.

Van Dyne, G. M., ed. 1969. The ecosystem concept in natural resource management. New York: Academic Press.
Vaughan, Michael R. 1975. Aspects of mountain goat ecology, Wallowa Mountains, Oregon. Master's thesis, Oreg. State Univ.
Virginia Commission of Game and Inland Fisheries and USDA Forest Service. 1972. George Washington National Forest, Dry River District. Unnum. map.
Walker, E. P. 1975. *Mammals of the world.* Vols. 1, 2. Baltimore: Johns Hopkins Press.
Waller, George R., R. D. Morrison, and A. B. Nelson. 1972. Chemical composition of native grasses in central Oklahoma from 1947 to 1962. Okla. State Univ. Agric. Exp. Stn. Bull. B-697.
Warner, R. G., and W. P. Flatt. 1965. Anatomical development of the ruminant stomach. *In* Dougherty, R. W., 24–38.
Wascher, Herman L., John D. Alexander, B. W. Ray, A. H. Beavers, and R. T. Odell. 1960. Characteristics of soils associated with glacial tills in northeastern Illinois. Ill. Agric. Exp. Stn. Bull. 665.
Wascher, H. L., B. W. Ray, John D. Alexander, J. B. Fehrenbacher, A. H. Beavers, and R. L. Jones. 1971. Loess soils of northwest Illinois. Ill. Agric. Exp. Stn. Bull. 739.
Weeks, Harmon P., Jr. 1974. Physiological, morphological, and behavioral adaptations of wild herbivorous mammals to a sodium-deficient environment. Ph.D. diss., Purdue Univ., Lafayette, Ind.
———. 1978. Characteristics of mineral licks and behavior of visiting white-tailed deer in southern Indiana. Am. Midl. Nat. 100(2):384–95.
Weeks, H. P., Jr., and Charles M. Kirkpatrick. 1976. Adaptations of white-tailed deer to naturally occurring sodium deficiency. *J. Wildl. Manage.* 40(4):610–25.
———. 1978. Salt preferences and sodium drive phenology in fox squirrels and woodchucks. *J. Mamm.* 59(3):531–42.
Weis, Paul L., and Paul K. Theobold. 1964. Geology of the Mont quadrangle, Kentucky. U.S. Geol. Surv. GQ 305.
Welch, Robert N. 1942. Geology of Vernon Parish. La. Dep. Conserv. Geol. Bull. 22.
Welles, Ralph E., and Florence B. Welles. 1961. The bighorn of Death Valley. U.S. Natl. Park Serv. Fauna Ser. 6.
Wenger, L. E. 1943. Buffalo grass. Kans. Agric. Exp. Stn. Bull. 321.
Westerhuis, J. H. 1974. Parturient hypocalcaemia prevention in parturient cows prone to milk fever by dietary measures. Cent. Agric. Publ. Doc., Wageningen, Netherlands.
White, Abraham, Philip Handler, and Emil Smith. 1973. *Principles of biochemistry.* 5th ed. New York: McGraw-Hill.
White, Robert G., Brian R. Thompson, Terge Skogland, Steven J. Person, Donald E. Russell, Dan F. Holleman, and Jack R. Luick. 1975. Ecology of caribou at Prudhoe Bay, Alaska. *In* Brown, Jerry, 151–87.
Whitman, Warren C., D. W. Bolin, Earle C. Klosterman, H. J. Klosterman, Kenneth D. Ford, Leroy Moomaw, D. G. Hoag, and M. L. Buchanan. 1951. Carotene, protein, and phosphorus in grasses of western North Dakota. N.Dak. Agric. Exp. Stn. Bull. 370.
Whittington, Charles L., and John C. Ferm. 1967. Geologic map of the Grayson quadrangle, Carter County, Kentucky. U.S. Geol. Surv. GQ 640.
Wickenden, R. T. D. 1945. Mesozoic stratigraphy of the eastern plains, Manitoba, and Saskatchewan. Geol. Surv. Can. Mem. 239.
Wiklander, Lambert. 1964. Cation and anion exchange phenomena. *In* Bear, Firman E., 163–205.
Wilcox, G. E., and J. E. Hoff. 1974. Grass tetany: An hypothesis concerning its relationship with ammonium nutrition of spring grasses. *J. Dairy Sci.* 57(9):1085–89.
Wilde, S. A., F. G. Wilson, and D. P. White. 1949. Soils of Wisconsin in relation to silviculture. Wis. Conserv. Dep. Publ. 525-49.
Wiley, James W., and Beth N. Wiley. 1979. The biology of the white-crowned pigeon. *Wildl. Monogr.* 64.
Willard, Bradford, ed. 1939. The Devonian of Pennsylvania. Pa. Geol. Surv. 4th Ser. Bull. G19.
Wilson, Alexander, and Charles L. Bonaparte. 1878. *American ornithology, or the natural biology of the birds of the United States.* Philadelphia: Porter and Coates.

Wolff, J. O. 1977. Habitat utilization of snowshoe hares (*Lepus americanus*) in interior Alaska. Ph.D. diss., Univ. Calif., Berkeley.

Wood-Gush, D. G. M., and Morely R. Kare. 1966. The behavior of calcium-deficient chickens. *Br. Poult. Sci.* 7:285–90.

Yoakum, J. 1958. Seasonal food habits of the Oregon pronghorn antelope (*Antilocapra americana oregona* Bailey). Int. Antelope Conf. Trans. 9:47–59.

Yoakum, Jim. 1967. Literature of the American pronghorn antelope. U.S. Dep. Inter., Bur. Land Manage., Reno, Nev.

Zawacki, April A., and Glenn Hausfater. 1969. Early vegetation of the lower Illinois valley. Ill. State Mus. Rep. Invest. 17.

INDEX